N O   D O W N L I N K

THE *CHALLENGER* CREW. BACK ROW FROM LEFT: ELLISON S. ONIZUKA, CHRISTA MCAULIFFE, GREGORY B. JARVIS, AND JUDITH A. RESNIK. FRONT ROW FROM LEFT: MICHAEL J. SMITH, FRANCIS R. SCOBEE, AND RONALD E. MCNAIR. (NASA)

NO DOWNLINK

A Dramatic Narrative about
the Challenger Accident and Our Time

CLAUS JENSEN

Translated by **BARBARA HAVELAND**

FARRAR • STRAUS • GIROUX

New York

Originally published in Danish under the title *Challenger — et teknisk Uheld*, copyright © 1993 by Claus Jensen
Printed in the United States of America
Designed by Debbie Glasserman
First edition, 1996

Library of Congress Cataloging-in-Publication Data
Jansen, Claus.
    [Challenger, et teknisk uheld. English.]
    No downlink : a dramatic narrative about the Challenger accident and our time / Claus Jensen ; translated by Barbara Haveland. — 1st ed.
        p.   cm.
    Translation of: Challenger, et teknisk uheld.
    Includes bibliographical references (p.   ).
    1. Astronautics — United States — History.   2. Aerospace industries — Social aspects — United States.   3. Challenger (Spacecraft) — Accidents.   I. Title.
TL789.8U5J46   1996   363.12'465 — dc20   95–33518   CIP

TO BIRGITTE

Contents

Preface

This is not primarily a book about either American space exploration or the space shuttle *Challenger* — although, involving as it does so much talk of astronauts, solid rocket boosters, high-pressure fuel turbopumps, firing rooms, and NASA Administrators, it might seem that way.

When I was growing up in Denmark in the 1950s, the Americans were known to me chiefly as the invincible liberators of Europe. I also knew what they looked like, thanks mainly to the well-thumbed editions of *Popular Mechanics* in the local library. They were rugged outdoor types, men of action in lumber jackets who were always improving their family homes with new and wondrous practical devices, such as radio-controlled garage doors, that were only dreamed about in Denmark. Very often — in fact, almost always — there would be a pretty woman standing in the background gazing admiringly at her handyman hubby.

A few years later, it did not surprise me at all that men such as these were capable of embarking upon a grandiose space program. I devoured every word on the Atlas rockets and the Mercury Seven in *Reader's Digest*, to the point where I knew the articles more or less by heart. Late one evening, some years earlier, my parents and I had watched the tiny dot that was *Sputnik* moving across a dark autumn sky. So I knew that the Russians were in the lead, but, like most of my friends, I had no doubt that the men from *Popular Mechanics* would soon catch up with and then pass them.

Later, I remember — as though it were a daily occurrence — how we sat expectantly for hours in front of a new, but flickering, black-and-white television; sat through countless holds in launch countdowns, until it reached the point where the experts in the Danish television studios were at a loss to find anything new to say.

But by the time *Challenger* exploded and crashed into the sea, my life had moved on and I had lost my uncritical faith in the United States and in men in lumber jackets. Spaceship launches were no longer the sort of events that merited the interruption even of children's television. Of course, in some corner of my mind, I knew that the space shuttle existed and that it was manned by astronauts. But I did not know their names — until, suddenly, the shuttle was the death of them and I discovered that we had been contemporaries.

Like most people, I accepted the disaster, recognizing that space travel was obviously a risky business and serious faults were bound to occur at some time or other. Having also read in the newspapers that the technical cause of the accident had been traced, I did not think much more about it.

But some years later, when I happened to come across Richard Feynman's in many ways shocking, but also surprisingly funny, personal account of his involvement in the Rogers commission, *What Do You Care What Other People Think?*, my interest and my curiosity became aroused. It almost made my hair stand on end to read what had happened to the NASA of my childhood, and I was taken in by the Nobel laureate's calculated and exaggerated humility as he acted the part, in James Gleick's words, of another "Holden Caulfield, a plain old straight shooter trying to figure out why so many other people are phonies."[1] This was an accustomed role for Feynman, but there is no doubt that the circumstances surrounding the *Challenger* disaster provided him with an especially fitting backdrop.

We have all experienced situations when some extremely complex and confusing set of circumstances, which we seemed unable to get to the bottom of, suddenly fell into place in a blinding moment of clarity. This unexpectedly dazzling light can, of course, be triggered by events in our private lives, but it can also strike an entire nation or whole continents simultaneously, in one immense, collective flash. In the United States and throughout the rest of the Western world, the sinking of the *Titanic*, the attack on Pearl Harbor, and the assassination of President Kennedy were such events. Each of these caused a whole generation to differentiate sharply, for the rest of their lives, between how things were before and how they were after that momentous occurrence.

Initially, the *Challenger* disaster might seem to be an event of this magnitude. And many Americans automatically likened the incident to their feelings concerning the murder of a young Democratic President in the early days of the space age. They all felt instinctively that more than a space shuttle had been lost in that sudden flash high in the sky. Throughout the nation, grief was instantaneous and intense. Many drew comparisons between the *Challenger* disaster and the earlier nuclear accident at Three Mile Island, and matters were not helped when *Challenger* was followed, just a few months later, by the Chernobyl accident.

President Ronald Reagan wasted no time in appointing a presidential commission to look into the affair. Suddenly, the perpetual flurry of activity within NASA was brought to a standstill. Thousands of documents had to be handed over, every desk ransacked, and every process laid bare, right down to the smallest and seemingly most trivial detail. The explosion cast a harsh and unforgiving light into the dimmest corners of this massive, sprawling organization and on innumerable subcontractors. But only for an instant. Soon merciful darkness descended once more.

Any violent explosion of such magnitude generates, first of all, a deep hush; this is followed by fear and a loss of orientation. Later come the attempts to understand the causes and get at the truth, along with a determination to prevent it from ever happening again. And finally, as time goes on, things gradually return to normal and everyone can breathe easily once more. I wonder if that is what has happened to the American space program. "The disaster" has unobtrusively become "the accident," or perhaps even — to use a NASA expression — "a major malfunction." The space shuttles are back in the air, and the whole affair was, by all accounts, probably just a bad scare.

Former President Ronald Reagan, who requested the Rogers commission's report, takes the view in his memoirs that, while the *Challenger* incident was certainly a grave matter, fortunately it did not take long to bring things back to normal. "The *Challenger* disaster was a catastrophe that bestowed pain and grief on all Americans. But after a lengthy and thorough investigation pinpointed the cause of the explosion, we picked up the space program, the shuttle flew again . . ."[2]

What Reagan here dismisses in two sentences I have devoted a whole book to. I feel we owe it, not only to the dead astronauts but also to all of the technicians and decision makers involved, not just to press on and forget, but to stop and try to learn a lesson from the tragic sequence of events that preceded the explosion. Unlike the former President, I do not believe that the cause is at all "pinpointable." What we are dealing with here is a kind of syndrome — countless factors, most of which are by no means unique to NASA or the space business, but which are to be found within all of the other large modern organizations and systems by which we are, to an ever-increasing extent, surrounded.

However, only very seldom are such large systems brought into the light and subjected to such close scrutiny. For this reason alone, we are almost duty-bound to take a good, hard look when, at long last, we do get the chance.

The creators of the American space shuttle considered it to be not only the world's most complex piece of machinery but also—and just as importantly—a symbol. And so it is for me.

In the wake of Vietnam and Watergate, it was a symbol of American recovery, a phoenix rising from the ashes, with flying colors and roaring rocket engines, attesting to American technological superiority and offering good reason for renewed national pride. The space shuttle proved that the American aerospace industry was also capable of producing high-technology equipment for peaceful purposes. It was the ultimate proof of what a large technological system could achieve for the good of all mankind, when the wheels of cooperation between the organization, the universities, and the political system were running most smoothly. This is why the shuttle became, and rightly so, a national icon.

And again, when it exploded in the Florida sky—almost before the wreckage had drifted down onto the Atlantic waves—it became a symbol; this time, though, in reverse. Now it symbolized vast contemporary forces that could no longer be contained; a huge technological structure crumpling under its own weight; abdication of responsibility, irresolution, and internal rivalry; and flight controllers who were suddenly out of control.

In short, the setup at NASA—that former model American organization—was turning out to be no different from what had been found so often elsewhere. In a flash, many universal and disturbing trends were drawn together by *Challenger*. By the end of the board of inquiry's investigation, the accident had become a symbol of forces within large systems, political organizations, and corporations which would, by all accounts, become more and more difficult to monitor and control in the future. It was a far cry from the optimistic Black & Decker world of *Popular Mechanics*.

I could have written a purely theoretical work about such general organizational trends, but I felt that the problem posed could be thrown most sharply into relief, and be most interesting for me personally, were I to adopt a concrete approach and concentrate on *Challenger*; turn every stone, study the technology, the corporate and political ethics, and allow the many leitmotifs presented by the affair to speak for themselves.

This book has been written in Denmark, based solely on the available written sources. There are several excellent books and articles on this affair — those by Malcolm McConnell, Richard S. Lewis, and Joseph J. Trento, to name but three. I have found useful information in all of them, which I have then tried to combine with the presidential commission's report and other sources to derive the lesson as I see it. I could have elected to go to America and visit Houston, Huntsville, and Cape Kennedy, to interview the players in this drama and talk to the bereaved families. But I did not. In my view, such a book ought to be written by Americans for Americans; instead, I have chosen to keep my distance and retain my perspective as a layman and an outsider.

I took my degree in Danish literature and history, subjects far removed from this one. So my background is by no means a scientific one — apart from *Popular Mechanics*, that is. But it is also the basic premise of this account that it is impossible to relate all the ins and outs of the *Challenger* story solely from a narrow technological standpoint. Space travel consists of much more than rockets and satellites. The American space program sprang from a political and cultural environment which has had a crucial effect on developments worldwide and upon which NASA has had a corresponding effect. And so a large part of this book deals not just with NASA but also with the United States in the years between *Sputnik* and the fall of *Challenger*.

The vast systems and organizations which the American government has brought into being to carry out missions in outer space have, over the years, gradually acquired a tremendous inertia and an obscure will of their own — and this has made it difficult for the politicians and the people to make sense of or influence developments. In this, as it happens, they resemble many other sectors of both public and private life — not only in the United States but in much of the Western world.

Traditionally, thanks to widespread specialization and rigid academic demarcation lines, the various political, economic, technical, and social elements are often kept separate in books like this one. But such distinctions cannot and ought not to be made; all of these factors are part of the story and together they constitute what I call the *Challenger* Syndrome. Often they are almost inextricably interwoven. Without this all-around view it is impossible to understand how those fatal seventy-three seconds could have had such an impact on a whole world; or why, in a wider perspective, they say so much about the realities of contemporary life.

Now and then, the assertion is made that there is no longer any point in appealing to the individual worker's own sense of responsibility, morality, or decency, when almost all of us are working within extremely large and complex systems, where the individual seems to be a totally insignificant and impotent cog. According to this perspective, there is no point in expecting or demanding individual engineers or managers to be moral heroes; far better to put all of one's efforts into reinforcing safety procedures and creating structures and processes conducive to ethical behavior.

One could, however, take the opposite stance and find in the *Challenger* story evidence to support the view that right now — at a time when the systems are threatening to set their own agendas and go their own way, without any apparent chance for human intervention and control — there has never been a greater need for individuals, within these systems, possessed of civil courage, reliable judgment, and personal integrity.[3]

There are also those who say that the realities of the modern world have grown too complex to be contained within just one narrative. That may be so. Nevertheless, I believe it is important for us to endeavor, for as long as is humanly possible, to tell the great stories of our time — and the story of *Challenger* is most definitely among them — in one account. Through my years with the space shuttle, I have become increasingly convinced that, if we are ever truly to understand our own day and age, then the story of *Challenger*'s fate is, sadly, not the worst place to start.

Claus Jensen
Aarhus, Denmark, February 17, 1995

NO DOWNLINK

A Major Malfunction

Bands of mist still hang in the hollows of the flat salt marshes. It is Tuesday, January 28, 1986, a beautiful day at the Kennedy Space Center in Florida. Seagulls cry in a clear blue sky. It is also a surprisingly chilly day. Florida is not used to frosty weather and this morning the temperature is four degrees below freezing. During the night it was even colder. In the morning sunlight, in an unreal, futuristic setting — amid concrete, steel, and state-of-the-art electronic equipment — the American space shuttle stands in readiness for its twenty-fifth launching. Like a tiny white fly, it clings to the side of the huge rust-red fuel tank.

This time, the ship being sent on a mission, which in NASA-speak has been code-named 51-L, is *Challenger*. This space shuttle has made nine previous trips into orbit and has returned safely every time, to either its landing strip in California or here in Florida.

As far as you could tell, after landing it was brushed off; a screw was tightened here and there; a wheel might have been changed. If it had landed in California, it was hoisted onto the back of its mother ship, an enormous jumbo jet, and returned soon afterward to its base at the Kennedy Space Center. There it was again raised into an upright position and the whole process could be repeated once more.

None of the other space shuttles — the sister ships *Columbia*, *Discovery*, and *Atlantis* — have made as many flights as *Challenger*. *Challenger* is known as a workhorse, the very epitome of the beast of burden that NASA has spent ten years — and $10 billion of American taxpayers' money — in developing.

The space shuttle is the world's first reusable spaceship. Quite in the spirit of a new age. In the words of President Reagan, it guarantees the United States "safe, routine, and cost-effective access to space." It is expected to do its job indefatigably and unobtrusively. Its launchings will not be particularly noteworthy, unlike the good old Apollo days, when the entire world held its breath for every liftoff. The whole idea of the space shuttle is that there should be no special reason for crossing fingers. There are those at NASA who actually call it, with some pride, "the space truck."

Gone also are the days when every child knew the names of all the astronauts; when even big, tough cops on Broadway were hard put to hold back the tears when John Glenn waved to the crowds from an open limousine, while at his side the Vice President looked on sheepishly. The space shuttle has brought about the democratization of the astronaut corps. The physical stress to which astronauts are now subjected is only a fraction of what the earliest astronauts had to withstand. Thus it is no longer necessary to use supermen in peak condition. Almost anyone can now get in on the act. And it is important for NASA to demonstrate that this is, in fact, the case.

On the launch pad, "the stack," as it is called in NASA slang, bears no resemblance to the familiar streamlined moon rockets. It is as amorphous and compact as an oil refinery, its individual components seeming to have been assembled in somewhat haphazard fashion. The pile as a whole weighs 2,000 tons and is as tall as a nineteen-story building. Throughout the night it has been bathed in the white glare of numerous xenon lamps. In this unreal light, as the night was drawing to a close, preparations were made for filling the enormous fuel tank.

The fuel for the space shuttle's three main engines consists of 143,000 gallons of liquid oxygen, chilled to −298° F, together with 383,000 gallons of even colder (−423° F) liquid hydrogen. If the temperature of these two liquids were to rise so much as a fraction of a degree they would evaporate and escape through the safety valves by means of which, at regular intervals, the whole setup is enveloped in steam. And it is for this reason that the pumping in of what amounts to 800 tons of liquid fuel is not begun until four and a half hours before the launch, with topping up continuing until the last minute. For the same reason, the fuel tank is also fitted with a thick insulating lining, like a gigantic thermos flask.

NASA once calculated what would happen if this external tank were to explode on the launch pad. They estimated that it would unleash an explosive force equivalent to that of 3,000 tons of TNT, or a bomb a quarter the size of the one dropped on Hiroshima. The blast would hurl even quite large chunks of metal a distance of 5,000 feet and shatter windows 13,000 feet away. It goes without saying that everyone within a radius of 1,000 feet would be killed outright, including, of course, the astronauts who would shortly be taking their seats alongside it.

One of the space shuttle's many astronauts was once asked whether he felt any fear in the moments before liftoff. He replied that anyone who did not feel just a little apprehensive in such circumstances had no real idea of the volume of energy which, for just a few moments more, would be pent up in harmless chemical form.

It may look as though this dominant feature, the external tank, is actually supporting the space shuttle. But this tank, in turn, is firmly secured to two slender white columns — their feet bolted to the launch pad. And it is these which ensure the stability of the entire construction. Each of these snow-white boosters contains 500 tons of solid fuel, which will provide the space shuttle with by far the greatest share of its booster power during its first two minutes of flight. They are the largest solid rocket boosters ever built, the first designed for reuse and for sending men into space.

Five hours before launch time ("T minus 5 hours" in NASA-speak) the technicians board the space shuttle to check that everything is ready for the crew and that all switches are in the correct positions.

It is probably not fair to say that no one nowadays knows the names of the astronauts on the crew. Presumably, a good many people are familiar with the name of at least one crew member: Christa McAuliffe, a thirty-seven-year-old high school teacher from New Hampshire and the mother of two children. She has been chosen by NASA from among 11,000 teachers — at President Reagan's suggestion — to conduct three classes from outer space. She is meant to stand as proof of the government's increased commitment to the field of education and also — from NASA's point of view — to show that the space shuttle has the capability to transport ordinary citizens safely into space and back again. NASA, which prefers to give acronyms to everything that has to be taken seriously, calls this part of the mission TISP (Teacher-in-Space Program).

McAuliffe is just one of the seven astronauts who will — approximately two hours before the scheduled liftoff and after having breakfast together — walk the short distance from their transfer van to the tower elevators. Astronauts no longer eat their last meal on earth before rolling cameras, but there is still something of a nostalgic ritual about the trip to the launching pad. It is only 8:30 a.m. and everyone shivers slightly in the cold. At their head, Francis Scobee, appointed as leader of the mission. He is forty-seven years old and has 7,000 flying hours to his credit, including time in Vietnam. He already knows *Challenger*, from a space trip two years ago. At his side in the cockpit he will have forty-year-old Mike Smith, who has also flown missions in Vietnam. There is another woman on board, apart from McAuliffe — thirty-six-year-old Judith Resnik, a career astronaut and an electrical engineer. She is also very attractive and NASA has not neglected to exploit her photogenic potential. She too has made a previous space trip on *Challenger*.

NASA has not only tried to demonstrate its breadth of vision by including both sexes. Mission Specialist Ronald McNair, for instance, knows quite well that he has most likely been selected because he is black, and this, he feels, puts him under a particular obligation. He is thirty-five and a trained physicist. He too has traveled into space on *Challenger*, in 1984. Thirty-nine-year-old Ellison Onizuka is also well aware that he is representing a minority. He is the grandchild of Japanese immigrants, brought up in Hawaii. Onizuka flew with *Discovery* last year.

Bringing up the rear is Gregory Jarvis, a forty-one-year-old electrical engineer, who has been designated as "Payload Specialist" — a newly created title born of the space shuttle program. Jarvis' task is to represent Hughes Aircraft Company, which is already a major NASA client, and might well — with a little active customer service — become an even bigger supplier of communications satellites for this "truck."

The seven crew members wave to the reporters and smile — the rookies a little diffidently (for them, this is far from routine), the four veterans with more poise. All are dressed in comfortable pale blue boiler suits, contrasting sharply with the clumsy space suits that hampered the movements of earlier astronauts.

The temperature is still just below freezing and up in the tower the wind, though light, is bitingly cold. The view across the flat countryside is stunning, and high over the sea, gulls can quite clearly be seen and heard. All over the steel structure of the tower, icicles, large and small, glitter in the morning sunlight. The astronauts hop up and down a bit while those in front of them are helped into their seats. This is by no means the simplest of operations, with the space shuttle in its vertical position. The NASA people who lend a hand have saved a little surprise for the schoolteacher. She is presented with a shiny polished apple. Smiling, she asks them to look after it for her until she gets back down.

It takes about half an hour to settle all seven into their seats. Front left in the cockpit, in the traditional airline captain's seat, lies/sits Scobee, with Smith on his right in the copilot's seat. Behind these two, Resnik and Onizuka respectively. Through the cockpit windshield all four can see the ice-blue sky above them. Now and again they may catch a glimpse of a colleague in a little gnatlike T-38 plane, checking out the wind conditions higher up, on the first stretch of the route to space.

On a lower deck, huddled close together with no view of the outside world, are the two novices, McAuliffe and Jarvis, along with McNair. The hatch is screwed down, the cabin is pressurized, and the crew are alone. That is to say, as alone as anyone can be when you know that everything you say is being transmitted to different control centers the length and breadth of America.

In one of the executive offices at the Kennedy Space Center, the astronauts' families have assembled to watch a video of the morning's preparations and to follow the launching. The youngest children soon find their attention wandering and begin playing on the floor. The grown-ups tell them that they are going to go up onto the roof once *Challenger* is finally ready for takeoff.

Meanwhile the countdown continues as scheduled and the space shuttle's navigational instruments are given their precise position in the solar system. In front of the pilot, the launching pad coordinates are glowing and the instruments register the fact that the apparently motionless stack is, nevertheless, constantly moving eastward at a speed of 895 mph, or 1,300 feet per second, by virtue of the earth's rotation. The crew have been through this many times in the simulator, but without the sounds. Now the metal creaks and moans as it contracts and then gives on contact with the supercooled fuel.

Obviously the countdown was not designed to increase suspense. The countdown is a huge checklist that monitors the thousands of details vital to a safe launching. At particular moments the countdown can be interrupted and the clock stopped while minor irregularities are corrected. Many of these points are checked by computers which can inform the operators should anything be amiss. Other details require the active participation of the crew, so that the moments before liftoff are (perhaps also for psychological reasons) completely taken up with practicalities. The crew are now linked up to the preprogrammed computerized sequences. This gradually begins to affect their way of talking. The tense banter slowly dies away, to be replaced by curt, businesslike reports. Emotions are concealed behind — and controlled by — test pilot jargon.

SCOBEE: *Control, this is* Challenger. *Commander's voice check, over.*

CONTROL CENTER: *Roger, out.*

SMITH: *Control, this is the pilot; voice check, over.*
Roger, out.

Twenty minutes before liftoff, the programs for the launch are keyed into the four main computers. These computers have the job of steering the space shuttle upward, through the earth's atmosphere, along its optimum trajectory, thus saving the shuttle from exceeding its stress limits. Even quite minor deviations from this trajectory could rip the spaceship to smithereens. Meanwhile, the computers are constantly being supplied with fresh data from the 2,700 different sensor points and are primed to abort at the first sign of the slightest irregularity.

At T minus 9 minutes a scheduled ten-minute break is taken to allow for any eventual delays. When the clock is once more ticking backward from nine minutes, the computers in the control center start the program that will direct the launching. Beyond this point, decisions are too critical to be left to human beings to fumble with.

T minus 7 minutes. The crew access arm retracts from the space shuttle, which now truly stands alone. All personnel in the area take cover in bunkers.

T minus 5 minutes. The auxiliary power units (APUs), which supply the electricity and hydraulic power to the space shuttle's nozzles and control surfaces, are switched on. The crew check that the hydraulic pressure level is up to the mark. The space shuttle now switches to its own power supply. Yet another lifeline to the outside world is cut.

> *Challenger*, this is Control. You are on internal power, over.
> Roger, out.

The words "Roger," "Go," "Affirmative" echo around the control desks. The big bell-shaped nozzles for the main engines are maneuvered into the position they will occupy during the launch.

T minus 2:55. The valves to the oxygen tank, which — right up to the last minute — have seen to the replacing of any evaporated gas, are closed and the tank is pressurized in preparation for liftoff. A moment later the same process is repeated with the hydrogen tank. The fuel pressure now builds up in the tank and pipes. Scobie checks that the three crew members on the lower deck realize they now have only two minutes to go.

At many spots along the coast, motorists have now pulled over to the side of the road — car radios on — to follow the launch. The specially invited guests on the viewing platforms — among them a Chinese delegation — get to their feet. The astronauts' children and spouses have now come out into the cold on the roof of the office building. The little ones are lifted up.

> Ninety seconds and counting. The 51-L mission ready to go.

T minus 25 seconds. The space shuttle's own computers assume control of the launch sequence. The astronauts settle themselves more comfortably in their seats. From now on there is nothing they can do. At this point, on previous missions, astronauts have often been shown to have a pulse of 150. Everything is on automatic. They are in the hands of the computers (and NASA). Even the captain, Scobee, can now do nothing other than monitor the systems. Human intervention during liftoff would be far too slow, imprecise, and, hence, hazardous.

T minus 10 . . . 9 . . . 8 . . . 7 . . . Things are now happening at a furious pace. Eight seconds before liftoff an enormous water tank sheds its full load over the launch pad to deaden the reverberation of sound waves, which could damage the ship. At T minus 6 seconds, at intervals of 120 milliseconds, the computer gives the go-ahead for liftoff to the three SSMEs (space shuttle main engines). The valves are opened. Under high pressure, oxygen and hydrogen are fed into the engines along thick umbilical cords. Once there, they vaporize, are mixed and electrically ignited. Some of the gases drive powerful turbopumps fitted with heavy rotor blades which — within just four seconds — are spinning at up to 40,000 revolutions per minute. The pumps compress the subsequent intake of oxygen and hydrogen in such a way that the combustion here is more efficient than that of any previous rocket engine. The whole structure is still bolted to the launch pad, while the computers compare the performance of the engines. Even at this late stage the flight can still be terminated.

This time, however, there are no problems. "There they go, guys. Three at a hundred," reports Scobie over the intercom. All three engines are functioning perfectly. They are all generating maximum power — altogether, as much as five jumbo jets during takeoff. The thrust from the three engines tilts the whole pile almost three feet to one side.

At T = 0, 11:38 a.m., as the stack bounces back into an upright position, the two rocket engines in the noses of the two booster rockets are started up. Immediately, the flames from these engines ignite the solid fuel down the length of the hollow rocket cores. In less than half a second the booster rockets generate a pressure that puts the main shuttle engines totally in the shade. Twenty-five extra jumbo jets at a stroke. At maximum output — 50 million horsepower — the eight explosive bolts, which are now having a hard time anyway holding back the 2,000 tons of machinery, are discharged. Unlike the main shuttle engines, the booster rockets cannot be switched off once they are in operation. From now on *Challenger* has no choice.

For a moment the whole stack seems to hover, motionless, on the cushion of smoke and flame from the rocket jets. Then, hesitantly, it makes up its mind. Slowly and majestically it takes off and leaves the earth behind at ever-increasing speed. The upper section of the tower is cleared with little more than a hair-raising yard to spare. Forced back into his seat, Mike Smith calls out to the others, "Here we go!" Moments later he adds, almost tenderly, to *Challenger*, "Go, you mother!"

On the ground the applause is drowned out, after a slight delay, by the thunderous roar of the launching. In line with tradition, launch control is transferred to the center at Houston as soon as the space shuttle has cleared the tower.

The journalists' press folders are full of impressive figures. During every second of the first phase, ten tons of fuel are expended. The main shuttle engines are thirsty. Collectively, they consume 70,000 gallons of liquid fuel per minute; 1,150 gallons are pumped through every second.

After sixteen seconds, *Challenger* goes into a graceful roll, which brings it onto its back, while its course alters from the vertical to a more oblique curve directed toward its eventual orbit. At this point too the output from the main engines is reduced for a moment, quite according to plan, while a particularly turbulent atmospheric layer is negotiated at reduced engine power. Neither the crew nor the space shuttle should be subjected to more than 3 g. The booster rockets also seem to be burning less fiercely. The whole stack sways and shudders as the gyro motors make continual adjustments to the five rocket nozzles, in order to stay on course.

The space shuttle computers keep an eye on the ship's flight path, its bearings, acceleration, speed, and direction. The ground computers are fed a steady stream of 3,000 updated readings per second from the space shuttle. Most of these pass straight into the data bases, to serve as the basis for later analysis of the mission. This continuous flow of data far exceeds what human beings could ever take in at a glance.

The control center in Houston concentrates on studying the data on such vital aspects as engine performance and the trajectory through the atmosphere. At a height of just under 18,000 feet, *Challenger* breaks through the sound barrier. Thirty-five seconds later comes a report over the loudspeakers: "Velocity 2,257 feet per second [1,538 mph]. Altitude 4.3 nautical miles. Downrange distance, 3 nautical miles." The crew's attention is focused on the computer screens, which show their speed and position in relation to the course calculated to bring them up into their orbit.

The spaceship's progress is followed at all times by long-range cameras, for which the clear, frosty air provides the perfect conditions. Today—with the aid of extremely powerful telescopic lenses—there should be a chance of seeing the jettisoning of the booster rockets, in just over two minutes' time, once they have used up the last of their 1,000 tons of solid fuel.

Again Smith gives vent to his elation: "Feel that mother go! Woo-hoo!"

Seventy seconds into the flight, Houston informs *Challenger* that engine output is again being raised to maximum. Scobee acknowledges receipt of this message: "Roger, go at throttle up." Everything is still going as planned.

And then it happens. After seventy-three seconds in the air, *Challenger* disappears in an explosion on the television screen. Smith's "Uh-oh!" is the final radio signal to be received before all communication is cut off. At that same moment, the range safety officer at Cape Kennedy sees the luminous point on the long-range radar screen split up into a large number of small dots. The commentator at Houston is not yet aware that disaster has struck and is still concentrating, as he should, on the information coming up on his visual display unit (VDU). Altitude and velocity figures still read as expected. Three seconds after the explosion he reports, unperturbed, "One minute, fifteen seconds. Velocity 2,900 feet per second [1,977 mph]. Altitude 9 nautical miles. Downrange distance 7 nautical miles."

On the ground, there are many who think that the booster rockets have simply been jettisoned a little early. "Where in hell's the bird? Where is the bird?" one technician shouts at another. The loudspeakers crackle: "We have no downlink." Bill Graham, head of NASA, has set his watch to indicate the precise moment when the booster rockets are scheduled to be jettisoned. He stares in disbelief at the television screen in front of him and then back at his watch. This is definitely not meant to happen. After a very long pause, the Houston commentator is back on the air, assuring everyone that the situation is being looked into very closely. In the engineers' dry jargon he concludes: "Obviously, a major malfunction."

All of the computer screens are now showing nothing but long rows of *s*'s. "Static"—no data from *Challenger*. On the radar screen two dots can now be seen forging steadily onward while the others fan out like the cascade of stars from a Roman candle. Even now, twenty-five seconds after the explosion, the remains of the space shuttle keep heading upward, undaunted. The two booster rockets — now released from their heavy burden — carry on, out of control until, after 110 seconds of flight, a radio signal detonates them.

Cameras from a host of television networks have all zoomed in on the big distorted Y in the heavens. In the frosty air, the trail of the exhaust forms a macabre imprint of the explosion and of the launch as a whole. The two branches of the Y can be traced back to the swell of the explosion, and below that, *Challenger's* course — while it was still unwittingly heading for disaster — can be discerned.

At the control center in Houston all eyes are now glued to the television screens. A few cherish the hope that the space shuttle will heave into sight and execute a standard emergency landing on the Kennedy Space Center runway. NASA has an acronym for that too. "RTLS," some voices cry out in desperation, as though the very technical jargon itself might harbor some hope. "Return to launch site." But most know better. The spectators on the platform, among them Christa McAuliffe's stunned parents, are hustled onto buses which will carry them away from the area. The children are led down from the roof by shocked adults, with no explanation given.

At McAuliffe's high school in New Hampshire most of the students had gathered in the auditorium to view the launching. Many of them had put on party hats and brought along noisemakers. When *Challenger* lifted off from its tower the din was deafening. When the explosion occurred the cheering merely reached new heights, until, sensing that something was wrong, one of the teachers yelled, "Shut up, everyone!" The school was hastily cleared of reporters, and shocked teachers had to begin the task of attempting to comfort the hysterical students.

In the White House, President Reagan is interrupted at his work by Vice President Bush: "Sir, there's been a serious incident with the space shuttle." National Security Adviser John Poindexter, borrowing from NASA, adds, "A major malfunction." Finally, the Communications Director, Pat Buchanan, clears his throat and tells him straight. "Sir," he says, "the shuttle has exploded." After a while Reagan asks, "Is that the one the schoolteacher was on?"

In Florida, later that afternoon, the "firing room" is sealed, so that no one can erase vital evidence. Numb key figures are primed for a short, preliminary debriefing.

Across the world, teleprinter bells are ringing. American television stations interrupt their programs to bring special broadcasts. Prior to this, only a handful of them had taken any great interest in the 51-L mission. It would have featured as a tailpiece in the evening news broadcasts. Now the explosion is shown over and over again, burning itself into the national consciousness in slow-motion replay.

An hour later, small pieces of wreckage are still falling out of the heavens, far out over the sea, while all over the Kennedy Space Center big digital clocks quietly continue to count the minutes and seconds of a mission which has long since ceased to exist.[1]

2. Heroism and Noble Sacrifice

That same evening, President Reagan was due to give his annual State of the Union address. His first impulse was to go ahead with it as planned, but advisers persuaded him to postpone it. Late that afternoon, he appeared on television and made a short speech about "the *Challenger* Seven." Speaking directly to the bereaved, the President said:

> Your loved ones were daring and brave, and they had that special grace, that special spirit that says, "Give me a challenge and I will meet it with joy." They had a hunger to explore the universe and discover its truths. They wished to serve, and they did. They served all of us.

Reagan followed this with a special message for the country's schoolchildren:

> I know it is hard to understand, but sometimes painful things like this happen. It's all part of the process of exploration and discovery. It's all part of taking a chance and expanding man's horizons. The future doesn't belong to the fainthearted; it belongs to the brave. The *Challenger* crew were pulling us into the future, and we will continue to follow them.

Thus, Reagan established the theme that was to echo in the media channels as solace for a stunned world. Exploration craves its sacrifices, but these are martyrs, who die in the service of a higher cause.

For Reagan's closing line, speechwriter Peggy Noonan had unearthed a sonnet ("High Flight"), written by a young Canadian Spitfire pilot, John Gillespie Magee, shortly before he was killed, at the age of only nineteen, in a dogfight over England in December 1941:

The crew of the space shuttle *Challenger* honored us by the manner in which they lived their lives. We will never forget them, nor the last time we saw them, this morning, as they prepared for the journey and waved goodbye and "slipped the surly bonds of earth" to "touch the face of God."[1]

Former astronaut John Glenn, the first American to orbit the earth, made this comment, in a voice that clearly betrayed his emotional state:

We used to speculate, the first group of seven, how many of us would be alive after the program. We always knew there would be a day like this. We're dealing with speeds and powers and complexities we've never dealt with before. This was a day we wish we could kick back forever.[2]

The following day *The New York Times* was able to report that people throughout the country had burst into tears, quite spontaneously, on hearing of the accident. In supermarkets, courtrooms, and schools, people who had never met before sought comfort from one another. The telephone networks jammed up, with everyone trying to call friends and family to discuss the explosion.

Many of the people whom the newspaper had interviewed compared their feelings to the day President Kennedy was shot. The same feeling of instantaneous nationwide grief, numbness, and disbelief was aroused. The author Ann Beattie described her experience:

A friend said it best, in a phone call around noon. He said, "Not again." Everything affects me as a writer. I was sitting alone writing and it pulled the rug out from under me. When you are a writer, in some ways, you space out when you're working. After the phone call, I just spent most of the day looking out the window and thinking.[3]

Another author, David Leavitt, said:

What seems to have impressed me most about this is that there was such an association with this space shuttle and children. Children in Oakland, N.J., actually built a cardboard space shuttle to replicate the mission. It was such a fantasy — this wonderful teacher on board, and we all remember the great love which one has for a teacher in our past. In certain ways, it brought the shuttle to a new symbolic height. What I'm feeling most profoundly, if it doesn't sound too hifalutin, is that there's a certain loss of innocence associated with it. It wasn't just any space shuttle. It's the one that became the children's crusade, a journey the children were taking into outer space.[4]

In Houston, contractor Jack W. Johnson watched the explosion over and over again on his television, switching from channel to channel. Right after the accident occurred he had sent his fifteen employees home. After a while he found he had had enough of sitting at home. He had to get into his car and drive out to NASA's Johnson Space Center. He told *The New York Times*:

> I just had to come here, I've lived in Houston for twenty years and I'm ashamed to say I've never been here. It did something to me. It was like a member of my family had been lost.
>
> People I've talked to today compare it to the day Kennedy was shot.[5]

Chuck Yeager, the celebrated test pilot, did not manage to soar to quite such heights in his reaction. To him the space shuttle was an experimental aircraft like so many others:

> I probably look at it a little different from someone who's never seen an accident such as this. I've seen accidents, a lot of accidents . . . but of course the shuttle is unusual because it receives a tremendous amount of exposure.
>
> It won't cause any great long-term effect on the space program. I see a need to look at all the information we have, telemetry data, photo coverage and the documentation of the vehicle and then find out what caused the accident and take corrective action and then press on. It's already happened, so there's nothing you can do about it. The object is to prevent it from happening again.[6]

In the Soviet Union the official reaction was restrained. After the announcement of the explosion on the radio, records by Glenn Miller and other American artists were played. This is 1986, and we are still in the days before the thaw in East-West relations; nevertheless, Gorbachev declared that he understood and shared the grief of the American people. As a gesture, the Russians intended to name two craters recently discovered by Soviet space probes on the planet Venus after the *Challenger* women, Resnik and McAuliffe. Soviet cosmonauts sent telegrams openly expressing their sympathy for their colleagues at NASA.

The Russian messages of condolence did not, however, fool the extreme right-wing American senator Jesse Helms. He was convinced that what they were dealing with was a Communist assassination attempt, and insisted that the whole affair be thoroughly investigated. Like an echo from the East came the voices of the Soviet hawks, linking the accident to the Star Wars program and the reckless impetuosity of capitalist society.

From the Vatican, Pope John Paul issued a statement in which he said that the accident had "provoked deep sorrow in my soul." The United States' loyal supporter British Prime Minister Margaret Thatcher struck a more patriotic chord: "New knowledge sometimes demands sacrifices of the bravest and the best," she said. "I just felt we saw the spirit of America and the spirit of the American people."

On Friday, three days after the accident, a memorial service was held at the Johnson Space Center in Houston, where the astronauts live and are trained. Six thousand NASA employees took part, along with congressmen and relatives of the dead astronauts. Earlier, the President and Nancy Reagan had individual meetings with the bereaved families, at which they were photographed attempting to comfort the astronauts' children. Afterward, Reagan made a brief speech outside, in which he again tried to express the feelings of the nation:

> The sacrifice of your loved ones has stirred the soul of our nation, and, through the pain, our hearts have been opened to a profound truth: The future is not free; the story of all human progress is one of a struggle against all odds. We learned again that this America was built on heroism and noble sacrifice. It was built by men and women like our seven star voyagers, who answered a call beyond duty.[7]

Over the heads of the crowd, at the close of the service, flew four T-38s, the astronauts' training aircraft. A perfect arrow formation would have required five planes, but on one side a plane had purposely been left out. In the Air Force this maneuver goes by the name of Missing Man, the formation for a lost pilot. Finally, one plane veered sharply away from the others and headed off alone toward the horizon — an aeronautic counterpart to the cortege in which a lone, riderless horse followed President Kennedy's hearse.

. . .

It was a week filled with great — and perhaps also pathetic — words, but there is no doubt that a great many people were deeply affected by the accident. More deeply than they might have anticipated. American citizens regarded the space missions as being closely bound up with national prestige, technological superiority, and admiration for the NASA engineers and scientists who had made the space adventure possible. The astronauts were still looked upon as heroes, even if not to the same extent as Glenn. The very fact of their having come down to earth brought them closer to ordinary people. It was not difficult to identify with McAuliffe and Jarvis. Both could easily have been living down the street.

Many cities organized special tributes. In New York, on Wall Street, the Stock Exchange observed a minute's silence. And the spotlights which illuminate the Empire State Building at night were switched off. The Ford Motor Company lost no time in canceling an advertising campaign in which their new Aerostar model was pictured in front of the space shuttle.

A week later, the major newsmagazines assessed the disaster in a somewhat wider context. In *Time*, one of the magazine's very best wordsmiths, Lance Morrow, set himself the task of finding words for the grief that had seized America. The recurring theme in his column is the shared sense of grief. Uncontaminated pain.

> [The *Challenger* disaster] inflicted upon Americans the purest pain that they have collectively felt in years. It was a pain uncontaminated by the anger and hatred and hungering for revenge that come in the aftermath of terrorist killings, for example. It was pain uncomplicated by the divisions, political, racial, moral, that usually beset American tragedies (Vietnam and Watergate to name two). The shuttle crew, spectacularly democratic (male, female, black, white, Japanese American, Catholic, Jewish, Protestant), was the best of us, Americans thought, doing the best of things Americans do. The mission seemed symbolically immaculate, the farthest reach of a perfectly American ambition to cross frontiers. And it simply vanished in the air.[8]

The trade periodical *Aviation Week and Space Technology* said in an editorial that the vital thing now was not to let paralysis set in:

> The measure of our resolve will be demonstrated by a decision to press on, to take corrective action and to resume the quest, in the same way explorers who have gone before were undeterred by serious setbacks.

The magazine expressed full support for NASA:

> The scientists, engineers, planners and technicians who form the NASA/industry team that made the marvel of the space shuttle a reality must be credited with striving to the limits of human endeavor to make the system as safe as possible. In the soul searching and second-guessing and calculated examination of data that will ensue in the weeks and months ahead, the basic premise has to be that the thousands of men and women who helped build and launch the shuttle gave it their best effort.[9]

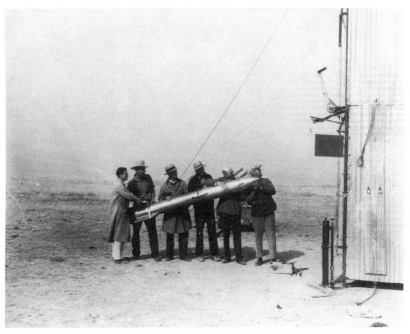

ROBERT H. GODDARD (SECOND FROM RIGHT) WITH TRIAL ROCKET IN 1932. (NASA)

3. *Some Useful Groundwork*

In point of fact, Isaac Newton said everything worth saying about rockets back in 1687, in his three laws of motion:

> First law:
> Every body perseveres in its state of rest, or of uniform motion in a right line, unless it is compelled to change that state by forces impressed thereon.

> Second law:
> The alteration of motion is ever proportional to the motive force impressed; and is made in the direction of the right line in which that force is impressed.

> Third law:
> To every action there is always opposed an equal reaction: or the mutual actions of two bodies upon each other are always equal, and directed to contrary parts.[1]

A firework sits at rest in its bottle. If we want it to move, it must be forced to change its state; a force must be impressed (first law). If we exert a lot of force on it, the "alteration of motion," or the firework's acceleration, will be significantly greater than if we were only to exert a little force (second law). Then again, the force might be so slight that the firework feels in no way compelled to abandon its state of rest.

The gunpowder inside the firework is what provides the force. It too will lie dormant and totally inert in its cardboard cylinder — unless, and until, we ignite it, in which case it will release its chemical energy and send hot gases rushing out of the base of the firework at high speed; and then the miracle occurs: as a reaction to this great speed, the firework is propelled in the opposite direction (third law).

The firework will shoot into the air, the gunpowder will burn out —

but that will not stop the firework. In accordance with the first law, it will carry on at its new speed, by virtue of its inertia, until affected by other forces. In this case, a combination of air resistance and the earth's gravity will soon curb its skyward flight and — by dint of the second law — relentlessly inflect its course in the direction of the earth's surface.

The third law is probably the strangest of the three, but it is one well known to all of us. If a child lets go of a balloon without tying a knot at the opening of the balloon, the balloon will whiz around uncontrollably for as long as air keeps rushing out of it. Similarly, if we let the shower-head fall into the bathtub, we run the risk of a thoroughly arbitrary soaking. Or, again, we are all familiar with the sight of a DC-9 aircraft sitting at the end of the runway with shimmering currents of hot air issuing from the two jet engines at its tail. Due to the effect of gravity on the huge aircraft, these jets are not yet powerful enough to shift its mass (first law). Only when the pilot supplies the engines with more fuel, by pushing the throttles forward, is the force of the jet streams increased, thus causing the aircraft to vibrate and strain at the leash (second law). Shortly after he releases the brakes, the plane — as a reaction to its exhaust — moves down the runway and takes off (third law). Were both of the aircraft's engines to shut off during its flight, the jet streams at its rear would peter out and — since those forces which gravity and air resistance have, at all times, been exerting on the plane are no longer being counterbalanced by the jet streams from the tail — the pilot will soon have to start looking around for a suitable landing site.

But as long as there is enough fuel for the engines, the plane can climb higher and higher, driven by its exhaust jets, until the air becomes "too thin." Hence the fact that standard passenger flights normally maintain an altitude of around 30,000 feet. This might lead one to imagine that the jet streams at the rear of the DC-9 need air against which to push in order to move the aircraft forward. But this is not the case. The aircraft moves forward because the airstreams from the jet engines are moving backward. It is as simple as that. Nor did Newton stipulate any more than this in his third law. The only reason the plane has to keep below a certain height is that its jet engines need the oxygen in the air for combustion. As the oxygen content diminishes with the altitude, the jet plane will run into difficulties. But if we brought along our own oxygen to feed the combustion, the altitude would no longer be a stumbling block. Then we would have an aircraft with, not jet engines, but rocket engines.

The same applies to the firework. If it were capable of reaching greater heights, it too would be faced with the problem of obtaining oxygen for the combustion of its gunpowder. It would also have to bring along its own oxygen, in some form, if it were to reach greater heights.

If it flew high enough, it would reach the ideal world of classical physics — the total vacuum — in which it would carry on, unimpeded, in uniform motion, affected only by the ever-weaker pull of the earth's gravitational field.

The propulsive force of both jet engines and rockets is traditionally measured in *pounds-force*. On takeoff, a jumbo jet produces approximately 200,000 pounds force to lift its 370 tons off the ground — more than half of what an Atlas rocket can produce when it gets going. When a rocket makes its stately takeoff from the launching pad, it might seem to be driving itself upward on its pillar of fire — as though it needs to push off from the reinforced concrete beneath it. But this is not correct. The rocket needs only a sufficient volume of gas particles to be thrust out of the rocket jets at its foot at a sufficiently high speed so that its mass can be compelled into changing its inert state and be sent, at altered speed, in the opposite direction from the gas particles. Preferably upward. Paradoxically, the rocket engines function better "without impetus" in the vacuum of outer space, where there are no air molecules to offer resistance to the exhaust gases; thus they are allowed to flow unhindered from the rocket jets and supply the rocket with an opposite directional force approximately 20 percent greater than that afforded on the earth's surface.[2]

The point about direction is not quite as simple as it sounds. A firework uses its long stick to shift its center of gravity so far back that there is a good chance of its shooting upward, but even this does not work every time. And we can certainly never be sure of exactly where our firework will land. Moreover, if the stick should slip off, the firework becomes totally unpredictable, just as the runaway balloon and the showerhead were. So the stick — or the low center of gravity — is not enough if we want to have any hope of steering the firework.

By looking at a walking stick we can tell that, with a little practice, it would be possible to balance the stick on one finger. But this requires constant minute and speedy adjustments of the finger, and sometimes even a bit of fast footwork, if the stick is to be kept in balance for more than a moment. In the same way, a long rocket "sits" on its rocket jets, which should, ideally, be capable of making adjustments in the direction of the exhaust if stability is to be maintained and the rocket steered with any degree of certainty.

The optimum rocket is, therefore, dirigible (self-correcting) and produces the greatest possible volume of exhaust gases at high speed. Conversely, it will have the least possible mass tied up in fuel and rocket
cylinders, allowing maximum acceleration to be achieved by the rocket's
combustion. Which is why multistage rockets are often employed. The
first and most dynamic stage accelerates the whole structure to a given
speed. The burned-out first stage is then jettisoned, leaving the second
stage with a greater initial velocity on which to work and less dead weight
to accelerate. With the third stage, the process is repeated with an even
smaller mass and with an even greater initial velocity.

The oldest known rockets can best be compared to fireworks, with a
gunpowder charge and a stick. The Chinese were familiar with them
more than 1,500 years ago. And in 1807 Copenhagen had the doubtful
honor of being the victim of history's first massive rocket attack when
the English bombarded the Danish capital with a hail of ten-foot-long
incendiary rockets. This method was developed by William Congreve,
whose efforts earned him a peerage.[3]

Using rockets for the drive into outer space was a concept that was
bound to bear fruit sooner or later, once Newton had established the
basic premises. But for a long time, authors such as Jules Verne and
H. G. Wells were alone in formulating their visions with any degree of
clarity. In 1903, however, a Russian schoolteacher, Konstantin Tsiolkovsky, published a book entitled *An Exploration of Cosmic Space by Means
of Reaction Devices*, in which he summed up his theoretical deliberations on space travel. He described weightlessness; devised formulas
for the volume of fuel required by a rocket of a given mass; recognized
the significance of the speed of the exhaust gases and, thus, realized the
inadequacy of gunpowder as a fuel. He predicted satellites and multistage
rockets and described liquid fuel — especially liquid oxygen and hydrogen — as better suited to breaking beyond the earth's gravitational field.
Tsiolkovsky also gave some thought to adjustable vanes mounted inside
the rocket exhaust, thus making steering of the rocket possible. Like his
colleagues in other countries, however, the Russian was given very little
support and was regarded as something of a dreamer.[4]

The American Robert H. Goddard, who was twenty-five years younger
than Tsiolkovsky, put his similar theories to the test in practical experiments with rockets. In 1919 he wrote A *Method of Reaching Extreme
Altitudes*, in which even at this early date he looked into the future,
outlining rocket flights to the moon, to photograph its far side. This

prompted *The New York Times* to retort contemptuously that Goddard evidently did not realize that a rocket was dependent upon "something better than a vacuum against which to react."[5]

On March 16, 1926, on his Aunt Effie's farm in Auburn, Massachusetts, Goddard launched the first rocket ever driven by liquid fuel. He did not, however, use liquid hydrogen, since the extremely low storage temperature it required put it beyond the bounds of practicality. Instead, he combined liquid oxygen with gasoline.

The rocket jet was hung from a slender stand built from steel tubing. To lower the center of gravity, the two fuel tanks were hung at the bottom of the stand. The hot exhaust gases would, therefore, stream down over the fuel tanks. Being a rocket pioneer was thus not without its risks; in fact, almost every rocket enthusiast was, at some point in his career, badly burned by explosions. Goddard had asked his wife to capture the experiment on film, and she did start the camera rolling as he ignited the fuel mixture with a blowtorch; but after seven seconds the film ran out, with the rocket jets still spluttering, not yet having generated enough power to send up the rocket. This had all the makings of a dud. It took a full twenty seconds for the "impressed force" from the rocket exhaust to reach the point where it was great enough to propel the rocket skyward. The rocket shot forty feet into the air, flew for two and a half seconds, and landed 168 feet away.[6] The experiment was a success. Goddard calculated its speed at approximately 60 mph. So liquid fuel did work in practice; it *was* more powerful than solid fuel — and not only in theory.

In the years that followed, Goddard continued his experiments, despite receiving only scant government support. He eventually held over two hundred patents, pertinent to all areas of future space research. His inventions included turbine-driven fuel pumps, which enabled the rocket's combustion to produce a greater thrust; he worked on self-cooling engines and gyroscopes which were connected to vanes inside the exhaust, making it possible to stabilize the rocket. After his death in 1945, the American government, which had not given him anything like his due while he was alive, had to pay his widow one million dollars in full and final settlement of all rights of patent.[7]

The third father of space research was the Romanian-born, German-speaking Hermann Oberth, who knew nothing of Goddard until 1922 and first heard of Tsiolkovsky's work in 1925. In 1923, Oberth published a book entitled *Die Rakete zu den Planetenräumen*, in which he presented a popular version of his theories and results within the field of rocket research.

1. At the stage now reached in science and technology it is possible to build machines that can climb beyond the limits of the earth's atmosphere.

2. The further perfecting of these machines will enable them to achieve such high speeds that — left to their own devices, out in the ether — they would not need to fall back down to earth and, what is more, would be capable of breaking free of the earth's force of attraction.

3. This type of machine can be built in such a way that men could (in all probability without any damage to their health) also take part in these ascents.[8]

This book was read by Wernher von Braun, at that time just thirteen years old, and in 1927, when Oberth formed the Verein für Raumschiffahrt and gathered together a band of amateur rocket enthusiasts at the so-called Raketenflugplatz outside of Berlin, von Braun was one of the first to join him.

The Germans

There are some old photographs from May 1945 of the young Wernher von Braun. He has just surrendered to American troops at Schattwald near the German-Austrian border and yet, even in this, Germany's hour of defeat, his smile is bright and self-assured. His left arm — in a cast, following a car accident — has set in an odd, crooked "Heil Hitler" salute; he is listening attentively to the American officers, like an obliging host faced with unexpected guests.

Everyone who met him soon abandoned their instinctive wartime distrust and found him charming and sincere. No one doubted his superior intelligence. Most did, however, find it hard to believe that this thirty-three-year-old man had sixteen years' experience working with rockets and was the person chiefly responsible for the technical development of the dreaded V-2 missile.

Von Braun's superior, Major General Walther Dornberger, found it considerably more difficult to adapt to the altered state of affairs. When an American soldier asked him a question related to the capitulation, he barked at those around him: *"Der Idiot soll Deutsch sprechen, wenn er was von mir wissen will!"*[1] A couple of days later, a member of the Allies' kitchen staff, a Pole, was prevented, just in the nick of time, from doing away with the sleeping German notables. Evidently he too found it hard to adapt to the new order.

But SS-*Sturmbannführer* Wernher von Braun was more flexible. During those first few days, he drew up a preliminary paper with the imposing German title of *"Übersicht über die bisherige Entwicklung der Flüssigkeitsrakete in Deutschland und deren Zukunftsaussichten."* In this paper, he states modestly that "the stratospheric A-4 rocket" — which he does, however, admit to be better known to the general public by the name *Vergeltungswaffe-2* (V-2) — should be regarded as no more than an imperfect harbinger of future possibilities.

> We are convinced that a complete mastery of the art of rockets will change conditions in the world in much the same way as did the mastery of aeronautics, and that this change will apply both to the civilian and military aspects of their use.

He does not, however, gloss over the costs involved.

> We know, on the other hand, from our past experience, that a complete mastery of the art is only possible if large sums of money are expended on its development, and that setbacks and sacrifices will occur, such as was the case in the development of the aircraft.[2]

In the same paper, Braun also predicts flights to the moon and to other planets.

By this time the Americans were beginning to realize what had fallen into their hands—four hundred German rocket experts on a silver platter. In next to no time, Allied intelligence agents were swarming around the top brass. Time and again—to their obvious annoyance—the Germans were questioned about the same elementary facts. They were particularly disappointed by the childish level at which discussion had to be conducted, and very tired of all the repetition. They had expected a more professional reception.

Furthermore, it soon became obvious that the Germans by no means considered their surrender to be unconditional. They saw it more as a prelude to negotiations between parties of equal standing. On behalf of all his colleagues, von Braun put forward his proposal of a three-year contract, as proof that the Americans' intentions were honorable, and that they were not simply going to take all they could get and then drop the Germans. Hence the extreme care taken by the Germans not to offer the Americans any tidbits too early on. They could always be put in the picture once conditions had been agreed upon.

The Americans were not, however, the only belligerent power on the Allied side. Other countries were welcome to submit their bids. But the Germans were in no doubt as to the order of precedence. As one of them said to an American interrogation chief: "We despise the French; we are mortally afraid of the Soviets; we do not believe the British can afford us, so that leaves the Americans."[3]

It was not only that the British could not afford them; they were not as open and straightforward in their dealings with the Germans as the

Americans were. Rather, their treatment of them was cool, formal, and correct. The fact that V-2 missiles had left 2,770 Britons dead and 21,000 wounded may well have had something to do with this.[4] Thus a German rocket expert might have had a hard time settling down in Britain.

Nor was there any need to, since, as early as the end of May, the Allied commander-in-chief, General Eisenhower, sent a telegram to Washington in which he advised that he had in his custody four hundred prominent scientists from the Peenemünde rocket research station, who were "anxious to carry on research in whatever country will give them the opportunity, preferably U.S., second England, third France." Eisenhower went on to say: "The thinking of the scientific directors of this group is twenty-five years ahead of the United States." He therefore recommended that the top one hundred scientists be transferred, together with all available drawings and a number of V-2 rockets, complete with attendant equipment—so that they would have some hardware to start with. Eisenhower, who was well acquainted with Washington bureaucracy, urged that no time be lost in making a commitment, saying: "Immediate action is recommended to prevent loss of whole or part of this group to other interested agencies." In closing, he also requested that America's attitude toward other German experts be reconsidered: "Suggest policy and procedure be established for evacuation of other German personnel whose future scientific importance outweighs their present war guilt."[5]

Actually, interrogation chiefs were hard put to detect any particular sense of guilt among the German scientists. On the contrary, they often encountered a marked pride in their technical achievements. To these Germans, the war had not been a nightmarish experience. Rather, it had presented them with a project close to their hearts, and provided them with ample resources for meeting rocket technology's challenges. For them, this far outweighed Hitler and Nazism. The V-2 was not primarily a weapon of mass destruction. This missile was, first and foremost, a forty-seven-foot-long rocket weighing thirteen tons which, with the aid of nine tons of liquid oxygen and alcohol, could reach a height of 260,000 feet and a speed of one mile per second. By means of a gyro platform, registering velocity, acceleration, and direction, it could control the vanes in its exhaust, thus making it possible to hit a target two hundred miles away, with an accuracy of within thirteen miles—a far cry from the 1930s and the amateurish rockets of Hermann Oberth's space society. The Army had stepped in and given Oberth and von Braun whatever they asked for.

In an unpublished article from 1972 von Braum offers what Michael

J. Neufeld characterizes as a "depressingly frank statement" regarding the attitude common among the engineers and scientists:

> Our feeling toward the Army resembled those of the early aviation pioneers, who, in most countries, tried to milk the military purse for their own ends and who felt little moral scruples as to the possible future use of their brainchild. The issue of these discussions was merely how the golden cow could be milked most successfully.[6]

No one else had ever managed to produce and control a rocket of this size — *that* was their true achievement. To get this far, a succession of seemingly insurmountable obstacles had had to be swept aside through intensive scientific and technological teamwork. For a long time their problem had been simply to keep the rockets going long enough to have any chance of discovering what went wrong soon afterward. Apart from the mangled wreckage of the rocket, they had recourse to just four separate radio readings from it — whereas the space shuttle, many years later, would need several thousand if a launch was to succeed. So it was necessary to fire off a great many rockets in order to solve even minor problems. Twenty rockets had to be sent up just to get the fuel valves to close properly — but then that problem *was* solved and they could move on to the next. This process went on for two years and involved two hundred trial shots.

During World War II these rocket enthusiasts had their own red-letter days. There was June 13, 1942 — the day on which the rocket was to be sent up for the first time — when personnel clustered around the launch ramp, keeping their various distances from it, depending on how much confidence they had in the venture or, possibly, on how much they knew of the risks involved. But on that occasion all went well. The rocket did go up; although, sadly, it came down in the Baltic only 3,000 feet away. So perhaps it would be fairer to say that the really big day was October 3 of that same year. Then, suddenly, everything fell into place; the rocket reached an altitude of 280,000 feet and, in just 312 seconds, covered a full 120 miles.[7] And then there was June 1944, when, for the first time, they were ready to fire on British targets.

One of the German scientists, Dr. Paul Figge, has described the enthusiastic mood within the group, even when things looked blackest for Germany, during the massive bombing raids on Berlin. According to him, their motto was: "Enjoy the war — the peacetime will be terrible!"[8] The mood was not altogether unlike that prevailing, at that same time, in Los Alamos, New Mexico, where the race was on to finish building the first atomic bomb. In both places, scientists were given all the financial support they needed; they were in daily contact with the leading authorities in their research fields and their days were completely taken up by solving scientific and technological problems, the tougher the better.

In both places the feverish activity of the scientists and engineers also lay at the heart of a massive technological network that had been set up in record time. Parallel to the work of the scientists, military experts had had to devise complex networks and systems for the procuring of adequate supplies and specially developed equipment. In both places, huge industrial plants scattered across the country were a prerequisite for their work and an integral part of it — although only a few of the top German scientists ever set foot in the underground plants where handpicked, technically adept prisoners from the concentration camps worked themselves to death on the rocket production line, once the V-2 missile had been perfected.

Now, however, the electric hustle and bustle of wartime had been superseded by a singular calm. A certain amount of exasperation was unavoidable. The Germans had a commodity for sale — know-how. So they did not take kindly to the blend of suspicion, contempt, and awe evinced by prospective buyers. One manifestation of this attitude occurred in the autumn of 1945, when the British asked for a demonstration of the V-2 launchings, wanting to document the procedure before the technique could be forgotten. They had "been looking forward, at long last, to studying the other end of the rocket's trajectory," as one of them remarked. To this end, two teams of German experts, together with their entourage of other ranks, assembled at Cuxhaven. The Germans were the master craftsmen, the English the apprentices; but in this case, every evening when the day's work was done, it was the apprentices who put their masters back behind lock and key.

The British also tried keeping the Germans on reduced food rations. This prompted an immediate strike and no time was lost in bringing the German rations back up to the same level as those of the British; a level which, however, the pampered Germans found to be considerably below the American standard. The victors also tried to impose a strict prohibition on fraternizing with the enemy, although this was not easy to enforce. Especially not with every rocket needing approximately four tons of alcohol and a fair number of the personnel believing that it would be quite absurd to shoot all of it sky-high. At any rate, tales are also told of some rowdier episodes, when the Germans tried their best to teach the British soldiers the words to *"Wir fahren gegen England."*[9]

But there must have been some alcohol left over for the rockets. At any rate, three were launched at the beginning of October, under the direction of Kurt Debus, who was later to assume overall technical responsibility for the launching of all manned American space flights prior to the space shuttle. All three functioned perfectly during launching and came down a little under five minutes later, in the North Sea, forty miles south of the Danish town of Ringköbing. The Germans had not lost their touch. The aforementioned Major General Walther Dornberger took part in the preliminary trials at Cuxhaven, before being moved by the British to London, not so much to draw on his expertise as to intern him for two years — his punishment for developing the V-2 missile system.

It was just at this time that the Americans transferred around one hundred rocket experts — with von Braun at their head — to the United States. Not, in this case, as prisoners of war but as salaried staff under contract to the United States government. Von Braun had not, however, managed to secure a three-year contract. Instead he had to content himself with six months, and the possibility of extending this by a further six. The Army was eager to tap the Germans' know-how but had no wish to enter into any kind of long-term commitment. Von Braun himself had to accept a salary of $750 per month, while every one else was paid at a lower rate. Several of the men had asked to bring their families with them, a request which was turned down flat by the Americans, partly out of regard for public feeling back home in the United States. After a bit of haggling, their families were billeted in a special camp set up in a barracks in Germany, where they received substantially better treatment than everyone else in that war-torn country — to the great disgust of local Germans.[10]

The Russians had not exactly been idle either. Both the former rocket station at Peenemünde on the Baltic coast and the V-2 factories at Nordhausen in the heart of Germany — where about 10,000 prisoners from the concentration camps had worked themselves to death as slave labor[11] — both now lay in the Russian zone. Quite a few rocket engineers of varying rank fell into Russian hands and were immediately presented with tempting offers from east of the Urals. With, moreover — at least to begin with — a higher salary than the Americans offered. Later, once they were caught in the trap, conditions grew tougher. The Russians had no illusions as to their postwar relationship with their former allies, and

American soil lay well beyond the range of their bombers. Rockets were most definitely a possibility that had to be explored. Nor did they have any illusions about "their" Germans. The latter were badgered and brow-beaten; they met with nothing but suspicion on all sides; and they were forced to instruct one team of Russians after another. If the Germans achieved results in the laboratories, these were immediately taken away and handed over to the Russian experts, led by Koroliev, their chief designer, for further development by the Russians themselves.[12] All in all, the Russians did not have much of a feel for the exceptionally creative work ethic and scientific teamwork which the enforced war work had produced both in Los Alamos and in Peenemünde.

The "American Germans" found themselves in the desert, at White Sands, New Mexico, not far from the spot where the first atomic bomb had been tested earlier that year. With them they brought the makings of a hundred V-2 rockets, which were to form the foundation of the American rocket program. But suddenly the Americans, unlike the Russians, seemed to be in no great hurry. They had ample access to bases within comfortable flying distance of the Soviet Union, if it should come to the crunch. And, also, there was a certain reluctance toward giving the Germans the impression that they were smarter than the American experts. So many a disgruntled letter was dispatched to the barracks in Germany. Von Braun described their status as that, not of prisoners of war, but "Prisoners of Peace."[13] It also came as a shock to him to see how stingy the Americans were with funds for rocket research during the immediate postwar period, at a time when the American public was crying out for cuts in military spending.

Little by little, however, the Germans found their bearings — even if the first "American" V-2, unveiled in 1946, did turn out to be a bit of a dud. To an ever-increasing extent, American industry was borrowing the services of German specialists, for shorter or longer spells. One of their main tasks was to advise against following procedures which had already been tried out and proven to be a waste of time, then steer research into more productive channels. There was no talk now of six-month contracts. After two years, just as soon as this option was open to them, many of the Germans sent for their families.

The language, too, had to be learned. Some set to work with charac-teristic thoroughness, spending entire Sundays in the cinema, in an effort to get a feel for the American vernacular. Unfortunately, when they tried to use what they had heard, it often came out sounding rather more hard-boiled than intended. Others struggled on alone, with their dictionaries. Often, they would draft their scientific and technical papers in German and then translate them into English word for word. This practice prompted one American to marvel at the fact that, in German, it seemed to be quite acceptable to say "My old grandmother a red cow with a bell round its neck has"[14] — and, at least to begin with, this was how they wrote and spoke English. The standard cliché of the fiendish scientist, advancing ominous theories in a thick German accent — a character brought most unforgettably to life by Peter Sellers in *Dr. Strangelove* — can be traced to those days.

The Germans' rigid organizational structure never ceased to amaze the Americans. Everyone had a title, which they guarded jealously, and respect for one's superior was the order of the day. Hard-and-fast chains of command were unavoidable and the clearly defined procedures which were laid down allowed for no deviations whatsoever.

In all, in the late 1940s, sixty-four V-2 rockets were sent up from White Sands, and a start was made on more daring experiments, with a newly developed second stage, which could be mounted on top of the V-2 rocket, thus increasing its range and speed.[15] Soon the desert would no longer be large enough to contain all of the crash landings. On one occasion a runaway missile came down in Mexico and caused a diplomatic incident. So, more and more often in the early 1950s, the launchings were transferred to the Atlantic Test Range at Cape Canaveral in Florida, where the scientists had the whole of the Atlantic in which to practice. According to the writer Tom Wolfe, the Germans thrived in the pioneer atmosphere of Florida, and after some successful launches, they permitted themselves the occasional well-earned break, let their hair down, brought out the *Glühwein*, and roared out lusty renderings of songs from the old days.[16]

Around the same time, the nucleus of the German contingent sallied forth from the desert to install themselves in Huntsville, Alabama, where — for the very first time — they truly felt at home. Some of the scientists had already been persuaded to join American military-industrial firms, but von Braun managed to keep a good number of the original team together and, at his recommendation, these were given jobs at the American Army's Redstone arsenal in Huntsville. The "everything-under-one-roof" approach from Peenemünde with its concentration of talented engineers and scientists proved extremely compatible with the U.S. Army's "arsenal system" of in-house research and development.[17] In Huntsville they settled in droves, in one particular neighborhood, and before long they were taking American citizenship and going on weekend excursions clad in lederhosen. Hermann Oberth and Major General Dornberger also joined their countrymen in the United States.

For the next twenty years, Wernher von Braun's world would revolve around Huntsville. In time, he, more than anyone else, would come to personify the American space program for the man in the street. Working from here, and plying regularly among Cape Canaveral, Washington, Houston, and Huntsville he — together with the men from Peenemünde — was to develop the first American ballistic missile, the rocket for the first American satellite and for the first American in space, as well as for the first American moon probe. And, at the pinnacle of his career, he was also to develop the rocket to beat all rockets, the Saturn V — tall as a thirty-six-story building — which gave a faultless performance on its trial run, filled the hearts of every American with pride, and in 1969 fulfilled its promise by sending the first Americans to the moon.

By then the Redstone arsenal had become NASA's Marshall Space Flight Center and created almost 40,000 jobs in the area.[18] The Atlantic Test Range had been renamed the Kennedy Space Center and there was not much pioneer spirit remaining in that high-tech, futuristic landscape.

5. The Russians Are Coming!

Every gun that is made, every warship launched, every rocket fired signifies, in the final sense, a theft from those who hunger and are not fed, those who are cold and are not clothed.

This world in arms is not spending money alone. It is spending the sweat of its laborers, the genius of its scientists, the hopes of its children.

The cost of one modern heavy bomber is this: a modern brick school in more than 30 cities.

It is two electrical power plants, each serving a town of 60,000 population.

It is two fine, fully equipped hospitals.

It is some 50 miles of concrete highway.

We pay for a single fighter plane with a half a million bushels of wheat.

We pay for a single destroyer with new homes that could have housed more than 8,000 people.[1]

These words do not come from some dyed-in-the-wool pacifist, nor yet from some Communist henchman, but from the President of the United States, Dwight D. Eisenhower, expressing his concern in a speech made in 1953, just a few months after his inauguration. No one knew better than Eisenhower that a strong military force was a necessity at a time when the superpowers harbored such mistrust of one another. But, to his mind, the development of new military hardware should never be an end in itself. The sole purpose of armaments was to safeguard fundamental American values — while spending as little as possible of the American people's hard-earned cash.

So it was vital that the President keep an extremely open mind when the arms industry or his own former colleagues on the general staff tried to sell him new wonder weapons. He was continually having to evaluate whether these would contribute to national security or merely enhance a sense of superiority which was already well enough developed. And all of these diverse projects also had to be tailored to meet the threat facing the United States. So reliable information from inside the sealed-off bounds of the Soviet Union would be of vital importance.

Eisenhower's predecessor, Harry Truman, had already made an attempt to bring expenditures for national defense down to the lowest warrantable level — which was why he had left von Braun kicking his heels in the desert for so long. Intercontinental missiles could wait. Instead, the atomic bomb had to be perfected; the international network of American military bases extended; and an effective force of strategic bombers put together. And, collectively, all of these measures should stand as a more than adequate guarantee of American national security.

As early as May 1946, Truman had, however, commissioned an independent report from the Rand think tank on the feasibility of sending satellites into space. Until now, all rocket launches had been of a ballistic nature, and space flight as such had been regarded as overly imaginative science fiction. But if a rocket could be made so powerful that, instead of traveling at the usual rate of 3,000–5,000 mph, it could be accelerated to 18,000 mph, then it would exert just the right amount of centrifugal force to keep the earth's gravitational pull in check, and would "fall" around the earth at an altitude of 700,000 feet. Round and round, without any recourse to motive power. Rand was to investigate the possibilities of such a venture.

In one of the very first interviews he gave in the United States, in 1945, von Braun had stated that the entire surface of the earth could be kept under observation from an orbiting rocket.

> The crew could be equipped with very powerful telescopes and be able to observe even small objects, such as ships, icebergs, troop movements, construction work, etc.[2]

The Rand report concluded that satellites "would undoubtedly be of great military value" and that they could be developed within the next five years. The think tank also predicted satellites being employed for scientific purposes: in the study of cosmic radiation, gravity, magnetism, and meteorology. With becoming modesty the think tank did, however, admit that it could not foresee every eventuality.

> We cannot envisage all of the uses and implications of space ships, no more than the Wright brothers could have predicted the B-29 bombings of Japan or intercontinental passenger flights.

Satellites would fire man's imagination and create as much of a stir as the dropping of the first atom bombs. They would be eminently well suited for reconnaissance work and communication. And above all, they could not be shot down.[3]

Nevertheless, in the interim, this project—like the intercontinental rockets—was filed away under "pipe dreams." On the face of it, it sounded like an expensive and risky business.

Little by little, this view was modified by the Korean War. Tight military budgets were, to some extent, relaxed, and the plans were brought out once more. Again, in the autumn of 1950, Rand was asked to provide a confidential report on satellites. This time the United States' need for spy satellites was emphasized. For the Soviet Union, taking stock of America's open society was relatively simple. It was quite another matter when it came to American opportunities for taking a look around the Urals or the Caucasus. So the very thought of American spy satellites could be expected to strike the suspicious Russians as an act of intimidation. Hence Rand's suggestion that the satellites' military potential be underplayed, with the emphasis being put instead on the peaceful uses of this "remarkable technological advance." Since overflying satellites could not be kept secret, it would be better to declare their (peaceful) purposes openly.

The legal aspect was particularly problematic. The United States would find itself in a rather awkward situation if satellite overflying were to be declared illegal under international law. So efforts were made to put the case for an upper limit to national airspace, similar to the twelve-mile-offshore territorial limits.

In the—highly confidential—opinion of the think tank, it would be best if the United States began by sending something in the nature of a trial balloon—an innocent experimental satellite—into orbit around the equator. This way, it would not infringe on Soviet territory, but would set a precedent for the unauthorized overflying of foreign nations.

These classified documents were inherited by President Eisenhower. They fitted in neatly with his need for accurate information on the Soviet threat, to enable him to keep a tight rein on rearmament. In March 1955, therefore, he approved the secret plans for the first American spy satellite, WS-117L. Space flight had, at a stroke, become a matter of national significance.[4]

The plans for intercontinental rockets were also taken out of mothballs—the decisive factor here being that both the Russians and the

Americans were developing the hydrogen bomb. Initial plans for the Atlas rocket involved a very large and very heavy atomic warhead, and counted on heretofore unattainable accuracy from a distance of 5,000 miles. The hydrogen bomb provided a more compact warhead and far greater explosive force — thus allowing for some leeway as far as accuracy was concerned. The target was bound to be obliterated anyway. At that time too came the breakthrough in C. Stark Draper's small inertia navigation systems, which made it possible — by monitoring direction, acceleration, and speed — for the rocket to "know" when it was over its target. Finally came the development of warheads fitted with heat shields capable of withstanding the descent through the atmosphere. The rockets themselves could be left to burn up once the warhead had been released.[5]

Eisenhower gave the green light for the Atlas in June 1954. He knew that it was going to be a terrifically expensive project, but he also knew that the Russians were hard at work. Not that he was pressing the panic button. He was banking on superior American technology, in terms of guidance systems, rocket engines, and warheads, to put the Russians in the shade. And intelligence service reports supported this view.

The announcement, in October 1954, of plans for an International Geophysical Year seemed heaven-sent. Every country in the world was invited to try its hand at launching a research satellite during 1957–58. Now the United States had a legitimate cover, one that might have been lifted straight out of the Rand report. No one could have any objection to a research satellite which was part of an international collaboration. The satellite could, in the most innocent of ways, open up outer space to other, less peaceful aims. Some other time.

But for the time being all energies had to be concentrated on the civilian project. They had to come up with a serviceable rocket. Naturally, the Eisenhower administration was well aware of the public relations mileage to be derived from the United States being first out of the gate. If this project was to be given top priority, as the nation quite rightly expected it to be, then they had only one option: von Braun's Jupiter C rocket, a more advanced version of the Red medium-distance rocket from 1953. The Redstone itself was, however, a direct descendant of the V-2 missile, and with *that* pedigree, all the stress being put on the highly peaceful nature of this satellite began to sound rather hollow. Nevertheless, the Jupiter rocket could have been tailor-made for the job in hand. In the summer of 1956 it flew 3,335 miles during warhead trials — a record that stood until the emergence of intercontinental rockets.[6]

Von Braun was more than willing, but they simply would not use him. The first American satellite could hardly be launched by former Nazis from Peenemünde.

And so they had to come up with the next-best solution. Work on the Atlas could not be delayed because of this research project. And besides, it was unlikely that the Atlas would be finished by 1957. So they opted for the American Navy's Viking, a rocket that had only been used for research purposes and, hence, had a decent civilian profile. No one seemed at all worried by the fact that it was not operational and would have to be fitted with three totally new rocket stages.[7] The new rocket was christened the Vanguard.

Von Braun was furious. He knew he could do it. But he was not allowed to. Instead a gamble was being taken on completely untested technology. It got to the point where inspectors had to be sent to Cape Canaveral to make sure that he did not make use of one of his own warhead trials to send the first satellite into orbit "by accident."[8]

The Germans grumbled a bit and looked around for other projects which might bring their employers, the American Army, into the space race. Optimistic scenarios depicted a time, far in the future, when there would be a need for huge rockets with much bigger engines and a resultant increase in propulsive force. The most ambitious of Army strategists visualized rocket-borne troops being set down behind enemy lines, and even an American Army base on the moon. And for undertakings of that nature, existing rockets, not least the Vanguard, were ridiculously inadequate.[9]

President Eisenhower was, however, playing his cards very close to his chest regarding the militarization of outer space. Old soldier that he was, he thought he had heard it all before. He listened with interest to the scenarios put forward by the armed forces and studied their technological shopping lists—all the while knowing that they would never get enough hardware to satisfy them and that turning outer space into yet another battlefield would only provoke a massive new arms race. He had reluctantly gone along with the Army missile programs, and he expected great things from the spy satellites, but he was also irritated by the fact that the armed forces seemed incapable of cooperating. The research work of one body would often end up being duplicated, and even triplicated, thanks to superficial internal rivalry—and at the taxpayers' expense.

The important thing, as Eisenhower saw it, was not to outdo the Russians but to achieve an adequate state of military preparedness; one which could be expanded—carefully and at an appropriate rate—if it became necessary. And the intelligence reports to which he had access suggested that the United States was under no threat whatsoever.[10]

Nor did Eisenhower believe it was so utterly crucial that the Americans should be first with a satellite. The emphasizing of civil objectives came before everything else. So, rather an unfinished rocket with an unblemished reputation than the Germans' military medium-distance rocket. No one wanted the Russians to get there first, but then again, it might not be such a bad thing. Then they would be the ones to kick the ball into overflight and, having done so, it would be hypocritical of them, later on, to dispute the Americans' right to do the same.[11]

Then comes *Sputnik*! The date is October 4, 1957, and all hell has broken loose. In actual fact, the Russians — under the leadership of that mysterious chief designer, Koroliev — have only done what they had always said they were going to do; and what the Americans themselves were working toward: launched a satellite to tie in with the International Geophysical Year. But that is not how the man in the street sees it. Because everyone knows the Russians are a race of backward farmers, that their country has been left in ruins by the Germans, and that just keeping the farmers' tractors going stretches the country's technological resources to the breaking point. And then they turn around and launch a 183-pound satellite, when the weight of the projected American model has to be kept to a scant 5 pounds so as not to overtax the rocket.

The way many Americans see it there is only one possible conclusion to be drawn from this: The American scientists are useless and lazy; the American Army is incompetent and far too sure of its own superiority; quite simply, the Army has been caught napping; and what about American technology? Isn't the quality of manufactured goods becoming poorer and poorer? You just have to look at them and they fall apart. The manufacturing industry calls it built-in obsolescence, but is it anything other than skimping, sloppiness, and shoddy workmanship? And what about the President, what does he actually do with his time? Play golf and make soothing, paternalistic speeches to the people instead of vigilantly guarding the nation's pride!

The public response, which bordered on the hysterical, might be difficult for us to understand today. Many saw America as being unprotected and vulnerable in the face of the Russian military machine. The United States had not had enemy bombers flying over its homeland during World War II and had never expected to find itself in such a situation. Now *Sputnik* was tracing its fine line across the American firmament. No threat in itself, but what did it mean? Impotence and defeat. Shame and humiliation. How was the United States supposed to take the lead in international affairs if the whole world was laughing at it? How was one supposed to look up to American science, the American educational system, American technology, when this sort of thing could happen?

And then they had to watch while the President appeared on television to say that this whole affair "did not rouse his apprehensions, not one iota,"[12] when everyone could see that this was nothing less than "technology's Pearl Harbor." The President said that he would have been much more concerned about panic-stricken initiatives, taken by people with no judgment, as a result of overheated public feeling.

Eisenhower did not feel that he had been sleeping on the job, but he was hampered by not being able to give the public all the facts. Work on the Atlas rocket was progressing well. The brand-new B-52 bomber had come off the production line on schedule. Hydrogen bombs were becoming increasingly smaller and more effective. The Navy was in the process of developing atomic submarines and the submarine-borne Polaris missile. The contract for the highly effective Minuteman missile had been signed. Top secret U-2 flights had been operating over the Soviet Union since the previous year and these reaffirmed his view of American superiority. And soon the spy satellites would be able to complete the picture. Succeeding generations have judged the last years of the 1950s to be among the most dynamic in American military research. It was at this time that the foundations were laid for every key weapon system produced over the next twenty years.[13]

Eisenhower believed that he had, quietly and single-mindedly, extended the American power base without going overboard financially. As far as he was concerned, prestige and propaganda were but secondary political factors. One had to concentrate on the facts — and the fact was that *Sputnik* posed no real threat to the United States. There was no cause for concern, he repeated in the days that followed and could not understand why the furor did not die down. And, most tragically of all, he had no idea what use his political opponents in the United States would make of *Sputnik*. Even though they knew better, thanks to the confidential information which he had passed to the leaders of the Democratic Party.

Twelve days after the *Sputnik* launch, on October 16, in New Mexico, the Americans sent up a slender Aerobee missile, which accelerated a small container full of steel balls to a speed of 34,000 mph. Thus, this container became the first man-made object to break free of the earth's gravitational field and go into orbit around the sun.[14] Now, you might think that this was just what was needed to calm the nation. But no. What the American people were looking for was an orbit around the earth.

Democrat Lyndon B. Johnson sensed that the tide was about to turn in his favor. Here was a matter of national importance, one that could be used to drum up opposition to Eisenhower, the Republican, in good time for the next presidential election. One of Johnson's advisers wrote to him, "The issue [*Sputnik*] is one which, if properly handled, would blast the Republicans out of the water, unify the Democratic Party and elect you President."[15] Moreover, the Democrats had a majority in the Senate, which should be exploited to the full. Even the Republicans could see how vulnerable they were. Two of them described it in this way: "We live in a political world, and no greater opportunity will ever be presented for a Democratic Congress to harass a Republican Administration, and everyone involved on either side knows it."[16]

In speech after speech, Johnson described how, on the first night of *Sputnik's* flight, he had gone outside, at his ranch in Texas, and suddenly that sky which was so familiar to him and which had so often made him feel at peace had been filled with ominous question marks. "I don't want to go to sleep by a Communist moon,"[17] he said later, and people knew what he was talking about. He also hit the mark with another of his metaphors. The Russians have now taken "the high ground"[18] around us, he said, and of course everyone knew what that meant. Thanks to all those Western movies, they knew the advantages of having your hideout up on a ridge, from which you could look down on the unsuspecting enemy. The Russians were skulking up there now, beeping away, while the Americans walked around down below, unable to do a thing about it. It was humiliating, it was embarrassing. A situation like this could do a lot of damage, and in Johnson's opinion, you would have to be mighty naive to think that the Russians were going to stop there.

> The Roman Empire controlled the world because it could build roads. Later — when it moved to sea — the British Empire was dominant because it had ships. In the air age we were powerful because we had airplanes. Now the communists have established a foothold in outer space.[19]

Moreover:

If, out in space, there is the ultimate position — from which total control of the earth may be exercised — then our national goal and the goal of all free men must be to win and hold that position . . .

From space, the masters of infinity would have the power to control the earth's weather, to cause drought and flood, to change the tides and raise the levels of the sea, to divert the Gulf Stream and change temperate climates to frigid.[20]

The blame for all this unbearable humiliation could be laid at the door of an incompetent administration. Eisenhower had done a good enough job in World War II as commander-in-chief of the Allied forces during the Normandy landings, but he was an old man now, and worn out. What was needed was some new blood. And it was high time.

LBJ was tired of hearing that the launching of an American satellite was, in fact, just around the corner. To the delight of his listeners he commented that no doubt — by the time it *was* at long last ready — it would be fitted with chrome tail fins and automatic windshield wipers.

There was something almost monomaniacal about his reiterations: "It took the Soviets four years to catch up with the Atomic Bomb. It took the Soviets nine months to catch up with the Hydrogen Bomb. And now tonight the Communists have established a foothold in outer space." Another Democrat commented lugubriously: "Let's not fool ourselves. This may be our last chance to provide the means of saving Western civilization from annihilation."[21]

Senator Styles Bridges, the senior Republican on the Senate Armed Services Committee, was not really in a position to denounce the present administration. Instead he made it quite clear that this turn of events meant that every ordinary American citizen had better start shaping up, and soon.

The time has clearly come to be less concerned with the depth of the pile on the new broadloom rug or the height of the tail fin on the new car and be more prepared to shed blood, sweat and tears if this country and the free world are to survive.[22]

And then Koroliev struck again. On November 3, Laika, the canine cosmonaut, was sent up in a space capsule weighing half a ton! Not only did this feat demonstrate that a living creature could survive in a weightless state; it also provided convincing proof of the Russian booster rockets' superior capacity. This time the Americans were almost too stunned to comment on such an achievement. Even Johnson could see that there was no point now in just lambasting Eisenhower. This disgrace touched every American citizen. The honor of the nation had been trampled underfoot. And so his pronouncements assumed a more statesmanlike character: "Well, I think the Russians are ahead of us; in some respects, I think, their achievements are rather remarkable, but there's no reason for panic because there's never any solution in panic. Progress is going to result only from bold decisions that are made by cool heads."[23]

In the winter of 1957–58, Lyndon Johnson capitalized on his objective view of the situation by arranging a series of congressional hearings intended to expose the present administration as being apathetic and incompetent and show that the Democrats alone possessed the required drive. There was no denying that seasoned government officials might find it rather hard to stand by and watch this kind of maneuvering, having — as they did — access to the actual, but often still classified, facts concerning the Democrat Truman's and the Republican Eisenhower's respective contributions in the area of defense. One of these, Oliver Gale, senior official in the office of the Secretary of Defense, described Johnson's shrewd tactics in his diary:

> The Johnson Committee has hit upon a very successful formula. They have set a series of hearings, a month apart, at which the Secretary and his people will report on the advancing state of our preparedness. They ask Defense to submit a week or so in advance a written report on what we are doing to catch up with the Soviets. . . . This — a highly classified document — is delivered by me to my opposite number, Jerry Siegel, of the Committee staff. Then, following the hearing, they issue to the press a report in which they urge Defense to do the very things we have said we are doing. The picture is clear: They are directing Defense, leading the nation in its frantic rush to reduce the state of peril, and we are gratefully — or perhaps reluctantly — doing as we are told.
>
> We go along partly because we have no choice, and partly because these are the same individuals who have to approve our military budget and we can do nothing but lose if we fight them.[24]

Meanwhile, ominous reports were rattling down the telex lines from American embassies across the world. Both the British Prime Minister, Harold Macmillan, and his Chinese counterpart, Zhou Enlai, had described *Sputnik* as "a real turning point in history." The Indian head of state, Jawaharlal Nehru, hailed *Sputnik* as a "great scientific advance," and opinion polls indicated that other Third World countries had been encouraged to pursue a neutral line between East and West. Cairo Radio was of the opinion that the new Russian booster rockets undermined the significance of "all kinds of pacts and military bases [and would therefore] make countries think twice before tying themselves to the imperialist policy led by the United States."[25] The word was that even the Europeans' faith in American technical superiority had taken quite a beating and that NATO's position too was weaker than it had been.

It did not help matters that, in November, American newspapers got their hands on the confidential Gaither Report, in which it was revealed that in the eventuality of an atomic war involving Russian missiles, the American people would be considered expendable, and that no one seemed able to conceive of any feasible form of defense — other than the obvious threat of retaliation. This report also estimated (wrongly) that as early as 1959 — in other words, in just two years' time — the Russians were likely to have over a hundred rockets at their command, all of them capable of homing in on targets in the United States.[26]

By December 6, 1957, everything was, at long last, ready for the launching of the specially adapted Vanguard rocket, carrying the first American satellite. For the very first time, American television viewers would be able to watch a live transmission of a rocket launch. For the occasion a kind of running commentary, in the form of a countdown announced over the public address system, had been devised — a ritual which was to become an integral part of space age procedure. It was now or never. Everyone held their breath. The rocket stood there, patient and unmoving; the very prototype of "a body at rest," enveloped in diaphanous clouds of steam, waiting for the start signal. At T minus 0 it belched out huge tongues of flame. Then, after teetering on the brink for what seemed like an eternity, the rocket slowly and ponderously rose off the launch pad — only, shortly thereafter, to — as it were — lose confidence in the third law, give up, and wearily sink back, before exploding in a vast sea of orange-red flames. Commentators hemmed and hawed and admitted that there were those who would construe this as being "a severe propaganda defeat" for the United States.[27] And many an American buried his or her face in their hands.

The newspapers were less diplomatic. The next day, wisecracking headlines outdid one another in their inventiveness. The ill-fated satellite was dubbed "Flopnik," "Kaputnik," and "Stay-putnik."[28] The Eisenhower administration could not even manage to launch "a small hunk of iron."

Members of the Russian delegation at the United Nations suggested, most politely, that the United States might be able to receive aid from the Soviet program for technical support to developing countries, to send "their little grapefruit" into space.[29]

In the wings, gnashing his teeth, stood von Braun; like an extra who has learned a role by heart and then kept his eyes peeled for the slightest sign that the play's leading man might be indisposed, thus offering him the chance to take over. Now his chance had come. Von Braun and his team wasted not a single second, and finally, on the last day of January 1958, they sent *Explorer*, the first American satellite, aloft, with the help of their trusty Jupiter rocket. For the first time in a very long while, Americans could hold their heads high, happily overlooking the fact that the ink was barely dry on von Braun's U.S. citizenship papers. The newspapers were full of articles on the Huntsville Heroes, who had raked the chestnuts out of the fire. In the nick of time.

The Birth of NASA

In January 1958 the American public opinion polls were kept very busy. The space race was the one topic on everyone's lips, and the nation was up in arms. As many as 82 percent believed that the United States was lagging behind the Soviet Union with regard to the development of advanced weaponry; 67 percent were convinced that Americans in general "had been too smug and complacent about our national strength"; 77 percent responded with a resounding yes to the statement that "*Sputnik* should make a difference in what we are doing about the defense of this country"; 61 percent were prepared "to pull in our belts and sacrifice for stronger defense."[1] Not only that, but a Dallas newspaper was of the opinion that democracy could learn a good deal from the dictatorship to the east: "Some advantages of tight totalitarian control will be helpful to our democratic process."[2]

No wonder, then, that a powerful arms industry — with the big aeronautics plants at its head — perked up. Very soon, there would be big money to be made from anything having so much as the faintest connection with outer space. And so, before very long, costly and ingenious space projects were pouring in thick and fast — all of them supposed to make the nation fit to meet the Russian challenge. The industry was by no means unhappy with newspaper headlines proclaiming the Missile Gap.

Events had taken the very turn that Eisenhower had dreaded most. An insatiable military force, fraught with infighting, had formed an alliance with the real heavyweights of the armaments industry. Together, they wooed a nation beside itself with panic. The Democrats were not exactly holding themselves back either. Everyone knew exactly what ought to have been done, long ago.

And yet Eisenhower's people had not been asleep at the wheel. American intercontinental rockets had been on the drawing board since the early 1950s and were now almost completed—following the investment of billions of dollars. Medium-distance rockets were already on hand. American intelligence had, moreover, a pretty accurate picture of what the actual threat facing the United States amounted to and saw no reason for undue concern. None of this did any good whatsoever. Eisenhower had underestimated the psychological factor; the loss of prestige entailed in being outstripped by the Russians. The American people had been set in motion by the "impressed Russian force" and were looking for instant and resolute action; they were looking for national leadership.

Eisenhower made a tentative opening move by acquiring a permanent scientific adviser, James Killian, Special Assistant to the President for Science and Technology. Well, at least it was a start. Then he set up an Advanced Research Projects Agency, with the power to coordinate all of the armed forces' various space projects and thus avoid any overlap.[3]

And finally he gave some thought to how he might use what he had learned from the International Geophysical Year to steer space research into civilian channels and out of the range of the Pentagon and Congress's hawks. He might even be able to kill two birds with one stone. For many years America's powerful aerospace industry had been having things far too much its own way in its dealings with the armed forces—who could easily get all their bills paid by defense-minded taxpayers. Now it might be useful for them to encounter some opposition from a civilian space organization; one which would, to a great extent, come under the control of the President and Congress.

They already had an organization known as the National Advisory Committee for Aeronautics (NACA). This had been formed in 1915, when aviation was still in its infancy, to carry out research—the results of which could be placed at the disposal of civilian bodies, the aeronautics industry, and the armed forces. It was an organization with contacts in all the right places and a reputation for being unbiased, if still a bit on the stuffy side here and there. Some of its members had still not entirely reconciled themselves to the shrill whine of jet aircraft engines, and yearned for real internal-combustion engines that sounded the way airplanes ought to sound. Others had, however, been quick to take up the challenges of rocket technology and to collaborate with the Air Force on supersonic aircraft and the experimental X-15 aircraft, soon to make its maiden flight.[4]

Breathing new life into this organization, granting it certain powers, and giving it overall responsibility for American space research would put a ceiling on the Pentagon, which would then have to concentrate on the armaments side of the rocket business and give up all thought of taking part in any future manned space flight. Eisenhower wanted, at all costs, to avoid any kind of arms race in outer space. Naturally the armed forces would oppose him, and they would be bound to bring up the fact that it had, in fact, been the Jupiter, a military rocket, that had saved the day when the Vanguard, the specially adapted civilian rocket, failed so miserably. So timing was of the essence. The Pentagon must be given no time to gather itself for a counterattack. Everything would have to be accomplished with one bold stroke.

This was where Democrat Lyndon B. Johnson came in. Johnson had the ability to play Congress as though it were a Steinway grand, and he was burning to make his mark on the space front. He immediately recognized the tremendous potential of such a civilian body. Better than anyone else, Johnson had grasped the importance of public opinion. The right moves made now could turn space research into an incredibly useful tool in American politics. If the wraps were taken off space research, at both the national and the international level, the right politician would have a tremendously powerful card up his sleeve; and, in any case, Eisenhower had only two years left in office.

But Johnson realized, too, that for Eisenhower and his people it was also a question of keeping a tight rein on the space budget and of putting the project where Eisenhower deemed it to belong: into a low, harmless orbit under the auspices of a special "Space Board." In the course of private talks, Johnson demanded that Eisenhower accord this body higher status and greater independence, as a National Space Council. Eisenhower made the objection that such a move would give the organization too much power, too much "momentum," which might prove impossible to administer politically. This was precisely the point which Johnson had been venturing, with all due respect, to make, and this Eisenhower was shrewd enough to realize. He also realized, however, that he could not do without the support of the leader of the Senate's majority party, and so he conceded the point.[5] In return, Johnson suggested that the President in office should always have the option of chairing the Space Council and, hence, exercising control through it. This suited Eisenhower perfectly and in no way worried Johnson. There would be other Presidents after Ike.

In April 1958, the National Aeronautics and Space Act was submitted to Congress. The Pentagon had been given only twenty-four hours in which to comment upon it. Later, Johnson was to say, with a wry grin, that the proposal "had been whizzed through the Pentagon on a motorcycle."[6] Before the proposal could be adopted, the projected Agency had grown into a veritable "Administration," credited with a more authoritative status than anyone in the White House had ever intended it to have. At one fell swoop, civilian space programs had been whisked to the top of the national agenda. The National Aeronautics and Space Administration (NASA) was born, and commenced its operations on October 1, 1958, under the leadership of a government-appointed Administrator, Republican T. Keith Glennan.

One of the masterstrokes of the NASA bill was that it was framed in such a way that the organization would not be hampered by all the snags that had soon rendered other public bodies sleepy and impotent. NASA was to be financed from government funds, but would have the same flexibility, and utilize the same management techniques, as top companies in the private sector. NASA's management were, at all times, free to discharge incompetent employees and replace them with better-qualified individuals. Top-notch staff could be transferred from other branches of the public sector on a permanent or temporary basis, and retirees with valuable skills could be pressed back into service. Right from the start, the idea was to create an elite organization with a high opinion of its own worth; and to make rigid hierarchical systems more pliant, thus allowing anyone in middle management to have easy and unimpeded access to the men at the top.[7] The mood was to be modern, free and easy. The collective enthusiasm of its employees was to be the driving force behind NASA's work.

The military were furious, but powerless. The armed forces felt that those damned civilians had swooped down and snatched territory which was rightfully theirs, before they could do anything about it. NASA was ruining their chances of fighting the Russians in outer space. But, as would soon become apparent, it was not that simple. If anything, the dividing line between the civilian and military sectors had been wiped out by the Cold War. It was quite conceivable that epoch-making civilian achievements could carry weight in the superpowers' struggle for a balance of power. Military strength could no longer be assessed in terms of gunpowder and bullets when such strength might just as easily manifest itself in great technological achievements founded on a well-developed educational system.

Viewed in this light, NASA, with its appealing civilian profile, might very well be regarded as having an even more effective "military" impact on the Russians than the Pentagon ever could. Moreover, using NASA as cover, the Pentagon could continue with the development of new intercontinental rocket systems and the building of spy satellites. With this in mind, General Bernard Schriever lectured Congress on the dangers of letting itself be taken in. No matter how strong NASA's civilian profile might be, the military aspect of space technology was going to account for between 75 and 90 percent of the business.[8]

On May 15, as an accompaniment to the congressional debate, the Russians launched *Sputnik III*, weighing in at just under one and a half tons. The CIA was no longer in any doubt. The Russians were working toward the launch of a manned space capsule, and they evidently had rockets capable of propelling such a capsule. Eisenhower still had not grasped the fact that it was possible for certain propaganda ventures to be accorded so much importance that such vast sums would be spent on them. For his part, he remarked, he had no intention of "hocking the family jewels" for the sake of outer space. Nevertheless, in August 1958, he was forced into granting an incipient NASA the authority to start making plans for an American manned space flight. Project Mercury had been given the green light.

In November 1958, the Air Force could present the Atlas, its intercontinental rocket. It seemed enormous, with three engines capable, collectively, of producing over 360,000 pounds-force and of reaching targets 5,000 miles away on the other side of the globe. One problem was, simply, that it was still not particularly dependable. Another was that—a year previously and with a minimum of fuss—the Russians had launched *their* intercontinental rocket, which NATO had christened Sapwood. This exerted a force of 1,100,000 pounds—three times that of the Atlas.

There was, in fact, a relatively simple explanation for this disparity, inasmuch as Russian nuclear weapons were still huge and unwieldy; and the Russians had not yet mastered the alloying techniques necessary for extremely powerful rocket engines. They were lagging far behind the Americans where sophisticated technical solutions were concerned. So the Russian booster rockets had to exert an enormous thrust, and this they achieved by means of clusters of more modest rockets which, when bound together, could accomplish the task. The American weapons, on the other hand, had been tailor-made for the far more compact American nuclear warhead.[9] But although this explanation seems both simple and obvious, it was not good enough for the American public or for many politicians. For the moment, at the start of the space race, they had nothing but military rocket technology to work from; manned space capsules were going to be very heavy indeed; a heavy weight required a powerful thrust; and there, paradoxically, the Russians—precisely because of their less advanced technology—were clearly leading the field.

7. NASA's President

In October, NASA had more than 8,000 employees, the majority of them recruited from the old NACA. And although a number of former military installations had also been taken under the civilian organization's wing, the Army Ballistic Missile Agency in Huntsville, Alabama, was, at least to begin with, exempt from this. Von Braun's people continued their work on powerful booster rockets designed to meet future military ends. The Army had not yet given up hope of colonizing outer space.

It did, however, have to abandon this idea after strong political pressure was brought to bear on it. All military rockets with a range of over two hundred miles would, in the future, come under the aegis of the American Air Force. And there was only one place where very large booster rockets would be required: in the civilian space program. In October 1959, Huntsville was unwillingly handed over to the newly formed NASA, and in March 1960 was renamed the George C. Marshall Space Flight Center, after the recently deceased former Secretary of State. This move brought NASA approximately 5,000 new employees.[1]

One of the people charged with the somewhat delicate task of inspecting the newly acquired premises on NASA's behalf and taking stock of current projects, equipment, and facilities has described his first encounter with the management style at Huntsville: "We walked into a conference room and saw all the Germans sitting all over the place. I sat down. And then von Braun comes in and everybody stands up and clicks their heels."[2]

As early as April 1959, the first seven American Mercury astronauts were presented to the press. The American public now had some wholesome, purposeful young men on whom to pin their space-age hopes. All seven signed contracts with *Life*, granting the magazine sole rights to their stories,[3] and NASA's top men beamed with pride. Right from the start, they were very much alive to the fact that public relations would prove to be the lifeblood of this civilian body.

The only problem was that the Atlas, the intercontinental rocket that was meant to form the cornerstone of the Mercury project, still had a habit of blowing up at the most inconvenient moments. For this rocket, a new and very advanced construction technique had been employed. To save weight, its walls were made of thin aluminum which also did service as the outer side of the fuel tanks. The warhead alone was reinforced. Quite a change from the heavy, robust steel structure of the V-2 rocket. The Atlas rocket was not even capable of bearing its own weight until its tanks were pressurized and able to stabilize the structure.[4] This meant that even very minor deviations from its optimum trajectory sent waves rippling along the whole length of this "soft" rocket—resulting, every time, in its ripping itself apart. Not much PR material to be gained there. Eventually the newspapers were reporting as fact that the United States was not capable of making rockets that worked.

By July 29, 1960, the people at NASA believed they had finally overcome all of their problems with the Atlas. They brought the astronauts and their families to Cape Canaveral, where, along with numerous guests of honor, they were to witness the first trial launch of the Mercury-Atlas, with a real space capsule fitted to the rocket tip—just the way it would look when the seven astronauts were really and truly on their way. The rocket rose up majestically on its three pillars of fire. Everyone applauded and tilted their heads back to watch as the rocket grew smaller and smaller. After about a minute, at a height of 30,000 feet, it ran into a particularly turbulent atmospheric layer and bid its stability farewell. It was promptly ripped apart like so many of its predecessors. Several of the guests of honor were actually terrified of being hit on the head by wreckage.[5]

While the work of solving the Atlas' problems continued, later in the year, on November 21, 1960—a demonstration was to be staged of the somewhat smaller Redstone rocket, which was supposed to launch the first American into space—not into orbit, but merely following a ballistic trajectory. Tom Wolfe has provided an unforgettable account of what happened.

Everything except an astronaut was on the launch pad. The dignitaries were all seated in grandstands, and the countdown was intoned over the public-address system: "Nine . . . eight . . . seven . . . six . . ." and so forth, and then: "We have ignition!" . . . and the mighty belch of flames bursts out of the rocket in a tremendous show of power . . . The mighty white shaft rumbles and seems to bestir itself—and then seems to change its mind, its computerized central nervous system, about the whole thing, because the flames suddenly cut off, and the rocket settles back down on the pad, and there's a little *pop*. A cap on the tip of the rocket comes off. It goes shooting up into the air, a tiny little thing with a needle nose. In fact, it's the capsule's escape tower. As the great crowd watches, stone silent and befuddled, it goes up about 4,000 feet and descends under a parachute. It looks like a little party favor. It lands about four hundred yards away from the rocket on the torpid banks of the Banana River. Five hundred VIPs had come all the way to Florida, to this goddamned Low Rent sandspit, where bugs you couldn't even see invaded your motel room and bit your ankles until they ran red onto the acrylic shag carpet—all the way to this rock-beach boondock they had come, to see the fires of Armageddon and hear the earth shake with the thunder—and instead they get this . . . this *pop* . . . and a cork pops out of a bottle of Spumante. (It was the original Project Vanguard fiasco all over again, except that it was worse in a way.) At least with Vanguard, back in December of 1957, the folks got lots of flames and explosion. It at least looked halfway like a catastrophe. Besides that, it was very early in the game, in the contest for the heavens. But this—it was ridiculous! It was pathetic![6]

The infant NASA had not exactly gotten off to the best of starts. And matters were not improved by the fact that in 1960 John F. Kennedy and Eisenhower's Vice President, Richard M. Nixon, were battling for the presidency. The space issue provided an obvious focal point for the campaign. And of course the Democrats tried to rub the Eisenhower administration's nose in any and every space fiasco. Nixon had already made a fool of himself the year before in Moscow, when he was to open an American trade exhibition and almost crossed swords with Khrushchev at a kitchen equipment booth, while the television cameras rolled. "There are some instances where you may be ahead of us; for example in the development of the thrust of your rockets for the investigation of outer space," said Nixon, clearly flustered, "[but] there may be some instances, for example color television, where we're ahead of you."[7] Now people the length and breadth of the United States had had occasion, more than once, to view American space fiascos—in color— and they had just about had enough.

.   .   .

John F. Kennedy was everything that Eisenhower was not. He was young, he was daring, he was an articulate intellectual and, above all else, he was champing at the bit in his eagerness to bring about a dramatic change in the course of events. Not only that; he was good-looking and had great television appeal. For the first time, hardened political commentators were faced with a situation where it did not seem to matter much what political messages the candidate might be trying to put across. Many voters came to his rallies only to experience this man's charisma, his "star quality," and perhaps to be among those lucky enough to take home his autograph. Kennedy was very much aware that he represented a new generation, which had fought in World War II and which was less cautious, less tightfisted than its parents' generation.

Eisenhower considered it of paramount importance that the state interfere as little as possible in social developments. Money was best left in the pockets of private citizens and private enterprise. In its zeal to take on the Soviet Union, the United States must not start leaning toward the Communist view of the state's role. Above all else, the United States could not be allowed to turn into a technocracy, with technological development being assisted by government investment. The state might make adjustments and modifications, it could give encouragement, but in any self-respecting capitalist society the responsibility for new technology and pioneering research had to lie with the private sector.

The situation was almost paradoxical. Congress and the American people were begging the executive body to assume more responsibility and broaden its sphere of influence. The President was desperately trying to opt out. The outgoing President at least.[8]

When Kennedy accepted the Democratic nomination in the summer of 1960, he renounced "the safe mediocrity of the past" and described the Eisenhower era as "seven lean years of drought and famine" which had "withered the field of ideas" and deprived "too many Americans of their will and sense of historical purpose." Now especially—at a time when, wherever you turned, there were the Russians, poised to spring— "good intentions and pious principles" were no longer enough; "soft sentimentalism" would no longer suffice. What was needed was "strong and creative leadership" and "critical and intelligent vigilance."[9]

In the final televised debate between Kennedy and Richard Nixon— who did not come across nearly as well on television—Kennedy took his opponent to task over the "Kitchen Debate." Nixon had not been right in what he had said back then, said Kennedy; quite the contrary, because we are living in an age when "rockets are more important than television." He was irritated by the American fiascoes and the resultant damage to the United States' international image. Nixon could only retort, lamely, that with such remarks Kennedy was not exactly helping to improve that image.[10]

As his running mate, Kennedy had selected (or rather, had forced upon him) Lyndon B. Johnson, the political tactician, who had had to abandon his own presidential hopes. Soon after his inauguration, by means of a legal amendment, Kennedy saw to it that, in the future, the Vice President would preside over the National Space Council, whose genesis Johnson himself had engineered two years previously. From now on, the Space Council would have to get used to the idea of having a chairman who did not simply display polite and absentminded interest, as Eisenhower had done, but who burned with all the intensity of a blowtorch over the space question. In all probability, Kennedy had also reasoned that this would give Johnson something to get his teeth into, and keep him from sticking his nose, and his ambitions, into other and more important issues.

On January 17, 1961, Eisenhower used his farewell address to make some points that smacked of anything but "soft sentimentalism." Instead, this man—who had, more than anyone else, represented the personification of the armed forces to a whole generation—issued an urgent warning against a new and dangerous force in American society, the military-industrial complex, which the political system had not yet learned to guard against.

> This conjunction of an immense military establishment and a large arms industry is new in the American experience. The total influence—economic, political, even spiritual—is felt in every city, every state house, every office of the Federal government. We recognize the imperative need for this development. Yet we must not fail to comprehend its grave implications. Our toil, resources, and livelihood are involved; so is the very structure of our society.
>
> In the councils of government, we must guard against the acquisition of unwarranted influence, whether sought or unsought, by the military-industrial complex. The potential for the disastrous rise of misplaced power exists and will persist.
>
> We must never let the weight of this combination endanger our liberties or democratic processes. We should take nothing for granted.[11]

Eisenhower also touched upon the way in which the status of scientific and technological research had changed during his presidency. All research had become more complex and more expensive, and an increasing proportion of this research was being financed out of public funds. Gone were the days of the little inventor in his workshop; the days when the blackboard was still an essential tool. Their place had now been usurped by computers, and "task forces" of scientists working in laboratories the size of football fields. Which was all well and good, Eisenhower felt, as long as it did not reach the point where "a government contract becomes virtually a substitute for intellectual curiosity." He had seen how even the private universities, formerly the fount of all independent research and pure scientific study, would now roll over and beg for lucrative private and public research projects.

In conclusion, Eisenhower said:

> Only an alert and knowledgeable citizenry can compel the proper meshing of the huge industrial machinery of defense with our peaceful methods and goals, so that security and liberty may prosper together.[12]

Privately, however, a disillusioned Ike believed that more and more areas of American society were being run along Soviet lines. The Cold Warriors and the Social Liberals had joined forces under the watchword: More power to the public sector. More technocracy. But they could not see the flaw in a strictly regulated, top-down system. And no one was more familiar with this flaw than Eisenhower—from his Army days. In the Army, each unit was told: "Don't worry about the big picture; if each unit does its job, the whole will take care of itself." But the sum total of all those duties, no matter how excellently discharged, might very well add up to one big fiasco. On the other hand, the spontaneity, the unpredictability, the failures, and the total arbitrariness of civilian life might seem like a hopeless muddle; but, taken as a whole, these chaotic building blocks might well form the foundations of societal success.[13]

Eisenhower obviously hankered after a bygone era. A sort of Norman Rockwell era, when the world was simple and intelligible; when relations among the government, the armed forces, private enterprise, and the universities were still clear-cut, with each individual body keeping itself to itself. The profound irony was, of course, that, during his presidency, Eisenhower had been instrumental in building and nurturing ever-greater technological systems, with his atomic programs and high-technology arms projects. In fact, it was during his time in office that the military-industrial systems became so impenetrable, so vast and powerful, that even the nation's head of state no longer felt in charge of developments.

Nor was it all that easy to determine who *was* running the ship, if not the President. For the first time, a situation seemed to have arisen in which vast technological systems could be driven by some kind of obscure self-will, which could not be precisely located, but which was inherent, rather, in the systems' own inertia.[14] The government could not spend such large sums on the setting up and expansion of powerful technological complexes without "impressing a force upon them" and, thus, setting in motion something which could not simply be halted at the drop of a hat. And, as with the systems, so with the rockets; the main problem was not how to make them bigger and more powerful; the real challenge lay in guiding and controlling their titanic strength.

"Before This Decade Is Out"

As 1960 moved into 1961, the NASA people were not feeling happy about their situation. In line with common practice, in November — immediately after Kennedy's election victory — the organization's government-appointed head, T. Keith Glennan, was asked to tender his resignation. Since then, they had heard nothing.[1] Nor had the two abortive demonstration launches done anything to lessen their feeling of insecurity.

Things went from bad to worse when it was rumored that Kennedy's scientific adviser, Dr. Jerome Wiesner of MIT, had drafted a memorandum which was very critical of the Mercury project. "By having placed highest national priority on the Mercury Program," he wrote to the newly elected President, "we have strengthened the popular belief that man in space is the most important aim of our non-military space effort."

> The manner in which this program has been publicized in our press has further crystallized such belief. It exaggerates the value of that aspect of space activity where we are less likely to achieve success, and discounts those aspects in which we have already achieved great success and will probably reap further successes in the future.[2]

Wiesner found it hard to see the scientific value in catapulting an astronaut into a gigantic arc, as if he were a cannonball, and then fishing him out of the Atlantic — or, for that matter, to have him orbiting around the earth, be it once or several times. He saw far greater prospects in unmanned space probes, sent far out into the solar system, along the lines of the Aerobee rocket in 1957.

Wiesner's views won widespread sympathy within American scientific circles. A study carried out among physicists at Berkeley showed that the great majority believed that military goals and propaganda were the vital driving forces behind the space project and that the nation was, to a great extent, wasting its money.[3]

To NASA, such views were anathema. Everyone there was prepared to pull out all the stops to put the first American astronaut into space. That was what the whole organization was working toward. For two years, all training programs and grants had been geared in this direction; and then this egghead, this intellectual, comes along and interferes with their laboriously generated momentum before the project can even get under way. NASA stepped up its trials program in a frantic race against time. Somewhere along the line, they had to get results, even if it meant skipping the occasional safety check.

On January 31, 1961, Ham — a thoroughly drilled and highly trained chimpanzee — settled himself in his Mercury capsule at the top of the Redstone rocket, the one that had popped its cork in November. Ham had been in hard training for just as long as the astronauts and undergone all the same tests. But, unlike his human colleagues, he had received electric shocks in the soles of his feet every time he pressed the wrong button. So now he knew his lessons perfectly. The rocket worked and Ham flew sky-high. Granted, he landed a two-hour sail away from the target ships and was in danger of drowning, but even so, when his capsule hatch was opened he was found to be in good spirits, lying back with his arms folded. According to the writer Tom Wolfe, it was probably the first time in a very long while that he had had a couple of hours to himself, undisturbed by men in white coats and electric shocks in the soles of his feet.[4] He tucked into a well-deserved apple with great gusto. Ham might just as easily have been Alan Shepard; basically there was no difference between his flight plan and that of a human astronaut. The chimp was presented to the world's press wearing diapers, so that no call of nature would be allowed to get the better of him in that situation. NASA breathed a sigh of relief as the cameras flashed; canceling Mercury now would not prove quite so easy.

Nor, as it turned out, was that what Kennedy had in mind. He understood Wiesner's scientific arguments, but politics was more than a science and, generally speaking, while the public could not have cared less about space probes, they lapped up every word printed in *Life* about the astronauts' wives. And there was a great deal more to NASA than just scientific research. Wiesner was apparently deaf to all the overtones, to the idea of populism in space technology, but Johnson and his boss definitely were not. The Americans still had time to be first with a man in space.

At long last Kennedy had appointed Glennan's successor as head of

NASA. The post fell to fifty-five-year-old James E. Webb, a man with extensive management experience in the oil industry and in the public sector, though with no particular technical acumen. Webb would go on to become a director who, to this very day, is still spoken of within NASA with almost religious fervor.

So what were they waiting for? Why hadn't the first American — with all due respect to Ham, the chimpanzee — been sent into space by now? The reason was that von Braun was not satisfied with the way in which his Redstone rocket had overaccelerated. Everything had to be just right. And so he wanted to have one more unmanned flight, at the end of March, before Shepard risked not only his own life but, in all probability — should an astronaut be killed — NASA's too. Webb saw his point of view and von Braun had his way, even though both the nation and the President were growing impatient.

But the Americans did not make it this time either. On April 12, 1961, the Russians were able to announce with pride that Yuri Gagarin had completed a full orbit of the earth, before landing safely at a predetermined spot. Not only had the Russians come first again; they had also skipped the intermediate, human cannonball stage and moved straight to putting a man into full orbit.

The House of Representatives was in an uproar. One member stepped up to the podium and spoke of how that morning he had woken up with "a lump of lead in his heart." Who today remembered who had been second to fly the Atlantic? Now they might as well forget the whole thing. Another said he felt that Americans were not "simply shocked and awed, they were literally stunned." For his own part, he might as well say it straight out; he felt "downright hurt" by recent events. James Webb was summoned to a hastily improvised hearing, to be told by Representative Fulton:

> Tell us how much money you need and we on this committee will authorize you all you need. I'm tired of being second to the Soviet Union. I want to be first.
>
> I would work the scientists around the clock and stop some of this . . . scientific business.[5]

Others too were infected by this mood, and vied with one another in their eloquence. In a voice like thunder, Representative Anfuso declared:

> I want to see our country mobilized to a wartime basis because we are at war. I want to see our schedules cut in half . . . I want to

see some first coming out of NASA, such as the landing on the Moon . . .[6]

An official at the Defense Department compared the legislature to "heavy drinkers leaping forward and falling over one another on hearing a cork pop."[7] NASA would have to make an incredible mess of things not to find itself up to its neck in funding.

In an editorial, *The New York Times* made the wry observation that Kennedy had perhaps been a shade overhasty in presenting himself as the young, energetic, and purposeful leader of a strong and progressive United States. The President, clearly upset, remarked cautiously that he had never concealed the fact that the situation might very well get worse before it got better.

But then, he had other things on his mind. Three days after Gagarin's flight, Cuban exiles, trained by the CIA and with American air support, landed at Cuba's Bay of Pigs, intent on removing Fidel Castro from power. From the very start, everything went wrong. Before long, all of the Cubans were either on the run, dead, wounded, or in Castro's custody. The entire operation seemed to have been poorly thought out and ineptly executed. And the responsibility lay with Kennedy.

He was in no doubt that he would have to act now if he wanted to preserve his credibility as President, but he was not sure how best to go about it. Of the three main feasible space ventures — the first satellite, the first man in space, and the first man on the moon — the United States had blown its chances at the first two. And the third was so costly, as well as being so demanding — from a technical and scientific point of view — that it almost brought him to his knees. On April 20, 1961, to gain time, he asked Johnson, as chairman of the National Space Council, to prepare a comprehensive summary that would provide the answers to the following questions:

1. Do we have a chance of beating the Soviets by putting a laboratory in space, or by a trip round the Moon, or by a rocket to land on the Moon and back with a man? Is there any other space program which promises dramatic results in which we could win?
2. How much additional will it cost?
3. Are we working 24 hours a day on existing programs? If not, why not? If not, will you make recommendations to me as to how work can be speeded up?
4. In building large boosters should we put our emphasis on nuclear, chemical or liquid fuel, or a combination of these three?

5. Are we making maximum effort? Are we achieving necessary
results? . . .
I would appreciate a report on this at the earliest possible moment.[8]

This was the word for which Johnson had long been waiting. He
wasted no time in summoning NASA people and scientific and financial
experts to a long string of conferences. NASA advised against the idea of
a space laboratory. The superior Russian rockets posed far too great a
threat to their chances of being first. On the other hand, the idea of a
manned moon landing was met with more enthusiasm, because the Russians were probably not very far ahead of the United States in that particular area and, not least, because the very fact that such a demanding
project would require much more than simply raw power from the
booster rockets meant that the United States could cash in on its superior
technological, scientific, and industrial potential. Such a landing could
conceivably be accomplished in 1967, if they now put everything they
had into the project. The year was by no means chosen at random. It
seemed very likely that in 1967 the Russians would want to find some
spectacular way of celebrating the fiftieth anniversary of the Russian revolution; one which would put the United States in the shade yet again.

Kennedy's economists were not at all concerned — in part, because
imminent completion of the intercontinental rockets would allow them
a bit more leeway with the budgets and also because it was by no means
certain that they were going to pursue Eisenhower's cautious economic
line on balancing the national budget and "cutting one's coat to suit
one's cloth." It might be that the country was in need of a powerful
injection of funds from the public sector, to kick-start an economic up-
swing. Even if it did mean that government finances would go into the
red for a few years.[9] Maybe it wasn't such a bad idea to hock the jewels
after all.

Kennedy would be able to reap the political advantage of these generous space grants, inasmuch as Congress would be easier to deal with
if its members were given some public money to take home to their
constituents. It was particularly important for Kennedy that he soften up
the leading Southern politicians and, by so doing, make it easier to get
them to cooperate on civil rights legislation. In addition, Kennedy would
find himself on better terms with the powerful aerospace industry, without having to feed the Pentagon hawks.[10]

Von Braun was tremendously heartened by the possibilities this presented. It would call for a booster rocket bigger than any that had gone
before. This was right up his alley, and he doubted very much whether

the Russians would be able to build such a rocket within the timespan specified. So, to the surprise of none of those present, he wholeheartedly endorsed the idea of a moon landing.

At long last, in May 1961 — midway through that long run of all-night conferences — Alan Shepard made his fifteen-minute space flight. The first American had traveled into space, and been weightless for just under five minutes. And although this really was not such an impressive achievement when set against Gagarin's full orbit, the American people went wild with pride.

Meanwhile, with no Americans as yet having executed their first orbit, experts were actually planning a trip to the moon involving a rocket of unprecedented proportions, with a booster rocket a hundred times more powerful than Shepard's Redstone rocket. The earth-to-orbit weightlifting capacity of the rocket would be increased 10,000-fold.

Johnson's professional instincts told him there would never be a better time for pushing something this big through Congress. And a look at the newspapers left one in no doubt that, even if certain sections of the population had previously had their reservations about the enormous space budget, they were now game for anything.[11]

Kennedy's people were also beginning to view the enthusiastic spirit prevailing within NASA as something that could affect the entire nation, if everyone was seen to be working together on an all but impossible task that would draw on the sum total of the country's intellectual, scientific, technological, and industrial resources. A lunar project would revitalize every section of American society; galvanize listless industries totally unrelated to space research; give rise to new, and as yet unknown, products; and modernize the educational system and the research work being carried out in the universities. In short, make the American people sit up and take notice and muster them around one contemporary undertaking; behind a young and dynamic Democratic President.

Seen in this light, NASA became something more than a technological system concerned solely with astronauts and rockets. It became closely associated with the young, innovative President and with important new trends in both American commerce and the American educational system. NASA became a symbol, a high-tech dream factory, which the media was going to love — and that was all to the good.

Space-age exploits provided dynamic proof of technological superiority in the relentless battle between technological superpowers. They were popular symbols of how the modern state had allied itself to science and pioneering research.

And if all this was possible, then money was no longer an issue. Congress, which had already been asked to increase NASA's funding by 11 percent following Kennedy's succession to the presidency, would now be asked to increase that amount by a further 44 percent. And all the indications were that it would say yes — if not yes please. But in the years to come a lunar landing was to call for increases to NASA's budget of approximately $5 billion a year, for at least three to four years.[12]

Not all government advisers were delighted by this new line of thought. Some considered a public deficit of the proposed magnitude to be politically acceptable, but only if channeled into more traditional, democratic public works which would provide employment for a large number of low-paid workers. This scheme smacked more of jobs for the highly educated and highly paid. Even Kennedy was not absolutely committed to the space question — a point which he impressed upon Wiesner. If Wiesner could come up with some other high-potential technological enterprise, of a civilian nature, which would have the same international impact, Kennedy was open to suggestions. For instance, he had the feeling that some dramatic new technique for the desalination of seawater could create the same kind of stir.

The academic world was still opposed to the idea, finding it hard to see how a moon landing could possibly lead to important scientific discoveries. The director of the Carnegie Institute's geophysical department conducted an opinion poll among his research staff. The result was 110 to 3 against Apollo.[13]

The presidential advisers could not, however, come up with another public program that could both kick-start the economy and issue a strong and aggressive signal to the Russians; a signal that would have the international community rallying round the United States and forgetting all about the Bay of Pigs affair.

And so, on May 25, 1961, Kennedy stepped up to the podium in Congress and declared:

> Now it is time to take longer strides — time for a great new American enterprise — time for this Nation to take a clearly leading role in space achievement which in many ways may hold the key to our future on Earth. I believe we possess all the resources and all the talents necessary. . . .
>
> Recognizing the head start obtained by the Soviets with their

large rocket engines, which gives them many months of lead time, and recognizing the likelihood that they will exploit this lead for some time to come in still more impressive successes, we nevertheless are required to make new efforts of our own. For while we cannot guarantee that we shall one day be first, we can guarantee that any failure to make this effort will find us last . . .

I believe that this Nation should commit itself to achieving the goal, before this decade is out, of landing a man on the Moon and returning him safely to Earth. No single space project in this period will be more exciting, or more impressive to mankind, or more important for the long-range exploration of space; and none will be so difficult or expensive to accomplish. In a very real sense, it will not be one man going to the Moon — we make this judgment affirmatively — it will be an entire nation. For all of us must work to put him there.[14]

Many members of Congress did not applaud this statement quite as heartily as expected. Not because they opposed it. It was just that the whole concept was so vast that you could not help but feel very small. What the President was proposing could be compared to his standing alongside an airplane in 1903 and promising to produce the supersonic Concorde by 1913 at the latest. On the evening news, on the other hand, his words met with nothing but enthusiasm. An ecstatic commentator announced:

As early as next year, Kennedy will award NASA a budget of 1.7 billion dollars. That's ten dollars per head of population. But that, surely, is fair enough for what is by far the greatest show on earth: the dramatic race to the Moon.[15]

In one stroke, all of NASA's worries were swept away. Kennedy was the ideal President. Their budgets would soon be tripled, and the workforce increased sixfold. Everything was possible. What had once seemed superfluous had become indispensable. Now they just had to get cracking.

2. The Astronauts

The selection of the first seven astronauts was not considered a big problem. The requirements were simple. Applicants had to be military test pilots with good references, be under five feet eleven inches in height, be college graduates and family men.[1] The American forces boasted plenty of personnel fitting that description. But doubts had already been expressed by certain decision makers at NASA as to whether such men would, in fact, be interested, since there was nothing very alluring about the prospect of exchanging a supersonic jet, over which the pilot had full control, for a space capsule. The very word itself! A capsule was a container. It carried a load. That load could be anything at all — a chimpanzee, for instance. But who would ever have thought of letting a chimpanzee test-fly a Starfighter jet?

At a meeting in Los Angeles in March 1958, shortly before the founding of NASA, a team of experts from the government, the armed forces, and the aeronautics industry had come to the unanimous conclusion that the right way of venturing into outer space would be by improving and upgrading the X-15 planes. More powerful rocket engines would make it possible for these aircraft to fly up there, with an experienced pilot at the controls, who could — after spending some time in space — turn the aircraft and bring it safely to a landing. At Edwards Air Force Base, where else? This air base was the home of the world's toughest pilots; from here, trailblazing test pilot Chuck Yeager took off to break the sound barrier for the very first time. This would be the only proper way to set about it, but, for this, more time would be needed, and everyone could see that time was running out.[2]

And so they agreed upon a stopgap solution, a "quick and dirty" approach, known as MISS, Man in Space Soonest. A capsule, or pod, would be launched — attached to the tip of one of the existing military rockets. And in this capsule they would put a man, who did not need to be a pilot. Why should he be, when all he had to do was act as a human cannonball or, to put it more politely, an "aeromedical test subject."[3] If he just lay nice and still and caused no trouble, the experts would do the rest and see to it that he landed, with an admittedly somewhat undignified splash, at some predetermined spot in the ocean.

With the advent of NASA, the MISS project was withdrawn from under the Air Force's wing, only to reappear a little later as the basic idea behind the Mercury project. More than 500 interested military pilots were carefully assessed and pruned down to 100. These then underwent extensive physical and psychological examinations before being reduced to "the Mercury Seven," who made their first, rather diffident, public appearance on April 9, 1959.

More than a year after the first astronauts had been selected, voices within NASA were still quite openly referring to the astronauts as "a redundant component," introduced into the system just in case any of the space capsule's instruments should, contrary to all expectations, fail to function. Other than that, it would be best if they kept their hands to themselves. What mattered most was that they be wired up to electrodes "from skull to rectum," as Tom Wolfe put it, and fitted with microphones, thus enabling those on the ground to form some picture of a human being's reactions during space flight.

Some NASA psychologists argued that, if this was what the job amounted to, then it was a mistake to opt for pilots, who had proved that they could cope with amazing levels of stress in their potentially lethal aircraft, precisely because they were capable of doing something; because they felt that they were in control of the situation and had only themselves to blame if anything went wrong. Instead, the psychologists suggested using the radar operator — the new man in the modern two-seater fighter-bombers, who simply hung on in his seat behind the pilot, with his eyes pinned unwaveringly on his radar screens, come what may. Such a man would make the ideal astronaut, having been trained for and, presumably, being psychologically predisposed toward finding himself under pressure without any chance of taking control of the situation.[4]

They even went so far as to consider various types of tranquilizers, which would ensure that the astronauts' test pilot side did not get out and start playing havoc with the flight. "An experienced zombie would do fine," as Wolfe writes. By the same token, a very large part of the training program was devoted to "deconditioning, desensitizing or adapting out of fears." Psychological theory had it that repeated exposure to powerful stimuli of a fear-inducing nature would gradually render the test subject insensitive to fear; hence he would not panic, or act inappropriately, or even just act instinctively when he ought to be keeping still.[5]

The fact that Eisenhower paid little heed to the astronauts was fully in keeping with this view of them as test subjects.[6] Likewise, it was hardly the President's job to sing the praises of other individuals — such as those brave souls who did such an excellent job of testing new types of ejection seats for the Air Force. They performed a valuable service, but that didn't make them special enough to warrant the attention of a busy President.

As we have seen, the public did not share this viewpoint. They had instantly identified the seven astronauts as being the incarnation of the best that America had to offer. They were courageous, modest, polite, young white men who radiated good health and had come a long way under their own steam. And they had nice families, unburdened by so many of the problems that had swamped other American families.

Already, during training, the distribution of power between the astronauts and NASA had undergone an imperceptible change. The intensive press coverage had increased the astronauts' status and, hence, their self-esteem. They were, for example, tired of hearing about "the capsule." At their insistence the capsule was first referred to as "the module" and later "the spacecraft."[7] They were also appalled to discover that the designers had no intention of providing the capsule with windows, which, from a structural point of view, were totally superfluous, besides posing a potential threat to the capsule's pressure sealing. Instead they had installed a little periscope for the astronauts' use. But the astronauts wanted windows, and so the designers had, grudgingly, to go back to the drawing board. They also wanted to have control of the rocket, and here the engineers had to bite their lips to keep from laughing at this instance of galloping megalomania. No man alive could control such a rocket. It was totally dependent on its automated systems, its gyrostabilizers and computers both on board and on the ground. Then along comes some astronaut and wants to take control!

NASA conceded another point and allowed the astronaut to operate the small reaction-control thrusters, which could alter the capsule's attitude in space during the weightless phase. These thrusters could be controlled by something that could, at a stretch, be taken for the joystick of a jet fighter. Now the astronaut could at least turn the capsule around, so that he had his back, and the rounded, heat-shield end, facing down the line of flight before the scorching return trip through the atmosphere, when everything would revert to automatic. The astronaut was also, reluctantly, granted the freedom to switch off the automatic controls before reentry and activate the retro rockets at the precise instant when the ground crew gave the word. And he was allowed to open the hatch himself, after landing, by means of detonator bolts.

Many engineers felt that life as they knew it had been turned absolutely on its head. At the top of the tree, as everyone knew, were the scientists, men who had hatched great theories, men descended in a direct line from Newton and the Laws of Motion themselves; then came the engineers, who strove to realize the scientists' theories, as far as was practicable; below them came the technicians, who kept an eye on all of these complex systems once they had been developed; and, at the bottom — the very bottom — was the test subject, no matter how brave he might be.[8] But now the test subject was actually starting to make his voice heard. And where did that leave you?

But the engineers too succumbed with astonishing rapidity as the patriotic response to the astronauts swelled. To the general public, they were far more than just guinea pigs; they formed the advance guard in a life-and-death struggle between the United States and the Soviet Union. And the way the American people saw it, it was up to the astronauts whether this fight could be won or whether the United States was to be consigned, for eternity, to the historical slipstream left by the Communist empire.[9] And although no one knew better than the engineers that this was by no means the unvarnished truth, they too were caught by the mood. More and more often, NASA workers and engineers would automatically put down their tools and applaud if an astronaut passed through the hall where they were working, on his way to yet another test.

This wave of sympathy for the astronauts reached its first peak, of course, in May 1961, when Alan Shepard became the first American to execute a flight through outer space — albeit for just 15 minutes and 22 seconds. And despite the fact that he had — with the aid of those hard-won manual functions — made a fair number of blunders.[10] Nevertheless, Shepard was a hero. The prototype of a true American. Ice cold under pressure.

He waited quite calmly, at the top of his Redstone rocket, for hours, while the countdown was halted temporarily several times, to sort out minor hitches. Waited — until the live sound transmission to millions upon millions of television viewers was interrupted for a moment — before telling the technicians that he couldn't hold out much longer. In a moment his bladder was going to burst. There was a long pause, while the static crackled over the public-address system. No one had allowed for such a problem arising on a fifteen-minute spaceflight. Going to all the bother of extricating him from his capsule, bringing him down from the tower, and accompanying him to the nearest toilet was out of the question. So, after hastily summoned experts had held a rapid, whispered conference, he was given a "Roger" to empty his bladder into his pressure suit—an act which led, shortly afterward, to all the temperature sensors giving out weird and hitherto unseen readings. But Shepard felt better. Coolly he told them they could now "light the candle" just as soon as they were ready.[11]

Shepard was received at the White House and his wife had a chance to see Jackie Kennedy at close quarters—the dream of every American woman at that time. He was also paraded in an open car through Washington, where tens of thousands of people turned out to cheer him, and later, in New York, he was treated to a ticker-tape parade. John F. Kennedy took a liking to Shepard. He had immediately recognized that he would find no better exponents of his political comeback than the astronauts. They were the very embodiment of the new America that Kennedy, at his inauguration just a few months earlier, had promised to build.[12]

In the days immediately after Shepard's flight, experts from NASA, the armed forces, and the Treasury drew up a memorandum intended to assist Kennedy in making his eventual decision on a lunar landing. It read:

> It is man, not merely machines, in space that captures the imagination of the world. All large-scale projects require the mobilization of resources on a national scale. They require the development and successful application of the most advanced technologies. Dramatic achievements in space therefore symbolize the technological power and organizing capacity of a nation. It is for reasons such as these that major achievements in space contribute to national prestige.[13]

From then on, the astronauts became an indispensable element in every PR exercise involving either the President or NASA. All in all, it was becoming increasingly difficult to differentiate between these two entities. NASA was Kennedy's tool now and the astronauts were Kennedy's personal friends.

But the one astronaut who was to deal the greatest blow to the nation's solar plexus was John Glenn. Glenn had it all. He sported a chestful of medals from his exploits as a pilot in the war over the Pacific and in Korea. In July 1957, the setting of a new coast-to-coast speed record brought him to the attention of radio and television stations nationwide, and he appeared on several quiz shows.[14] So Glenn was the only one of the seven whom people felt that they already knew. Besides, he was a practicing Christian and never neglected to emphasize that, without God, we human beings were nothing. Glenn rapidly developed into the astronauts' press spokesman. Alone among them, he had the knack of always saying the right thing at the right time. More than any of the others, Glenn personified such all-American virtues as piety, hard work, respect for the flag and the family.[15] And he spoke of these things, and of duty, without the hint of a blush. Every other American was bound to be shown up as an opportunist, a grasping, loose-living con man, eternally pursuing vacuous pleasures, when compared with John Glenn.

This is why, on February 20, 1962, 100 million people sat mesmerized in front of their television screens as John Glenn prepared, after several postponements, to take off on a spaceflight that would put the first American into orbit around the earth.

During an earlier attempt, less than a month before, John Glenn had demonstrated the astronauts' new power in no uncertain terms, when his house was surrounded by broadcast vans from the various television networks — all anxious to record his wife's reaction to the anticipated launch. Journalists and television technicians trampled the front garden into mire in their efforts to get close enough, while Annie Glenn sat behind closed shutters, scared out of her wits — not so much because of the spaceflight as because she had a stammer and did not want to find herself stuck, stuttering, in front of the entire nation.[16]

The situation was complicated further by a telephone call announcing that Vice President Johnson, who had long felt overlooked by Kennedy, wanted to drop by and watch the launch with Annie. And just to be on the safe side, he had instructed the three nationwide television networks to be in attendance. As an act of courtesy, his limousine would park in a side street until Annie was ready, and he could make his entrance with the appropriate pomp and circumstance. But Annie Glenn was not ready, and she was in no mood for any visit from the Vice President of the United States. Lyndon Johnson fumed with rage and asked his people to bring some pressure to bear. He had no intention of being made a fool of by any astronaut's wife.

When — with that day's launch postponed yet again — Glenn climbed down from his bed of thorns, he was asked by some very agitated NASA officials to call home immediately and talk some sense into Annie. Keeping the chairman of the National Space Council waiting might not be such a wise move. A bewildered Glenn rang home to find out what all the fuss was about. He had to hear very little of the story before giving his wife his full support:

> Look, if you don't want the Vice President or the TV networks or anybody else to come into the house, then that's it as far as I'm concerned, they are *not* coming in—and I will back you up all the way, one hundred percent, and you tell them that. I don't want Johnson or any of the rest of them to put so much as *one toe* inside our house![17]

James Webb too tried to talk to Glenn but was told that this was way out of line. Webb had never been spoken to in this way by a subordinate before. He contemplated having Glenn replaced, but horrified advisers persuaded him not to. If he took to doing that sort of thing, he could forget all about the other six astronauts. So Johnson had to drive home again. John Glenn had spoken.

At the end of February they pulled it off. Television viewers watched as the Atlas rocket, functioning perfectly for once, made its stately ascent, with John Glenn on board. They gazed, entranced, at the grainy pictures in which the rocket seemed to be moving in a series of sudden jerks— as it occasionally outdistanced the camera's powerful telescopic lenses and left the cameraman struggling to locate it and bring it back into the picture.

Glenn was accelerated to a uniform motion; one which was barely distinguishable from a state of rest, even though he was moving at a speed of five miles per second. This motion would continue in the vacuum of outer space—as stated by Newton—until the speed was altered by small retro rockets. Then the capsule's trajectory would no longer pass over the earth's surface, but would, instead, home in on the predicted target area. On its way down through the atmosphere its speed would suddenly become perceptible again, as it was converted into unbelievably high temperatures on the outer side of the heat shield, just inches from John Glenn's back.

When, during the flight, some doubt arose as to whether the heat shield had worked itself loose—which would mean that Glenn could be incinerated in the atmosphere on his downward journey—no one made any further moves away from their television sets or their new transistor radios for the duration of the almost five-hour-long orbit. The heat shield turned out not to have come loose after all. The nation breathed a sigh of relief when radio contact was once more made with John Glenn, after his trip down through the atmosphere. Then the Mercury capsule, complete with its unscathed astronaut, appeared, dangling safe and sound beneath its parachute; and not long afterward Glenn was standing on the deck of the USS *Noa*, where jubilant marines painted footprints on the deck at the spot where he had first set foot.[18]

This tribute was only the beginning. Johnson bore no grudge and, on welcoming him home, called him "one of history's great pioneers," while others spoke of the United States' "new Columbus." Kennedy made the trip to Cape Canaveral personally, to shake his hand and pin a medal to his chest. Unlike Shepard, Glenn did not have to go to the President. Three days later, in Washington, 250,000 people stood waiting in torrential rain to greet him. And there, in the seat of government, he was accorded the unique honor — normally reserved only for the President or prominent foreign heads of state — of addressing the two houses of Congress. Glenn told the assembly, "I still get a lump in my throat when I see the American flag passing by." And he asked them to look up to the balcony and salute his parents and, not least, Annie, without whom he would never have made it. By then — all around the auditorium — even the oldest and most cynical of politicians were openly weeping. Some days later, this "aeromedical test subject" received the biggest welcome in New York's history, with four million people also having difficulty in holding back the tears along the fifteen-mile route.[19]

10. NASA's Finest Hour

John Glenn's flight made it easier to bear the fact that the Russians and the mysterious Koroliev struck again and again; and that, the year before, Gherman Titov had made not three but seventeen orbits of the earth — thus traveling 436,000 miles, or approximately the distance to the moon and back. Americans also had to live with the fact that, in August 1962, the Russians had maneuvered two cosmonauts, each in his own space capsule, to within four miles of one another; and accept that the Russians had sent up a female astronaut who had, on her voyage through space, covered a greater distance than all of the American astronauts put together.[1] The main thing, they convinced themselves, was that the United States no longer screwed up all the time, and that the nation was now working toward a clearly defined goal.

NASA had never had it so good as in the years from 1961 to 1966. The organization had a definite, clear-cut goal; it was continually being supplied with new equipment and awarded additional grants; it was protected from its military rivals and it had accomplished a task of great national importance. In 1961, for the first time, this civilian organization spent a fraction more than the armed forces on space projects, in 1962 somewhat more again, in 1963 double the amount, in 1964 and 1965 over three times as much![2] Eventually, having been increased fifteenfold, NASA's budgets amounted to around $5 billion a year. And the workforce expanded accordingly. There was money enough to purchase all the necessary expertise; to attract the sharpest minds from the worlds of science and technology, and to equip the laboratories and technical facilities which they requested.

America's Apollo project is a textbook example of what is known as "visionary leadership." NASA had a vision and a time frame and that was just about all: Put a man on the moon and bring him back safe and sound. At the time when this vision was formulated, John Glenn had not yet been put into orbit around the earth and the most powerful

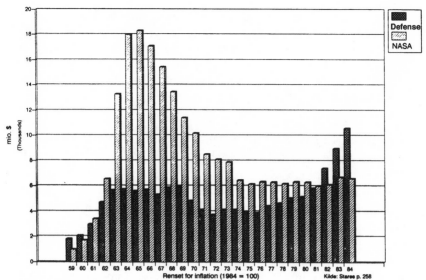

EXPENDITURES OF THE U.S. ARMED FORCES AND NASA'S SPACE PROGRAM. (STARES, 1985)

rockets at NASA's disposal (Atlas) exerted a force of twenty times less than that required for a trip to the moon. At that point no one knew quite how they were supposed to develop these rockets or how a lunar landing was to be effected.

One could be forgiven for thinking that such an incredible gap between reality and vision would leave those involved feeling unnerved and paralyzed, but, in fact, only at very weak moments was this the case. Most of the time, the most striking feature was an all-consuming commitment to the task at hand. Tom Wolfe has described the times when workaholics had to be forced to go home, to take a break. During his interviews, he heard how NASA staff were on a perpetual adrenaline high, and how veterans of those days felt that they never again experienced that same level of intensity. Everyone lived and breathed for one thing, and one thing only, during this period, and they experienced things normally experienced only by men during wartime — when a moment's inattention could cost lives. People worked their fingers to the bone on their own little segment of the project. They had never had it so good or felt more alive.[3]

Joseph A. Raelin has observed that it is a characteristic of highly educated technicians, engineers, and scientists the world over that they hate the idea of their time and their education not being exploited to the full.

> Professionals, by training and inclination, want to stretch their in-
> tellectual abilities to the limit. They desire, particularly early in their
> careers, to be involved in their work, and to make use of every
> minute. They also want to identify with the whole product or service
> on which they are working. They want to be able to see their con-
> tribution.[4]

During these years NASA had no trouble satisfying these requirements.
Every ounce of energy and every idea was needed if this project was to
succeed.

According to Raelin, there are four elements which the typical highly
educated professional needs to have fulfilled if he is to give his best.
First, he must have autonomy. Management is welcome to define the
aims, but should leave the ways and means up to the employee and not
keep looking over his shoulder. Professional people prefer to control
matters themselves and to have plenty of leeway. Hard-and-fast job de-
scriptions are highly unpopular. Second, there has to be a challenge.
Preferably a need to achieve something that has never been done before;
something calling for often unexpected combinations and applications
of the wide variety of skills at the employee's command. Third, the professional likes to think of his job as being meaningful. And if the work
happens to be socially significant or of a morally irreproachable nature,
then so much the better. Fourth, the work should be performance-based.
The employee does not object to being subjected to assessment and
criticism, but not until the results are in. And how these results were
arrived at is no concern of management.[5]

The above-mentioned observations would suggest that many of those
highly educated specialists whom NASA was eager to recruit had a somewhat ambivalent attitude toward the concept of management. They preferred to work independently, or at least to be given a great deal of
freedom. And such freedom could be facilitated by a corporate identity
that subtly set a standard or norms for acceptable behavior.[6] And from
the beginning NASA had a strong corporate identity. The familiar logo
cropped up everywhere. Colors were bright and the laboratories and
workshops sparkling clean. All the indications were that this was a place
where precision work was carried out. The organization was renowned
for its expertise and for being in the forefront of research. It was an honor
and a mark of distinction to be made part of the NASA family, and it
raised one's standing in the eyes of former colleagues. Who wouldn't
want to work for a winner? No new NASA employee needed a boss to
tell him that he was expected to do his best. In such surroundings he
would never dream of doing anything else.

As a bonus, he would find himself working with high-technology equipment that was used for benign purposes. This was civilian work of the same high level as work on the Pentagon's lethal, secret systems. But unlike his "secret" colleagues, the NASA employee had no difficulty in feeling unreservedly proud of his work, and could discuss it openly with his wife and his friends.

A former NASA chief, Thomas O. Paine, put it like this:

> I think we all tend to operate at about one percent of our capacity. But, by God, when you get into a wartime situation . . . or into something like the moon program, then you begin to operate at maybe 110 percent of your capacity. It really brings out the best in people, having tough and challenging things they have their heart and soul in. And that's the kind of operation NASA was. So even the ordinary people at NASA began to be giants in that era.[7]

But, of course, NASA did have a management structure, functioning at all levels. It was a new, modern style of management. Kennedy's idea of NASA as the driving force behind changes to the more rigid norms within American industry began to have an effect. Whereas many American companies still regarded the workforce as lazy and irresponsible and in need of constant supervision and, hence, believed that clear-cut hierarchical chains of command, punishment and reward, and, above all, respect for one's superiors were essential, NASA — like the smartest companies in 1960s America — was moving in quite a different direction, and making a success of it.

At NASA, you worked in teams; subsidiary projects were run by means of worker participation and often there was no way of telling the difference between workers and management. The bosses ate in the same cafeterias as the engineers and technicians; anyone could raise a problem with their department head while waiting in line for lunch, and very often have it sorted out there and then, before they reached the cash register. Project leaders at NASA understood that you could not boss around highly educated men but you could win their respect by showing a personal interest in their projects. It was also crucial that no one be punished for failures or mistakes, but that these be seen as an indication that the person concerned was taking risks, and that he was pushing the bounds of what was technologically possible. And, above all, the workers had to feel free to admit their mistakes, rather than sweep them under the carpet for fear of being punished. At a later date any mistake might prove to be disastrous.

Translated into management jargon, at NASA they were experimenting—as had, unwittingly, been done at Los Alamos and, to a lesser degree, at Peenemünde—with management techniques in line with the theories advanced by the idiosyncratic Human Relations movement as far back as the 1930s and 1940s:

1. An organization is not a machine but a social system made up of people with all their hopes, fears, envy, desires, and needs.
2. Noneconomic rewards have as much bearing on productivity as economic rewards. Friendship and collegiality are important parts of the job.
3. Productivity cannot be achieved simply through the direction of management but requires concurrence by the worker group in the form of social norms.[8]

These principles had arisen from the working methods adopted by scientists in great research laboratories which could trace their roots all the way back to Thomas Edison's day. In such places, keeping tabs on people and giving orders would get you nowhere. Results could be achieved only by letting the scientists have their head and encouraging their curiosity—in whatever direction it happened to take.

Around 1960 these principles were taken a stage further by Douglas McGregor in his so-called humanistic management movement, in which he mocked the views of more traditional executives, whose basic outlook he dubbed the X theory. This presupposed that all workers had to be forced into doing anything, and their work carefully monitored—quite in keeping with Frederick W. Taylor's and Henry Ford's scientific management theories from early in the century. Against these McGregor set his own Y theory, which presupposed the opposite: that all workers were creative and were prepared to shoulder their share of mature responsibility, if given the chance. All that was required was a management bold enough to work toward a situation where the employees became involved in decision-making processes that concerned them and had the confidence to admit their mistakes and take risks without the fear of being called to account. And that the management ventured to permit informal communication between workers and management with no weight being placed on formalities or any outward show of respect.[9]

In other words, the organization had to have the courage to promote a spirit of equality within the company. People had to have a respect for one another that did not stem from titles but came from an appreciation of each person's different abilities, all of which were vital to the work as a whole. And after a successful launch, the most telling managerial ploy might be to light some fat cigars and get drunk with the workers, thus communicating the message: We've done this together, and we all have good reason to be proud. To coin another management term, NASA was practicing professional accountability. The management trusted its specialists; the relationship was in fact very similar to that found between a layperson and an expert.[10] Many NASA workers found it an intoxicating experience to see a public company run this smoothly when, for so long, they had been used to anything that so much as smacked of government control being sluggish, stuffy, and impenetrably bureaucratic. It struck them as amazing that a government body could actually be galvanized by patriotism, enthusiasm, and commitment to such a degree that their associates in the private sector regarded NASA's achievements with barely concealed envy.[11]

Because, of course, NASA was not alone in this enterprise. The great majority of the workforce that eventually became involved in this space-age adventure were in the private employ of big aerospace companies which had previously worked almost exclusively for the Pentagon. These companies now had to compete fiercely for civilian NASA's favors, under conditions which they found much stricter than those they had been used to with the armed forces. They had to resign themselves to relinquishing their absolute control of production processes. NASA set great store by the fact that it commanded the highest possible technical expertise throughout and, thus, was in a position to keep close watch on its suppliers — in whatever field — and could, under the heading of "contractor penetration," assure itself of independent quality-control checks — often to the great annoyance of these suppliers.[12]

The Apollo program drew upon the services of more than 20,000 private companies, at all levels — although half of all the funds awarded to NASA in the years up to 1966 went to the four big boys of the military-industrial complex, North American, McDonnell Douglas, Grumman, and Boeing.[13] Only 10 percent of NASA's funding was retained within the organization.

And over all of this presided a President, John F. Kennedy, who had great faith in NASA, and a chief executive, James Webb, who could contact the President personally at any time — often to the indignation of Lyndon B. Johnson, who certainly thought he deserved better. Even Congress seemed anxious to please and not disposed to count the pennies — particularly not after Albert Thomas of Texas, the powerful chairman of the House of Representatives select committee that had sanctioned NASA's budgets, succeeded in having NASA's Manned Space

Flight Center located in Houston, a move for which most people had trouble in finding any other decent justification.[14]

John F. Kennedy was carried away by the new technological possibilities represented by NASA, and regarded the organization as a reflection of a society which no longer had to be held back by economic restrictions, rigid professional demarcations, and old party lines. It *was* possible for the public and private sectors to work together. Bosses and workers had buried the hatchet. Everyone willingly waived their own special interests in order to collaborate on a bigger, common project which would benefit the whole nation, or as he was wont to say in more impassioned moments, the whole of mankind. Concord *was* possible, if only the goals were high enough. In October 1963, Kennedy announced:

> As we begin to master the potentials of modern science we move toward a new era in which science can fulfill its creative promise and help bring into existence the happiest society the world has ever known.[15]

And on November 21 he struck the same chord of technological optimism in a speech made in San Antonio, in which he said:

> Frank O'Connor, the Irish writer, tells in one of his books how, as a boy, he and his friends would make their way across the countryside, and when they came to an orchard wall that seemed too high and too doubtful to permit their voyage to continue, they took off their hats and tossed them over the wall and they had no choice but to follow them.
>
> This nation has tossed its cap over the wall of space, and we have no choice but to follow it. Whatever the difficulties, they will be overcome. Whatever the hazards, they must be guarded against.[16]

The next day, Kennedy was assassinated and NASA had lost the President who had, more than anyone else, impressed the force that put the space organization into orbit. From now on it would have to get used to running under its own momentum.

Jim Webb's NASA

James Webb had been Kennedy's choice as leader of NASA, and Webb more than anyone was to miss the unconditional support of NASA's President. He ran NASA from the days when the organization could draw on the services of 75,000 employees in its own laboratories, in the universities, and with private subcontractors, until — in the summer of 1965 — the payroll peaked at almost half a million — whether directly or indirectly employed. Webb was the man who forged the management techniques necessary to the success of this colossal undertaking; a man who could not help but wonder to what extent these methods developed by NASA might be applied to other fields.

In 1968, while he and NASA were finalizing preparations for the lunar landing, he was unexpectedly fired by a volatile President Johnson, who was by then sinking deeper and deeper into the quagmire of Vietnam. His dismissal came as a shock both to him and to NASA, but it did present James Webb with the time — all of a sudden, and after seven years in the hot seat — to take stock. He recorded his experiences in a book entitled *Space Age Management: The Large-Scale Approach*. In this, he is revealed — not unexpectedly — as a most dynamic character, harboring no doubts whatsoever as to his own or NASA's worth. He quotes — possibly with a nod in Johnson's direction — a comment made by C. Stark Draper in 1965 that expresses the optimism also to be found in Kennedy's last speeches:

> Pioneering technology, which starts from purely mental conceptions and leads to valuable results without precedent in the real world, is among the most important ingredients of human progress. Modern civilization is particularly dependent upon continued contributions from technology as it advances toward ever-higher levels of achievement. Individuals and organizations capable of initiating

and leading the new technological developments essential to this
advance are rare and must be encouraged and supported as a simple
matter of self-interest on the part of society.[1]

By and large, Webb's statements are imbued with a passionate belief in
science and technology as the key to progress. NASA taught him that
everything is possible, if only sufficient energy and resources are invested
in accomplishing a task. Hence his fascination with large-scale enterprises: the great social reforms of the New Deal, which created jobs for
the unemployed and supplied cheap power to underdeveloped areas; the
building of the Panama Canal; the Manhattan Project, which produced
the atomic bomb, and the even more anonymous project which developed military intercontinental ballistic missiles. In every case, Webb as-
serts, these projects were preceded by chaotic crises, with the public in
a panic and the authorities not knowing which way to turn. Conflicting
solutions abounded — all of them pandering to ill-concealed special interests. Congress held hearings and instituted inquiries, none of which
did any good. And in every case the answer had been to give one man
or one authority the power to cut through the special interests and solve
the problem.[2]

Webb was obviously fascinated by wartime situations, when ordinarily
stringent procedures and precautionary measures are relaxed. Then everyone pitches in and the day is saved in an astoundingly short space of
time. That is how he sees NASA. The organization was able "to take a
caterpillar and make it into a butterfly, when we had never seen a butterfly."[3] The task assigned to them was an unnerving one; and again and
again, in the years just after *Sputnik*, their rockets came to grief. But they
persevered, to win through in heroic style with "by far the largest program
of its kind ever undertaken by a democratic society in peacetime,"[4] because they did not waste time talking. Instead they just got the job done.

What Webb admires most about NASA's structure is the way in which
Congress merely defines the goal the nation wishes to achieve. It is left
up to the space organization to suggest what methods to adopt — and
these Congress will then endorse, a year at a time. The rules are very
simple. If NASA does its job, everything will be just fine; if not, the purse
strings will be drawn tight. Hence, the head of the organization's primary
task is to ensure the continued support of Congress; this is especially
important because with such a project a very long time can elapse between the conception of an idea and the actual end product. In the
meantime, Congress has to be kept receptive. And while working toward
that final objective, it does no harm to produce some impressive intermediate results in good time for the annual budget negotiations.

Proudly, James Webb quoted from one of Adlai Stevenson's last public statements, in which he noted, with great enthusiasm, the capabilities we now have. Science and technology, he said, are making the problems and activities with which we are so concerned today

> irrelevant in the longer run because our economy can grow to meet each new charge made upon it. It will stagnate only if we do not ask enough. This is the basic miracle of modern technology. This is why it is, in a real sense, a magic wand which gives us what we desire. Don't let us miss the miracle by underestimating this fabulous tool.[5]

Difficult as it may be nowadays to comprehend such an optimistic view of technology, this was a perspective shared by many people in the mid-1960s. Webb himself believed that the lessons learned from NASA could prepare us "on an almost fixed time schedule [to] meet new needs or effect desired improvements in our situation as a people and as a nation,"[6] as well as "to ensure continuing progress toward the great goals we set for ourselves long ago"[7] and "the means to achieve, for ourselves and for others, levels of being that we have hardly dared dream of before."[8] All that is needed is for the right people to be given a free hand; for management, workers, and government to work together, for both the private and the public sectors to abandon unprofitable rivalries, and, finally, for academics and scientists to climb down from their ivory towers and enter wholeheartedly into the creation of a modern technological society — for the common good.

Although it is never said in so many words, one does not have to delve far into Webb's book to realize that he is itching to lick the *whole* of "USA Incorporated" into shape. He regards NASA as a visionary agency, creating "revolution from above."[9]

It is visions such as these that have prompted modern writers like Dale Carter to detect totalitarian traits in NASA. In his book *The Final Frontier* he uses the term "the Rocket State" to describe this image of a conflict-free society emphatically dominated by well-oiled technology. This is no old-style totalitarian society — the kind that functions by virtue of the exercise of power and subjection to authority. Here, the methods are much more subtle. PR and momentous symbolic events can combine to produce "voluntary totalitarianism": "The willful sacrifice of personal and factional interests to a greater public good."[10] Outer space (and NASA) seems so pure, so unblemished by all the things we human beings have done to one another in the past. Seen in this way, NASA becomes a divine society, promising redemption at a time when people have lost their faith in God. And the astronaut

> embodied those universal American values of piety and hard work, of family and flag. Whether in military service or atop Army and Air Force launchers, he "maintained a sense of discipline while civilians lived by opportunism and greed."[11]

The more insecure, powerless, and insignificant ordinary people feel, the greater their need to see the astronauts as pure heroes and role models. Dale Carter has described typical NASA employees as "the guys with crew-cuts and slide rules, who read the Bible and get things done."[12] Wherever he went in 1969, the writer Norman Mailer kept bumping into these men, and found them to be "true Christians, gentle, helpful, replaceable, and serving on a messianic mission."[13] Just the kind of employees that many American employers would have given their eyeteeth for during those years.

In his book, James Webb sets out the ten key factors which normally typify an endeavor of the Apollo program's nature and scale:

> *First*, they are ordinarily undertaken as a result of a significant change in the environment — social, political, technological, military, or other — that raises a new and urgent need or presents a new opportunity. . . .
> *Second*, interaction between the environmental situation and large-scale endeavors is a continuous and often turbulent process. . . . This means that for each large-scale endeavor there is a critically important need for continuing feedback of information regarding the environment and, at the same time, sufficient flexibility in organizational structure and management processes to enable the enterprise to "ride out" unexpected turbulence.
> A *third* very important feature of large-scale endeavors is that organizing, administering, reorganizing, and administering the reorganized structure provide the key to the effectiveness and usefulness of such endeavors, rather than the invention of entirely new machines or processes. . . .
> *Fourth*, large-scale endeavors do not generally require new organizational and administrative forms, but the more effective utilization of existing forms. . . .
> *Fifth*, all large-scale endeavors have a number of unusually complex managerial requirements. They depend on men with special, often unique, skills, men trained in a variety of disciplines, men

who by their very nature raise unique problems for management. Communications are of unusual importance, particularly communications related to the collection and use of feedback. Enormous quantities of data are indispensable, but this in turn creates a special problem of unwanted data, or "noise," and possibilities of confusion. . . .

Sixth, another common denominator of large-scale endeavors is the necessity of a continuing "critical mass"[14] of support. There must be enough support and continuity of support to retain and to keep directly engaged on the critical problems the highly talented people required to do the job, as well as to keep viable the entire organizational structure. As with an airplane, the initial support must be adequate to attain the equivalent of "flying speed," and support must continue at a tempo and at a level adequate to maintain the equivalent of an efficient flight path. . . .

Seventh, large-scale endeavors are increasingly concerned with the utilization, and often the development, of advanced technology and the application of new knowledge. Most require doing something for the first time and have a high degree of uncertainty as to precise results.

Eighth, all large-scale endeavors have important secondary and tertiary effects beyond those associated with the prime objective. These alter the environment and significantly impact events generally . . . produce social and sometimes political consequences of lasting importance.

Ninth, a number of intangibles characterize the large-scale endeavor. They are essentially investment enterprises representing a willingness of a group or a society to give up resources in hand for future returns that may be long delayed in realization. They are ordinarily unusually challenging from the standpoint of their uniqueness, their promise, or the urgency of their need, and they tend to appeal to creative instincts on the part of supporters and participants. Conversely, as new ways of doing things, they are prone to generate resistance or dissent.

Tenth, large-scale endeavors, whether public or private, invariably loom large in the public eye. They are subject to constant watchfulness on the part of supporter and opponent alike. Today they must operate under the glare of TV lights, not at times of their own choosing, but when someone else wants to look them over. They tend also to be subject to a double standard. When anything goes wrong, there is a rush for the seven-power glass and the microscope. Mistakes are heavily taxed.[15]

It is evident from this list that Webb had no time for laissez-faire management. Even though engineers on the shop floor at NASA might think that they had plenty of room for maneuver, and regard management as very open and accommodating, Webb is quite conscious of the need for monitoring the state of play throughout the organization. He is the cook, with seven different dishes on the stove at once—all of which have to be ready before the guests arrive. Such an undertaking is far too great, too complex, *not* to have a strong central leadership, capable of coordinating all of the NASA center's efforts.

Webb was not afraid of the power contained within these vast technological systems or of taking a hands-on approach to them. Quite the opposite—the power fascinated him. Put the right man on the bridge, and that power can be tamed—so long as he realizes that he will have to concentrate on a great deal more than the technical details. He will have to be on the alert, and ready to react like greased lightning, twenty-four hours a day.

To Webb, a large-scale endeavor is like a rocket. It is unbelievably powerful, but is also totally unstable if left to its own devices and not monitored many times per second, from thousands of sensor points— thereby enabling management and Houston to react to the slightest variance in the system, before it can develop into a disaster. Only when the rocket is safely in orbit can they relax.[16]

Webb's list further underlines why NASA, from the outset, had put press coverage high on its list of priorities. Attention of this sort is the lifeblood of the organization. Through their politicians, the American people have placed very large sums at NASA's disposal; sums which the nation could have elected to spend on other things. And these same people need to be given some return for their money—not just at the end of the day but also along the way. The people's interest will have to be stimulated and kept alive for eight years, before the first man sets his foot on the moon. This calls for a management with a nose for public relations and a well-developed sense for political legwork. The Congress select committee has to be cultivated. Politicians wanted a slice of the cake for their own state, or at least a visit from a couple of astronauts now and again.

Webb had his second big confrontation with John Glenn when the astronaut told him that he was no longer prepared to go traipsing all over the country just because some damned congressman needed to be photographed with him. He felt that he had more important things to attend to. Once more, Webb was infuriated by Glenn's arrogant defiance, but above all by the fact that he could not see how, from a management point of view, traveling around the country *was* the most important way of using John Glenn during this phase, when maintaining the momentum was all-important.[17]

Repeatedly, in James Webb's list, the point is made that an organization the size of NASA is not divorced from the society in which it lives and that it cannot turn its back on it. Sooner or later, the society's problems will become the organization's problems. And so it does not surprise him when, in the late 1960s, NASA is faced with growing criticism for not doing enough to help the problems of the inner-city slums; not building better schools; not supporting developing countries; not solving the world's food shortage; and not bringing about peace in Vietnam.[18] None of the above fall naturally within NASA's mission — no matter how loosely it might be interpreted — but that is not how the general public saw it. NASA too has a responsibility — if only because it looms so large on the American horizon. And Webb goes along with this view. Modern managers must accept the fact that it is no longer enough just to concern themselves with their company's performance. They also have to consider

> school dropouts, crime rates, the prevalence of poverty, the number of university graduate students, effectiveness of government policies, incidence of group violence, and even indexes of the nation's willingness to act like a great power on the international scene. Managers must look at these things not as concerned bystanders or innocent victims or charitable outsiders willing to lend a helping hand. They must look at them as elements of their own jobs, elements as vital to their success as managing internal conflict or providing psychological props for their subordinates.[19]

This last quote of Webb's reflects the fact that at this time — in 1968–69, during his enforced retirement — he is well aware that American society is not what it was ten years before when NASA was founded. Eisenhower's moderate fifties have been replaced by the ranting, raving sixties, Vietnam demonstrations, and fighting in the slums. And in the midst of all this, NASA was about to fulfill the promise of an assassinated President: to send a man to the moon and bring him home again, safe and sound.

12. *System Error*

To the general public, NASA seemed, at long last, to have got the hang of its highly developed technological systems. Rockets no longer exploded on the launch pad; astronauts went up, came back safely, and were picked up by Navy ships, all according to plan. The whole operation seemed to get more and more smooth. Television networks covered the various missions with increasing expertise, thanks to journalists who were specially briefed by NASA, and who also acquired a stronger physical resemblance to NASA personnel with every visit to Houston or Cape Canaveral. They started sporting crew cuts and even began to talk like the men at NASA, adopting a dry, clipped way of speaking which put the accent on efficiency and made it clear that allowances had been made all along the line for any problems that might arise, and that such problems would be solved in a professional manner.

But on closer inspection, execution of the individual missions proved to be far from perfect. On the Mercury 7 flight in May 1962, when Scott Carpenter became the first American since John Glenn to be put into orbit, he was so carried away by the sunrises and sunsets which he witnessed every hour and a half that he fell behind in his extremely tight schedule. His concentration was further impaired by a defect in his space suit, which caused the temperature inside it to rise almost to the boiling point. In addition, because of a flaw in the space capsule's automatic stabilizers, he had to make constant manual adjustments.

This manual firing of the small reaction-control thrusters drained the precious fuel reserves which were supposed to last until it was time to maneuver into position for reentry; the situation was worsened by the fact that when Carpenter switched back to automatic, he forgot to turn off the manual system. For ten minutes both systems were using up precious fuel reserves. Because of this the capsule was not squarely aligned for its descent, increasing the risk of its burning up. Not only that, but Carpenter was five seconds late in discharging his retro-rockets and had to release his parachute manually — the automatic system having cut out due to lack of fuel. Carpenter survived, but he landed 250 miles away from his designated splashdown point. For forty minutes television viewers thought that he had been killed, and Walter Cronkite, television's normally imperturbable newscaster, was seen to weep on-screen.[1]

NASA was not happy with Carpenter's performance. The operator on board the space capsule had made one blunder after another. The old debate over automation versus manual control flared up again. Scott Carpenter soon got the message, and handed in his resignation. But if we look more closely at what actually happened, we can see that it cannot all be ascribed to operator error; in several instances this much-maligned astronaut actually intervenes and saves the situation when the automatic controls have failed.

In May 1963, on Mercury 9 — the last Mercury flight — following a breakdown in the automatic system, Gordon Cooper had to execute a complete manual return through the atmosphere, and managed the maneuver beautifully. He landed four miles inside the target area — a degree of accuracy the automatic controls had never been able to achieve.[2]

As an intermediate stage between the Mercury program and the far more exacting Apollo program, NASA slotted in the Gemini project, which would entail two-man-operated space capsules practicing docking in space and doing the groundwork for space walks, both of which were prerequisites for the trip to the moon. But these flights were not without their hitches either.

Gemini 5, in August 1965, involved the first experiment with fuel cells designed to convert oxygen and hydrogen into electrical energy which would power the space capsule. Two hours into the flight, the pressure in this system dropped alarmingly and the astronauts had to survive for twenty-four hours in very low gear, until the fuel cells, quite inexplicably, started functioning again. To make matters worse, before landing, wrong coordinates were keyed into the capsule's computer and it missed the target area by ninety-five miles.[3]

Gemini 6 — due to be sent up in December 1965 to practice formation flying with a sister ship — had its launch terminated two seconds after the activation of the rocket engines, because a plug had fallen out of its socket. By its very nature, such a termination — with highly volatile fuel running through every pipe and hose — is extremely risky. NASA rules dictated that in such a situation the astronauts should not hesitate to activate the Abort button, which would catapult them away from the primed rocket while there was still time.

Walter Schirra and Thomas Stafford knew that this would destroy the capsule and scrap their mission, so they did nothing, while NASA firemen rendered the rocket safe. Later, NASA praised their courage and their resolve. They had lived up to their training. Even when their lives were seriously threatened, they had stayed calm, not stuck rigidly to security procedures, shown presence of mind, and saved the mission — which could be carried out a couple of days later.[4]

On Gemini 8, in March 1966, the first successful docking of two objects in space was achieved when the two astronauts, Neil Armstrong and David Scott, coupled their spacecraft to an Agena rocket, sent up for this purpose. (The Americans were now undeniably ahead of the Russians. The latter were, as time went on, undertaking increasingly desperate missions — for instance, when they squeezed three astronauts into a capsule designed for one.[5]) Not long after docking with the Agena stage, however, the entire combined setup started spinning uncontrollably, in its weightless state, at a rate of one revolution per second. The astronauts, believing the problem to originate from the rocket stage, uncoupled themselves from the Agena, but the somersaulting became even worse and there was a definite risk of a disastrous collision. Only intervention by the astronauts — the operators — prevented a disaster. By making use of the reentry system ahead of time, Neil Armstrong managed to stabilize the capsule and execute an emergency landing. Later it was discovered that a little thruster had become stuck in such a way that it had been impressing upon the spacecraft a constant force great enough to put the entire structure into a spin.[6]

The first space walks were similarly dogged by errors and irritating snags. The suits overheated when the astronauts were working in them in space. The Plexiglas helmet visors steamed up, so that the astronauts could not see a thing; and when, later, the helmets were treated with a demisting chemical, this stung the astronauts' eyes and made them run, so they still had trouble seeing.

All these stories notwithstanding, there is no denying that during these years the American space program was going from strength to strength. The dockings were a success. The computers worked out the requisite trajectories, making it possible for the radar systems — operated by the astronauts — to lock on to their targets. As time went on, the astronauts also mastered the technique of space walking. In fact, behind every single account there lies a success story in which either the operators or the ground crew are seen to improvise and save a mission which would otherwise end in disaster. NASA *did* have what it takes and, as the space capsules gradually became more and more complex, the astronauts were far from being superfluous. But these stories also illustrate that—even with the concerted efforts of a whole nation—such advanced and complex technology cannot be developed in such a short time without serious and unanticipated flaws inevitably cropping up, right, left, and center.

On Friday, January 27, 1967, at the Kennedy Space Center, 1,100 people were in the process of practicing for the first Apollo mission.[7] At the top of the empty Saturn 1B rocket, the three astronauts—Virgil "Gus" Grissom, Edward White, and Roger Chaffee—were lying on their backs in their molded reclining couches. To Grissom and White, who had sampled earlier models, the new Apollo capsule seemed incredibly spacious. Chaffee was a rookie. It was 6:30 p.m. It had been a long day, full of simulated countdowns and checklists. The instruments in the control room showed each astronaut's steady pulse. The electrocardiograms were recording only faint muscular activity. Similarly, the gimbals inside the spacecraft's guidance and navigation section showed only slight movement. Suddenly, telemetry from the spacecraft showed a momentary power surge in AC bus 2. At the same instant, all the instruments went wild and the voice of an obviously distraught Gus Grissom was heard announcing over his microphone that the atmosphere inside the capsule—which consisted of 100 percent pure, pressurized oxygen—was on fire. "We've got a bad fire . . . let's get out . . . we're burning up!" The rescue team did not waste a second, but it took them almost six minutes to get the hatch open, by which time it was too late. After a look inside the burned-out interior of the capsule, a NASA official had to report over his walkie-talkie that he did not want to describe over the radio what he had seen.

This was not the first fatality the American space program had suffered. A year earlier, two Gemini astronauts died as they tried to land their T-38 jet in bad weather at the McDonnell plant, where they were to see the finishing touches being put to their space capsule. They struck one of the production hangars and crashed. But of course, to the general public they were not real astronauts. Neither of them had as yet been in space, and a plane crash was not a real space disaster. The crew of the

Apollo spacecraft was quite another matter. Grissom had been the second American to be launched into space. He had run the full course — from Mercury, through Gemini, and now Apollo. Ed White had been the first American to walk in space, in June 1965. This was the first true NASA disaster, and the nation was stunned — even when told that the three dead astronauts had to be regarded as the tragic, but nevertheless necessary, price of scientific progress of this magnitude.[8]

The spotlight was at once turned mercilessly on NASA, as Webb had predicted, and the organization had to permit the representatives of the press to view the gutted Apollo capsule. A board of inquiry was immediately set up to look into the matter. This board was comprised, for the most part, of NASA officials — a fact which gave rise to a fair amount of criticism from outside. On the other hand, only at NASA could one find experts who knew about these things. In all, the board of inquiry drew upon the expertise of 1,500 specialists on 21 different panels, all of them determined to get to the bottom of things.

Afterward, in its 3,000-page report, the board of inquiry could reveal details of serious faults and deficiencies, both in the capsule structure and in the production finish. The bulk of the criticism was aimed at NASA's chief supplier, North American Aviation. As early as 1965, as many as 20,000 major and minor defects had been detected in equipment supplied by North American to the Apollo program — 220 of them in the capsule's air-conditioning system alone.

Objects were found in the capsule that should most definitely not have been there — including a stray socket from a set of socket wrenches. And despite scrupulous documentation, no one could say who had left it there. Worse still was the discovery of Teflon cable insulation — which would flake off if one so much as stepped on the cables, or if it came into contact with a sharp edge.

As many as 113 "significant" items of structural work had not been completed, on delivery of the capsule, and the necessary documentation for several of the completed tasks still was not forthcoming.

Throughout the capsule's life, it had also been plagued by faults in the soldering of the aluminum cooling system pipes which caused them to leak flammable cooling fluid. Tests proved that even a slight twisting of the pipes during installation could produce cracks in the joints. Specifications dictated that no wire should be positioned within three-quarters of an inch of a cooling pipe, but tests showed that glycol in the cooling mixture could be ignited from as much as four inches away.

Attention centered on the wiring — the accident having been caused by a short circuit. The board of inquiry found several places where hatches and panels had come into contact with sheaves of wiring, jamming them up against sharp metal edges; places where astronauts could not avoid treading on wiring and thereby damaging it. They also unearthed what were described as "rats' nests" of wiring. Tangled masses of wire which seemed to have been run quite haphazardly to their eventual home in the space capsule. In several spots, it was discovered that wires were much longer than they needed to be and that they had been "looped back and forth to take up the slack." Elsewhere, the statutory wire color-coding practices were not always adhered to. The space capsule alone contained 15,000 different wires — in all, fifteen miles of wiring to unravel.[9]

In its defense, North American said that originally all the wiring had been neatly tucked away, but that NASA's people kept coming up with alterations and adjustments — new specifications which inevitably created confusion. A single drawing might have undergone as many as 100 corrections. At the same time NASA was nagging at its supplier to get a move on — with the result that long wires were left as they were, rather than taking the time to calibrate the capsule's sensitive instruments to meet the somewhat lower electrical resistance of a shorter wire.

But that was not the only kind of problem discovered by the board of inquiry. North American had 8,000 employees working on the NASA project in California, and approximately 1,000 at the Kennedy Space Center in Florida. Their plant had won the $2.5 billion contract outright, but — in keeping with standard practice — auxiliary projects were farmed out to 12,000 subcontractors. Added to which, the aforementioned corrections kept pouring in from the various NASA centers. As far back as 1965, one of NASA's in-house committees had become so frustrated by endless delays and foul-ups at North American that they instituted a thorough inspection of the factory. Among other things, they found that the workers were too imprecise, too slow and apathetic, and that lines of communication between the main supplier and the subcontractors were indecipherable — just as communication between NASA and its main supplier was hampered by their mutual mistrust. So not everyone had been infected by the NASA spirit.

The board did not spare NASA either. They had been far too lax in their quality control. It was described as irresponsible not to have come up with a carefully engineered wiring system capable of being fitted onto a three-dimensional jig. Instead the wires had been fixed to a two-dimensional stand, and during the final installation the wires had been twisted and bent so much that even at this early stage they had suffered damage. Just as irresponsible was NASA's failure to realize that fire-retardant materials could easily become flammable if "soaked" in pure, pressurized oxygen for any length of time. And finally, it was not acceptable that it should have taken so long to open the hatches, and that it had been almost impossible to open them from the inside during tests carried out by the board.

Oddly enough, in its examination of those NASA personnel who had been present on the launch pad, the board encountered what might almost have been relief at the accident. The pressure of work had been so great, and on so many occasions serious accidents had been so narrowly avoided, that many had long been expecting disaster to strike, from one quarter or another. They could not keep it at bay indefinitely. Now it had happened and there was a sense of the air having been cleared — as though after a violent thunderstorm.

But this was not how the American people saw it, once the findings had been made public. The nation had been given the impression that NASA knew what it was doing; that it was infallible; that the organization was comprised of vigilant, courageous men who would step in the moment anything threatened to become at all risky. They had also been given the impression that every process was safeguarded by fail-safe mechanisms that could be activated within milliseconds, if need be, and avert any danger before it could become serious. Instead, what they were now seeing was an organization under intense pressure to deliver the goods and implement, in an unwarrantably short space of time, a program so demanding that sooner or later a disaster was bound to occur.

NASA's case was not helped by the fact that it had lied about the sequence of events surrounding the accident and had maintained that the three astronauts had died instantly — when hundreds of technicians had heard them, over the intercom, screaming in mortal agony.[10]

Naturally, NASA immediately set to work to incorporate the board's recommendations into the first Apollo flight, but the organization was never the same again. Nor was James Webb. Again and again, during the ensuing twenty-month interval, he was hauled over the coals by congressmen who lectured him as to what they would have done if it were *they* who were going to send a man to the moon. The tone adopted was, all of a sudden, peremptory, uncompromising, and contemptuous. To the man who had, more than anyone else, personified NASA's successful endeavors to solve almost insurmountable technical problems, one congressman could say, "The level of incompetence and carelessness we have seen here is just unimaginable." Unable to believe his own ears,

Webb could only reply softly, "The Saturn V is a very large and complex machine."[11] Later, he appealed to fellow Democrat Walter Mondale, who curtly informed him "that he intended to ride this disaster for every nickel's worth of political mileage he could get out of it." He also said, as one witness reported to Webb, "that he didn't give a hoot in hell about him or the space program."[12]

As a modern systems analyst, Charles Perrow is not the least bit surprised by the above accounts. It is his conviction that wherever human beings operate, sloppy work is done, defective products are turned out, warnings are ignored, and dangerous shortcuts taken by people who would not think twice about lying to save their own skins. That's just how people are and, in Perrow's view, it would be as well to make allowances for this when developing any new form of modern technology.[13]

Subcontractors will try to cut corners wherever they can when big money is at stake. Even the most thorough quality-control inspector is powerless to stop this. NASA's quality-control inspections were extremely thorough; more thorough than anywhere else in American industry. But they still could not avoid a serious accident.

Of course, it is quite normal for someone to leave a tool behind at the spot where he has just been working, then not be able to find it later on. It is normal for someone to be distracted for a moment and be five seconds late in pressing a button. Every day, in any average business, thousands of similar mistakes and oversights occur, and no one worries about them, because no harm is done. Who is going to report a colleague for forgetting to switch off a light in his office before going home, when they could easily do the same thing themselves another day?

The difference with space travel and other high-technology human activities, however, is that normal mistakes can have very visible and abnormal consequences. Hence, it is not normal for that five-second delay to result in missing a target area by 250 miles. And if that light switch happens, instead, to be the manual switch for the thrusters which will ensure the right approach to reentry through the atmosphere, rather than staying eternally in orbit because of a lack of fuel, suddenly it becomes fatal not to switch it off. Which is why astronauts are in training for months prior to their launch. But mistakes do still occur. Not because the astronauts are careless—they are far more conscientious than the average human being—but simply because little slipups such as these are part of the human condition.

Designers do, of course, try to anticipate the possibility of such errors by building in fail-safe systems, but strange as it may seem, these fail-safe systems often give rise to new risks; and in any case they only serve to complicate an already complex system even further. Many of the mistakes made in space, and on land, occur simply because complex components start to interact with one another in totally unforeseen ways. Unforeseen because it is no longer possible for anyone to assess all the likely permutations for interaction inherent within such complex systems.

Let us, with Charles Perrow, take a look at one famous, and simple, systems breakdown: Gus Grissom losing his capsule. This happened on the second American spaceflight, in July 1961, when Grissom—like Shepard—was to make a brief flight into space without going into orbit.[14]

When Grissom came down in the Atlantic, he began—quite according to plan—to tidy up a bit inside the spacecraft and jot down his instrument positions, before being picked up by a Navy helicopter. He spoke to the helicopter over his radio, asking them to hold off for a minute, until he was ready. He was still, understandably, a bit flustered and nervous after his trip into space. He then gave the helicopter the all clear to come in, as planned, and attach the hook to the spacecraft before he opened the hatch.

Suddenly the helicopter pilot saw the detonator bolts, which the astronauts had insisted on having fitted to the system, send the hatch cover flying off into the distance. As yet, the helicopter did not have a hold on the capsule, and this rapidly began to fill with water. Grissom, who had presumably bumped into the button while getting ready (although he denied this till his dying day), dived over the lip of the hatch and away from the capsule, which the helicopter had now hooked.

Soon, not only the capsule but Grissom too was on the verge of sinking, in his supposedly unsinkable suit. In his haste he had forgotten to close off the opening to which the capsule's oxygen hoses had been connected. Struggling to seal this opening, before going under altogether, he swallowed a fair amount of seawater and was growing frantic with fear. More and more often, the heavy suit was dragging his head beneath the waves. The helicopter had to abandon Grissom in an attempt to recover the waterlogged, and very precious, capsule. In no time, the waves were washing over its wheels.

A reserve helicopter had, however, been standing by to see to the astronaut in case of emergency. It passed low over Grissom while the skin divers on board took pictures of him. Many a time, during training, the astronauts had splashed about in their space suits and now Grissom looked just like any elated astronaut, bobbing up and down and waving.

The crew of the second helicopter were then distracted by the sight of the first helicopter almost being pulled under by the flooded capsule — although the helicopter was now winning the battle. Finally, the capsule was hauled free of the waves, water cascading out of it. Then a light came on in the cockpit, indicating that the capsule's weight exceeded the helicopter's lifting capacity by 25 percent, and there was nothing to do but let go of the precious cargo. Grissom's capsule, *Liberty Bell 7*, sank at a depth of 15,000 feet and was lost.

The second helicopter eventually caught hold of Grissom, who was dragged underwater for thirty feet before being pulled to safety in a condition later described as "incoherent." To cap it all, the overload light in the other helicopter was later found to be out of order. So the helicopter could have lifted the capsule after all.

This example involves a large number of fail-safe systems, and all of the individuals involved were intent only on doing their best. And yet the hatch cover is blown off thanks to a fluke accident, the oxygen intake has not been closed off, the people in the helicopter misread the astronaut's waving — interpreting it according to what they expected from such a situation. Another dramatic incident, occurring close by, creates a distraction, as does a fault in the overload indicator. Taken singly, each of these mistakes is manageable, but the unexpected interaction between them almost leads to the loss of both capsule and astronaut. Moreover, it is a fail-safe system — the pyrotechnic devices for the immediate release of the hatch — which instigates the catastrophe.

Even the most banal of incidents can easily trigger a sequence of events which is anything but banal; and in this case the systems involved are not even connected, and we are still on earth. In outer space the surroundings are a great deal more hostile; added to which, the capsule's systems are extremely complex, closely linked, and highly interactive.

Charles Perrow describes convincingly how the risk is increased throughout when systems shift from being linear to being complex.[15] In a linear system, such as a standard manufacturing process, the individual machines are spread across a wide area, each taking care of its own distinct stage of the process. Should a machine break down, the fault will very rarely affect the other machines, and often the damaged component is easily replaced. More often than not each lever governs one distinct function, and factory staff will have little trouble in pinpointing where the fault has occurred and grasping the nature of the problem, even though they may not be able to fix it then and there.

If, in addition, the factory is arranged in such a way that it is possible to cope with a breakdown in production at one point without having to bring the whole production process to a standstill; if the sequence can be reorganized slightly, perhaps to achieve the same result in some other way, or another machine brought into service — one capable of fulfilling almost the same function — then, according to Perrow, the system is both linear and loosely coupled. Such a system will seldom present major problems.

Conversely, if the system is tightly coupled, allowing no room for delays; if the sequence is inflexible and there is only one or very few correct ways of reaching the objective; if there is no chance of replacing components, apart from the programmed, built-in fail-safe systems, then one could soon find oneself in trouble should a fault occur anywhere along the line. Before long, such a fault is sure to paralyze the entire system.

If, apart from being tightly coupled, the system is also complex, with all of its components situated in the immediate vicinity of one another, and linked up to one another this way and that; if auxiliary systems are sending feedback to the main system, which is also linked up to other auxiliary systems; if there are a great many buttons to press, all looking alike; and if fluctuation of the instrument needles can be caused by several different factors, then one is not just in a fix, but pretty certain to experience sooner or later what Perrow calls a "normal accident."

At shorter or longer intervals, tightly coupled, complex systems are bound to break down in such a way that even their designer will have difficulty in figuring out what is going on. And often, later, the designer will claim that the operator was to blame for the breakdown. After the fire in the capsule too, people at North American thought that the fault could have been caused by Grissom kicking a bunch of wires, thereby triggering a short circuit. Wasn't he the one who messed up so badly that time when a space capsule sank? But one could point to just as many other instances where a designer has created a system calling for superhuman — and, hence, computerized — control at all levels, and where the system breaks down anyway, requiring the intervention of an alert operator — if, that is, the designer has made any allowance in the system for such intervention.

It goes without saying that this phenomenon is not unique to tightly coupled technical systems. Where vast technological systems create links between thousands of subcontractors in an intricate, interactive organizational structure so complex that even its head office has trouble obtaining an overall view; when what at one level are just trifling errors can mean the difference between life and death at another, then, sooner or later, these systems too are bound to experience "normal" catastrophes, as was the case with the fire in the capsule.

Space-age systems require supermen among their designers, manufacturers, quality controllers, and operators. And in Perrow's experience, there are no supermen. Which is why space missions and many other modern high-technology processes (such as aviation, air traffic control, nuclear power stations, the manufacture of chemicals, and genetic engineering) are so risky. And the modern world is constantly being presented with more and more such systems.

13. The End of a Decade

By the time James Webb was writing his book in 1968, NASA had, over the two previous years, already had to bid farewell to 140,000 employees. One-third of the workforce had become superfluous now that the Apollo system had been developed and produced and all that remained was to put it to the final test. NASA's budget showed a corresponding decrease. Back in 1966 a somewhat less cooperative Congress decided that in 1967 Webb ought to be able to make do with half a billion dollars less than he had requested. This trend had continued, until by now NASA's budgets had been reduced by a quarter and, as a consequence, Webb had been obliged to draw on NASA's internal resources by cutting back on projects not directly contributing to the lunar project.[1] This included closing down NASA's Long Range Planning Office. Long-range planning had now become a luxury.

It is apparent from the tone of Webb's book that he was proud of the vast sums of money which NASA had administered, but he also feels impelled to legitimize space expenditure and put it into perspective. Thus, he emphasizes the point that NASA's total funding from 1958 to 1968 amounts to *just* two-thirds of military spending in a single year, *just* 6 percent of the defense budget for the same ten years, *just* 3 percent of public expenditure over the same period, and *just* 0.5 percent of the gross national product. On average, the annual space budget has constituted *just* a third of Ford's or Esso's budget, and in 1966 alone General Motors earned as much as had been allocated to NASA during its first ten years.[2]

No self-assertive company talks like that. What shines through this almost manic recital of comparisons — intended to play down the financial significance of the sums spent on space research — is the sense of an organization under pressure. An organization that is beginning to doubt itself, even as the press is informing the nation that the necessary communications equipment for the moon shot alone would cost more than the entire Mercury project; and that—taking everything into account— the lunar module has cost fifteen times its weight in gold.[3]

The American people no longer looked up uncritically to NASA. In fact, the way things had been going, it may be that they found it hard to look up to anything or anyone at all. In the summer of 1967 the black ghettos of Detroit erupted. For five days even a combined force of 15,000 police officers and National Guard troops was unable to gain control over the situation, as gangs went on the rampage and people looted shops and started fires. From the air, the city seemed to be at war. When the smoke died down, 41 people were dead and 300 injured; 1,100 buildings had been completely gutted by fire and 5,000 of the city's poorest citizens had been left without a roof over their heads.[4]

In April, Martin Luther King was assassinated in Memphis. The one man who had, more than anyone else, symbolized black people's non-violent fight for civil rights. The same man who in Washington in 1963 made his famous "I have a dream" speech, in which he said, "The Negro lives on a lonely island of poverty in the midst of a vast ocean of material prosperity; he lives in exile in his own land." Now many blacks had abandoned their nonviolent principles and were rioting and plundering their way through 125 American cities—among them Washington, where burning buildings could be seen just three blocks away from the White House. Before order could be restored, 20,000 regular soldiers and 34,000 National Guard troops had to be called in.[5]

Two months after King's assassination, Robert Kennedy was also killed, shortly after launching himself into the battle to succeed Lyndon Johnson as President. It seemed as if all of America's hopes and ideals were to be crushed in just one year.

Meanwhile the Vietnam War was at its height, and it was becoming more and more difficult for the American people to accept the fact of young American men returning home in body bags; difficult to see any point in the war at all. Every evening, on their televisions, they were treated to more grisly pictures from the battlefields, pictures which only served to undermine morale even further. Within four years, Lyndon B. Johnson, who had won a landslide victory in the election following Kennedy's death, became the most hated of American Presidents. Wherever he showed his face, young people would shout, "Hey! Hey! LBJ! How many kids did you kill today?" and there was a general feeling of relief when he announced in March 1968 that he would not be seeking reelection.

In the autumn of 1968, young Americans demonstrated outside a

Democratic Party convention in Chicago — an action which emphasized just how hard it was for the established political system to get through to modern youth. As a joke, the Yippies threatened to put LSD in the city's water supply. They also announced that they had arranged for certain particularly charming, smartly dressed activists to seduce delegates' wives.[6] The Chicago police did not appreciate this brand of humor and responded with unheard-of brutality, lashing out in blind hatred at the pampered middle classes, as though they were the most deadly plague on American society.

When one of the Democratic politicians in the convention hall, Abraham Ribicoff, stepped up to the microphone and asked Chicago's Democratic mayor, Richard Daley, to put a stop to the police "Gestapo tactics," Daley screamed from the floor, "Fuck you, you Jew son of a bitch . . . Go home!"[7] It had not yet dawned on Daley that his day had come and gone. Suddenly the powermongering methods of the old politicians, methods fostered in smoky back rooms, could no longer take the glare of the television spotlight. On the other hand, there was no sign of a new Democratic leader capable of uniting the nation as Kennedy had done.

To the nation's dismay — and as if to rub salt into the wound — Jacqueline Kennedy married a short, old, ugly millionaire, the Greek shipping magnate Aristotle Onassis. If it had been up to the nation, she would have spent the rest of her life as the beautiful, grieving widow of a President who was now gradually becoming a myth.

Nor did the word "technology" exercise the same power as it had done for Kennedy and James Webb. The same aerospace companies which had supplied high-technology equipment to NASA are also supplying the Army in Vietnam. The word "technology" acquires ominous overtones and is girded by suspicion. Moreover, in Vietnam this technology seems to be useless. The world's mightiest military power, with the world's most advanced weaponry, was — the way the public saw it — being worsted by little men in pajamas who had never heard of infrared sights or heat-seeking missiles. American youth, which should — by virtue of age alone — have been all in favor of anything new, instead feels drawn toward a markedly antitechnology left wing. Technology is synonymous with the system, and you have to be wary of the system.

So, though Kennedy's aim with his space program may have been to establish the superiority of American technology, it is this very superiority which is now being discredited. And while it may have been his aim to make the people forget the Bay of Pigs, instead the people are now weighed down by the tragedy of Vietnam. The Russians too — who may have wanted *Sputnik* to divert the world's attention away from what was going on in Hungary after 1956 — have not only lost the space race; they have also lost every shred of international sympathy following their brutal entry into Czechoslovakia in August 1968. Both superpowers had lost international credibility.

This was how things stood as NASA prepared for a landing on the moon. As Dale Carter points out, the organization's space capsules could be seen as a sanitized American reality. A small, self-sufficient society equipped with all of the newest, most ingenious products of modern industry; inhabited by nice, clean-cut young people who put duty before comfort; and, above all, a world free of all the problems that were chafing at American society. Borrowing from Thomas Pynchon's *Gravity's Rainbow*, Dale Carter describes the Apollo module as a kind of "Happyville," floating high up above "Pain City."[8]

When, at Christmas 1968, Apollo went into lunar orbit for the first time, and its three astronauts became the first to see the earth as a round and fragile globe suspended amidst the icy pitch-black of outer space, it was thoughts akin to these that filled their heads. And what impressed the viewers, far more than the technological achievement, was the fact that, from their distant lookout point, the astronauts could put the world's problems into perspective. On Christmas Eve, NASA's expensive technological communications system relayed Frank Borman's tinny voice reading a prayer on behalf of all mankind, after which the astronauts took turns reading extracts from the account of creation in the Book of Genesis, while Lyndon Johnson sat before his little television, listening with tears in his eyes:

> Give us, O God, the vision which can see Thy love in the world in spite of human failure. Give us the faith, the trust, the goodness in spite of our ignorance and weakness. Give us the knowledge that we may continue to pray with understanding hearts, and show us what each one of us can do to set forth the coming of the day of universal peace.[9]

Johnson had only a few days left in office before making way for Richard M. Nixon, who was now about to bring the Republicans and conservatism back to the White House. Nixon was reputed to be no friend of NASA. He regarded it as Kennedy's creation and hence a symbol of everything that was to be done away with.

But Nixon would never have dreamed of stepping in and calling a halt to the Apollo project this late in the day. He too could see how Apollo 8 had caught the imagination of the nation. Outer space still held a fascination, at a time when not many illusions were left. In 1969, the writer Norman Mailer could not make up his mind as to whether the lunar project was "the noblest expression of the Twentieth Century or the quintessential statement of our fundamental insanity."[10] Like almost everyone else, he was experiencing both sentiments in equal measure.

Wernher von Braun was now facing the culmination of his entire career. Initially, he had been against the approach that had been decided upon for the moon landing. He would rather have built an enormous Nova booster rocket, powerful enough for a direct shot at the moon. This would, however, have required a colossal thrust—and NASA was not prepared to pay for that. Instead an indirect approach had been chosen, with a three-stage rocket putting the astronauts into orbit around the earth. Only when that had been achieved would the rocket's last stage accelerate the spacecraft away from the earth's gravity, en route to the moon. Here, the Apollo capsule and the lunar module would go into orbit around the moon; two astronauts would transfer to the lunar module and attempt a landing, while the third remained in orbit above the moon's surface. After spending some time on the moon, the two spacecraft would be reunited and the Apollo capsule would bring them safely back to earth.

This approach still required a huge rocket, one with its roots in the kind of rocket von Braun had had in mind while still working for the American Army. Doubtless von Braun was quite happy that the days of using Air Force Atlas and Titan rockets for manned spaceflights were over. Just as he had been approached when it came to sending up the first satellite with the Redstone rocket, so in January 1962 it was to him that they turned when it was a matter of maximum, brute force. Over the next five years, von Braun developed the Saturn V. It was taller than a thirty-story building, and at the start of the launch it would, all told, weigh 3,000 tons and exert a thrust of 7.5 million pounds—a force equal to that of 136 of the old V-2 rockets or, to draw a more peaceable analogy, 85 Hoover Dams—while carrying a 45-ton load to the moon. This rocket would not be marred by the faintest military connotation; instead, it would be the very symbol of civilian NASA.

Not everyone at the Marshall Center in Huntsville had ever really come to terms with the creative, civilian note struck by NASA. Discipline and precision had remained von Braun's guiding principles. He had disregarded complaints that he relied too much on his German colleagues and would not appoint Americans to top management posts. Gradually, however, the Americans had gained admittance — though only after having served a strict apprenticeship that taught them to appreciate von Braun's principles. At the Marshall Center they had good reason to be proud of their achievements. The Redstone, the Saturn I, and now the Saturn V had all suffered significantly fewer teething troubles than the Atlas, for example. The Redstone boasted a reliability rate of 98 percent, better than that of most new jet fighters. Von Braun's technological solutions, which leaned more toward the conservative than the advanced, paid off. This fostered an overweening team spirit which could, on occasion, ruffle a few feathers at the other NASA centers.[11]

For instance, Webb was annoyed to discover that von Braun not only intended to develop the Saturn V at Huntsville and produce the first test models there; he also expected — totally in accordance with his own hands-on approach and with his experiences in Peenemünde — to be directly responsible for the construction of subsequent models. Von Braun backed down on this point only after a direct order entrusting the actual rocket production to the aerospace giants.[12] Personnel at the other NASA centers suggested, caustically, that if it had been up to von Braun, he would have launched Neil Armstrong and the others from Huntsville.

Von Braun was also delighted by the fact that President Kennedy, the year before his death, had been shown models of the rocket and had used them as the basis for a speech made at Rice University in Houston in September 1962:

> My fellow citizens. We shall send to the moon, 240,000 miles away from the control station in Houston, a giant rocket, more than 300 feet tall — the length of this football field — made of new metal alloys, some of which have not yet been invented, capable of standing heat and stresses several times more than have ever been experienced, fitted together with a precision better than the finest watch . . .[13]

The comparison with "the finest watch" caught on and was used again and again in the years to come. Norman Mailer mentions the way in which many factory managers idolized von Braun, regarding him as the high priest of the manufacturing industry.[14] Whereas they had endless problems with their employees and the quality of their products, he could get his people to manufacture a product measuring over 300 feet in

height—a product of which, as an American, one could feel proud—
with the precision of "the finest watch." And could assemble it in such
a way that it gave a faultless performance on its first test in 1967; used
up its thirteen tons of fuel per second in approximately 150 seconds,
without faltering; switched to the subsequent stages—which, as an innovation, utilized a powerful combination of liquid oxygen and liquid
hydrogen as fuel—following Tsiolkovsky's recipe to the letter. And everything functioned exactly as it should.

At one of the press conferences held prior to the "proper" launch in
1969, a murmur ran around the room when an East German journalist
asked a question *in German*. Of course, everyone knew that von Braun
was German, but they had all agreed that it was only polite to disregard
this fact. He was an American now, of whom anyone could be proud;
then along comes this Commie . . . Von Braun hesitated for a moment,
then translated the question into English, made a comprehensive reply
in English, then translated his reply into German. At the end he smiled
wryly and said he regretted to advise the Japanese journalists that he
could not offer them the same service. After that—thanks to his fine
sense for the sort of quotes the press needed—he was once more holding
the whole room in the palm of his hand.

When, for example, he was asked how he assessed the importance of
the lunar landing, he declared without hesitation: "I think it is equal to
that moment in evolution when aquatic life came crawling up on the
land." Several members of the press literally got to their feet to applaud
this remark.[15] He had done it once again.

The evening before the launch, on July 16, 1969, Norman Mailer
attended a party held to pay tribute to von Braun's achievements.
Von Braun's mentor, seventy-five-year-old Hermann Oberth, was also
present—the only surviving member of the old rocket pioneer trio, Tsiolkovsky, Goddard, and Oberth. At the party, von Braun spoke of what the
next day was to bring:

> What we are seeking in tomorrow's trip is indeed that key to the
> future on earth. We are expanding the mind of man. We are ex-
> tending this God-given brain and these God-given hands to their
> outermost limits and in so doing all mankind will benefit.[16]

It was like listening to Kennedy all over again.

Every man achieves his own greatness by reaching out beyond himself, and so it is with nations. When a nation believes in itself as Athenians did in their golden age, as Italians did in the Renaissance, that nation can perform miracles.[17]

And at this point, there must have been those in the company who could not help but think of von Braun's original homeland, as he hastened to add:

Only when a nation means something to itself can it mean something to others. We are truly faced with the brightest prospects of any age of man. Knowing this, we can watch the launch tomorrow with a new dimension of hope. We can cheer the beginning of a new age of discovery and the new attainment that spans the space distances and brings us nearer to the heavens.[18]

Wernher von Braun could get away with employing such high-flown phrases. Gone were the V-2 rockets and the slave labor from the concentration camps, gone was the poverty of the slums, gone was Vietnam. All that existed was the bright future which space travel and NASA had helped to create.

14. The Eagle Has Landed

And so, at long last, it happened. What Kennedy had promised the nation at the start of the decade. Just six months before the sixties ran out. On July 20, 1969, Neil Armstrong brought his NASA boots cautiously down into the soft dust of the moon and uttered those famous words: "One small step for a man, one giant leap for mankind."

After the twenty-month hiatus that followed the fire in the Apollo capsule, NASA had been kept incredibly busy. It was as though the dead President's words were nipping at the organization's heels. As though NASA was afraid that if it did not make it before the sixties turned into the seventies, then — as a space organization — it might as well throw in the towel. After Apollo 8, flights followed so close on one another's heels that there was barely time to record the findings of the previous flight properly — much less incorporate them into the next mission before it was due to take off.[1]

Nevertheless, the Saturn V rocket had given a faultless performance and everything had proceeded according to plan. Perhaps to the faint disappointment of one lady among the spectators at Cape Kennedy who had, earlier in the day, told *The New York Times,* "It's so thrilling, maybe the engine will explode!" Kurt Vonnegut had been proved right. Just a few days earlier he had written that any exhibition which involved the genuine risk of someone being killed was always a surefire hit with the public.[2]

It must have seemed a source of wonder to the slightly overweight, sunburned millworker whom Norman Mailer saw at the launch.[3] He had worked with machines all his life. He had learned that machines can be as unreasonable as people, but also that you can reach the heart of them through hard work, patience, and perseverance. Eventually you could almost feel that you understood them, perhaps even take a nap beside your machine with a contented smile on your lips. But this new "machine" was something else again. With all those other machines, he had plunged his own oily fingers down into them until he knew every bolt, every oil pipe, with his eyes closed. And in this way alone could trust grow up between him and the machine. No one in their right mind took a nap alongside a Saturn—a machine with millions of parts, on which a million fingers had worked. How much evil, error, and deception in a million fingers? Now he was going to see this machine in action, and Mailer writes:

> He will see a world where machines are king and he does not know whether to cry from pride or the all-out ache that he does not really comprehend the new machinery.[4]

In London, four days later, Queen Elizabeth stayed up all night to watch the lunar landing "machine" in action. Pope Paul VI peered at the moon through his telescope with an eagerness that would have amused Galileo. Nixon sat "glued to the television for hours," even though this project had been Kennedy's brainchild. Together, they heard the crackling radio link from the lunar module's "downlink," when, after a four-day journey, it floated low over the darkened surface of the moon.[5]

EAGLE: *Lights on. Down 2½. Forward. Forward. Good. 40 feet, down 2½. Picking up some dust. 30 feet. 2½ down. . . . Forward. Drifting right . . . contact light. Okay, engine stop. ACA out of detent. Modes control both auto, descent engine command override off. Engine arm off. 413 is in.*

CAPCOM*: *We copy you down, Eagle.*

ARMSTRONG: *Houston. Tranquillity base here. The Eagle has landed.*

CAPCOM: *Roger, Tranquillity, we copy you on the ground. You've got a bunch of guys about to turn blue. We're breathing again, thanks a lot.*

EAGLE: *Thank you.[6]*

Just prior to this moment, the lunar module's automatic landing system had been heading toward a rugged crater the size of a football field that was strewn with very large boulders which would have damaged the module if Armstrong had not taken control and steered the *Eagle* to a smoother patch. Again the operator had saved the mission and, hence,

---

* Capsule Communicator: Usually an astronaut acting as a link between the spacecraft and the control center directors and technicians.

his own life and that of his buddy, Edwin Aldrin—although he almost used up all his fuel during this critical maneuver. Flight Director Gene Kranz, who had been looking forward to this moment for months—even years—and who had mercilessly driven his staff to the point where none of them would put a finger wrong, sat as though stunned, staring vacantly into space. It was some while before he came to, raised his fist, and brought it thudding down onto the control panel.[7]

The next day, *The New York Times* saw all of its 950,000 copies snapped up, and recognized this feat—as Wernher von Braun had done—as a new stage in the development of mankind. The paper really went to town, describing the "urge . . . deeply inscribed in man's psyche" that drove him to seek fresh challenges. It called up the names of Lindbergh and Columbus and reminded its readers that Armstrong himself had, during an interview, said that each astronaut was "ultimately driven skyward by the nature of his deep inner soul." They were required to do these things "just as salmon swim upstream."[8] The newspaper was to be reprinted, unaltered, a few days later as a souvenir.

Robert H. Goddard's widow, Jan, who had tried to film the first launch of a rocket using liquid fuel many years' earlier, told one newspaper that her husband would have beamed like a ray of sunshine. "Because it was always his dream to send a rocket to the moon." Neil Armstrong's wife, Ann, was asked by the media whether this was the greatest day of her life. Living up to the very best tradition, as set long ago by John Glenn's Annie, she replied, "No, that was the day when I married Neil."[9]

Workers in countless different industries the length and breadth of the United States swelled with pride at the thought that they had been a part of something remarkable. An almost sacred commission. For once, they had been involved in making something that worked, and it had been seen in an unforgettable way by the largest television audience in history, supposedly a quarter of the world's population.

The astronauts on the moon received a daily news summary so that they could hear how the media had described their exploits and learn what else had been happening back on earth. NASA did, however, practice a mild form of censorship. It did not include in its summary for that day the news that—on the evening before John F. Kennedy's visionary resolution was finally to be realized—the Democratic majority whip, Edward Kennedy, the last of the Kennedy brothers, while under the influence of alcohol, had driven over the side of a bridge near Chappaquiddick. He had then left the wrecked car in the water and only after nine hours had elapsed informed the police that they would find the body of a pretty, blond twenty-eight-year-old secretary inside the car. This incident finally ended the Kennedys' chances of making a political comeback.

As far as NASA was concerned, that was all to the good. The organization's feelings for Edward Kennedy had long hovered around the freezing point. Two months earlier, at Clark University — Goddard's alma mater — and in the presence of Mrs. Goddard, John F. Kennedy's brother had had the nerve to say that it would be only right for future space research grants to be put, instead, toward fighting poverty, hunger, pollution, and homelessness.[10] At any rate, in his opinion, space research grants ought to be weighed against other national priorities.[11] Such a man was not to be taken seriously, and now he had brought himself down.

Vice President Spiro T. Agnew proved to have been cast from a different mold, when he declared, ecstatically (though perhaps somewhat feebly), that before the end of the next decade the United States should be sending men to Mars.[12]

Richard Nixon spoke on the telephone to the astronauts from his office in the White House:

> I just can't tell you how proud we all are of you. For every American this has been the proudest day of our lives. And for people all over the world, they too join with Americans in recognizing what a feat this is. Because of what you have done, the heavens have become a part of man's world. And as you talk to us from the Sea of Tranquillity, it inspires us to double our efforts to bring peace and tranquillity to earth. For one priceless moment in the whole history of man, all the people on this earth are truly one. One in their pride in what you have done. And one in our prayers that you will return safely to earth.[13]

This was more a carefully phrased official speech than a spontaneous telephone conversation, but by now Nixon was personally moved and impressed by what was happening on the moon. In an unprecedented move, on his inauguration he had allowed the NASA director appointed by Johnson — Webb's successor — to retain his post. Later, this director, Thomas O. Paine, had no illusions about the reason for this largesse:

should disaster strike en route to the moon, the only heads to roll would be those appointed by Nixon's predecessor.[14]

When it came to another symbolic gesture, however, that generosity of spirit was singularly lacking. A number of key people at NASA felt that it would be appropriate for the astronauts to be picked up by the aircraft carrier USS *John F. Kennedy*. But furious Nixon supporters adamantly vetoed this suggestion.[15] So, instead, it was on the deck of the USS *Hornet* that the astronauts were able to view their President, through the little windows of their quarantine trailer, after a successful return trip.

Now it had reached the point where the President did not just receive the astronauts in the White House or at Cape Kennedy. He put to sea to welcome them home.

By this time, television viewers were, ever so quietly, growing tired of the moon. There was no mistaking what an achievement it was, but among themselves they were heard to mutter that, after a while, the flickering television images had become boring. Not that it wasn't amusing to see how they bounced and jumped about like big Michelin men in the weaker gravity of the moon, but once you'd been watching it for five minutes, so what? Wise guys said that the whole thing had been recorded in some out-of-the-way gravel quarry on earth, with some primitive 1950s cameras. Nobody would be able to tell the difference anyway.

In James Webb's book, published that same year, he wrote:

> It is entirely possible that an endeavor that had been strongly and enthusiastically endorsed on all sides at its inception and was making good progress towards its goals might suddenly find itself in trouble as a result of changes completely beyond its control. An important factor may involve changes in the public mood — changes in basic public attitudes towards the goals being sought. Congress is highly sensitive to such changes, and as endeavors become more and more complex, a greater and greater degree of confidence and trust is required to maintain essential levels of support.[16]

The American people had certainly not turned their backs on the space program. They were very pleased with what NASA had been able to accomplish. But now the show was over and the nation was making its way out of the theater. And out on the street, Pain City awaited.

In 1969, a million people demonstrated against the Vietnam War, which was still raging, despite Nixon's hopes for "tranquillity." A shocked nation heard of the twenty minutes when American soldiers were over-taken by a bloodlust that left over one hundred old men, women, and children dead in the Vietnamese village of My Lai. Charles Manson was jailed, together with some of his followers, after the brutal ritual murder of Roman Polanski's pregnant wife.

A month after the *Eagle* had landed, 40,000 young people gathered together for a weekend of "Love, Peace, and Understanding" at the Woodstock festival and heard Jimi Hendrix wring the familiar strains of "The Star-Spangled Banner" from his guitar in a manner that must have made it clear to everyone that, to a large proportion of the American youth that had grown up during the sixties, American values did not hold the same meaning as they did for their parents.

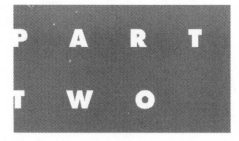

PART
TWO

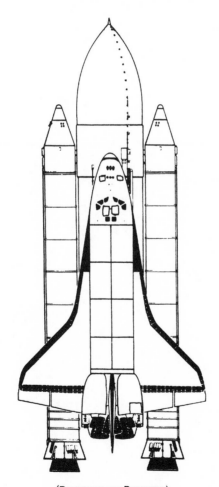

(Report to the President)

15. Say Again, Please

Most people today are probably under the impression that the Americans made only one landing on the moon. The pictures of Neil Armstrong from the summer of 1969 have imprinted themselves on our collective retina as NASA's ultimate triumph. But Apollo 11 was scheduled to be succeeded by Apollo 12, 13, 14, 15, 16, 17, 18, and 19. The astronauts had been training for years.

However, people had lost interest. Naturally, NASA still issued detailed press releases, and international journalists still visited Cape Kennedy and Houston, but in the living rooms of America, boredom was setting in. NASA's smooth-running, technological show had lost its fascination — perhaps precisely because everything was going so well. The reporters' press kits told them, down to the last second, just when the first stage would be jettisoned, when the Saturn rocket would set course for the moon, and when the lunar module would plant its fragile spider legs in the moon dust. This in itself was an impressive achievement — to be able to predict so accurately what would happen 247,000 miles away. But you have to be a scientist to delight in predictability.

At press conferences, the astronauts noted, with some disappointment, that journalists were struggling to come up with questions on topics that their readers had not heard all about already. Also indicative of this trend was the much smaller turnout in April 1970 for the launch of the third lunar expedition with Apollo 13 — just a tenth of the number that had flocked to Cape Kennedy to wave goodbye to Neil Armstrong. Astronauts Jim Lovell, Jack Swigert, and Fred Haise noticed the difference, but were of course too professional to let it affect them.[1]

This time, too, according to the television news broadcasts, things would follow the usual procedure: First the astronauts entering the command module — which, for this trip, had been named *Odyssey*. Then up, and into orbit around the earth a couple of times; then a fresh firing of the Saturn rocket, and off to the moon. *Odyssey* would then be uncoupled, and would turn around and unhook *Aquarius*, the lunar module, from the rocket stage, which could now be dispensed with. On arrival at the moon, deceleration and, again, a couple of orbits above the moon, before *Aquarius* would transport two astronauts down to the lunar surface. After thirty-six hours on the moon, *Aquarius* would take off once more and link up with *Odyssey*, which had spent all this time circling in a low lunar orbit, steered by their buddy, the third astronaut. And finally, back to earth in *Odyssey*. Clean well-oiled NASA routine.

The astronauts did everything just as they had done it many times before in the simulator on earth, and the equipment responded perfectly. One rocket engine did cut out a bit early, but the others compensated for it as planned and they made a bit more headway. Orbit: OK. Fresh firing of rocket engine: OK. Acceleration: OK. On course for the moon: OK. Linking up with the lunar module: OK. The crew even found time for a couple of television transmissions, during which the nation was shown around both the command module and the lunar module. For the moment, the pressure was off, and the crew could relax a bit, before having to be on their toes again, on arrival at the moon.

And just then, after fully two days in space, the impossible happened. A bang. Vibration. Suddenly large sections of the instrument panel lit up like a Christmas tree. Warning sirens.

SWIGERT: *Okay Houston, we've had a problem here.*

CAPCOM: *This is Houston. Say again, please.*

LOVELL: *Houston, we've had a problem. We've had a main B bus undervolt.*

CAPCOM: *Roger. Main B undervolt. Okay, stand by, 13, we're looking at it.*

One of the two main cables—the "main buses"—to the spacecraft's power supply had registered a sudden drop in voltage; exactly what had happened with the capsule fire on earth, but this time 205,000 miles away from home.

It was difficult to make any sense of the instrument readings. There were 250 different gauges and readings. Some showed a short circuit in the hydrogen system. Others showed the oxygen containers as being full to overflowing. Other needles shot up only to subside again. Radio contact with earth was lost—and then restored. The spacecraft lurched from side to side. The astronauts' pulses rose from 70 to 130. (The ground crew's pulse was not recorded.) Everyone was frantically intent on locating the fault in the electrical system. It had to be there somewhere. It had to be some minor fault, one that could be rectified, and thus enable the mission to proceed.

On earth, at first it was assumed to be an instrument failure, the sort of thing they had come up against before. Or a flaw in the downlink to earth. When the warning lights for the electrical system suddenly and inexplicably went out again, everyone breathed a sigh of relief. The problem had solved itself. And the bang might well have been caused by one of the astronauts bumping rather clumsily into the wall.

Three minutes later, two of the spacecraft's three fuel cells suddenly went out of commission. These cells converted liquid oxygen and hydrogen into electricity, which then powered all of the vital systems — including the navigation systems, the thrusters, the oxygen supply, the ventilation, and the production of water for drinking and for the cooling system. Again a host of warning lights blinked on the control panel. Only one cell left. When Lovell checked the reserves in the no. 2 oxygen tank, he could not believe his eyes. The needles were right down on zero. The tank was empty — 275 pounds of liquid oxygen gone. But this gauge had given wrong readings before now, and fortunately — partly for safety's sake — there were two oxygen tanks. When, stomach sinking, he glanced across at the no. 1 tank, he could actually see the needle slowly dropping. Still, no one could explain the relationship between the drop in voltage, the bang, the vibration, and the problems with the electrical and hydrogen-oxygen system. Naturally, the astronauts had been subjected to every imaginable kind of systems breakdown in the simulators, but there had always been an unwritten rule against throwing multiple, unrelated, simultaneous failures at them. Start doing things like that in a system consisting of thousands of components and you would never finish training.[2] So it was only reasonable to give the astronauts training in the one- or two-point failures that might occur. But Apollo's multiplicity of failures had never heard of this unwritten rule; and in this the Apollo 13 accident resembled all the other "normal" accidents described by Perrow.

Through the window, one of the astronauts saw a vaporous cloud emanating from the spacecraft. The force of this would be enough to cause them to lurch about. The men on the ground at Houston still clung to the theory of an instrument failure, but they were not on board and their bodies were not feeling the ensuing effects. Understandably, there is a great reluctance to admit that a carefully planned mission may have to be abandoned — not to mention the almost inconceivable thought that the lives of American astronauts could be in danger.

Charles Perrow has analyzed several high-technology accidents and on every occasion he has noted the typical behavior associated with system accidents:

1. Initial incomprehension about what is indeed failing.
2. Failures are hidden and even masked.
3. A search for a *de minimus* explanation, since a *de maximus* is inconceivable.
4. An attempt to maintain production if at all possible.
5. Mistrust of instruments, since they are known to fail.
6. Overconfidence in ESD's (emergency safety devices) and redundancies, based upon normal experience and smooth operation in the past.
7. Ambiguous information is interpreted in a manner to confirm initial (*de minimus*) hypotheses.[3]

All of these characteristics can also be found here. Added to which, in this case, time is running out and the situation is potentially disastrous.

Only very slowly did the true picture emerge: A short circuit (drop in voltage) in one oxygen tank had led to an explosion that had blown out the whole of one wall of the service module. The antennas which maintained contact with Houston had been badly damaged. The fire had immediately been extinguished by the vacuum outside, but the damage was irreparable. The second oxygen tank was indirectly linked to the first through the fuel cells—which meant that it was not actually a reserve system at all. So, after the initial explosion, *its* oxygen reserves too would inevitably seep away.

The spacecraft would very soon be reduced to a lifeless mass, since the one remaining fuel cell would, of course, be able to supply power for only as long as it still received oxygen from a tank with a gauge now rapidly dropping toward zero.

At the last moment they succeeded in isolating an oxygen surge tank in the command module in such a way as to save a little under 4 pounds of the total 275 pounds of oxygen in the two tanks. An attempt was also made to shut off every piece of power-sapping equipment which might reasonably be dispensed with. By this time it had become obvious to everyone that the crew's lives were in danger.

Almost simultaneously, those on earth and in space struck upon the idea of using the lunar module as a kind of lifeboat. *Aquarius* had its own supply systems, for use during its time on the moon. The spacecraft's oxygen supply allowed for the fact that 20 percent of the oxygen in the airlock would be lost with each moon walk, and there also had to be enough oxygen for the backpacks worn by the astronauts on the surface of the moon. The question now was whether the spacecraft, and supplies designed to accommodate two men for between 33 and 35 hours, could stretch to three men for 85 to 100 hours.

Oxygen, water, and electricity would have to be strictly rationed. Under normal circumstances, with all systems running, *Aquarius* would consume 50 amps. At its lowest anticipated "tick-over" point, the spacecraft consumed 20 amps. But if the batteries were to last, consumption could not exceed 15 to 17 amps. The consumption of cooling water would have to be reduced correspondingly, from 6 pounds to 3 pounds, if enough drinking water was to be left for the long journey home.

This solution was possible only because the explosion had taken place after a successful linkup between the command module and *Aquarius* and prior to the lunar landing. If the explosion had occurred early on in the mission, or not until the return journey, it would have meant certain death for everyone on board. In the first case, because maneuvering of the command module would have been out of the question, and hence the linkup with the lunar module would have been impossible. In the latter case, either because supplies on *Aquarius* would have been used up on the moon or because separation from *Aquarius* would have been completed and the homeward journey already begun.

So the crew huddled together inside the lunar module and shut off all systems in the Apollo module itself. The ground crew were having to operate at lightning speed. Lead Flight Director Gene Kranz was working feverishly, bringing all of his vast store of experience to bear and improvising from one minute to the next, without ever stopping to wonder whether a rescue was even feasible. Forty men were assigned to the task of staying constantly one jump ahead of events and, if necessary, trying out fresh solutions on the simulator. Throughout the United States, the subcontractors who had supplied the various systems called in their key people, in some places providing them with a police escort.

The checklists were now hopelessly inadequate. For instance, standard procedures presupposed that the command module would supply the power for starting up the lunar module. There wasn't much chance of that now, though. Eventually, someone found an earlier checklist showing how the lunar module might activate its systems by means of its own batteries. This list would be read out to the astronauts, who could accomplish the task manually within two hours. But by then there would be just fifteen minutes' power supply left in the command module. And so a greatly abbreviated emergency checklist had to be improvised on the spot.

During preparations for the moonshots, allowance had been made for the sort of emergency situation that might conceivably arise on the long stretch to the moon. The computers had calculated that if the powerful rocket engines at the back of the service module were ignited long enough to reduce the spacecraft's speed by around 4,200 mph, the earth's gravity would — in line with Newton's principles — take over, swing the rocket around in a gentle arc, and bring it back to earth, where it would land in the ocean some sixty hours later. However, no one dared even to contemplate starting up the rocket engines of the crippled service module. And the lunar module's rocket engine had nothing like the necessary power. It was designed only to slow down the much lighter module in the weaker gravity of the moon.

So there was no choice but to leave the spacecraft, for the time being, at the mercy of the "impressed force" already present. This would also give them some desperately needed time to consider the problems of the landing. "Newton" would bring the spacecraft around to its original target in just under twenty-four hours, at which point the moon's gravity would spin it around, some seventy-five miles above the surface of the moon, and send it back toward earth. The only problem was that, left to Newton alone, the spacecraft would miss the earth by 46,000 miles on its return trip. So again they had to improvise, and work out how to use the engine which had been designed for takeoff and landing on the lunar surface to impress the spacecraft with a force capable of adjusting its course. After some frenzied calculation, the answer was arrived at: first, five seconds at 10 percent of full power, then up to 40 percent for 30.4 seconds. Then, according to the computers, it shouldn't be too far out.

Another problem was presented by the carbon dioxide from the astronauts' breath. The filters in *Aquarius* were not designed to cope with three men for such a long time. So the filters from *Odyssey* would have to be utilized, but these were not compatible with the equipment on *Aquarius*. Again the ground crew had to experiment with cardboard and adhesive tape, to see whether the filters could be connected to the system by using the tubes from the astronauts' space suits, which would no longer be needed. This too paid off. Other personnel looked into the astronauts' own suggestion that they might stretch out their supply of cooling water with their own urine.

Cardboard and urine! The people at Houston and in outer space had come a very long way from the media's image of NASA. There was no sticking to manuals and carefully worked-out emergency procedures now, because no one had been able to predict a situation such as this. Instead they invested all of their common sense, resourcefulness, and inventiveness in an almost hopeless race against time.

In Charles Perrow's terminology, they endeavored to make this extremely complex, tightly coupled system more linear in form and more loosely coupled. Automation was replaced by manual intervention. They tried, too, to get the best out of the "operators" on the spot, as well as the colossal backup systems on earth.

And back on earth, the American people awoke to a new day and the sudden realization that spaceflight was no longer predictable and deadly dull. The lives of those boys in space were at stake, and there was no guarantee that even almighty NASA could save them. President Nixon asked to be kept up to date on the situation, and Georges Pompidou, the French Premier, offered the assistance of his country's navy on the high seas.

The astronauts themselves suggested that the backpacks for the space suits might be adapted in some way for use in collecting water from the command module. The men on the ground got together with the backpack supplier to work on this idea.

Not surprisingly, circling the moon meant a great deal to the crew. It must have been a bitter blow not being able to carry out the landing which they had been practicing for years, but from this point onward they were heading back toward the earth, 247,000 miles away. For the three days remaining, there was nothing left to do but bide their time, curled up in semidarkness and in severely cramped conditions, while the temperature slowly dropped to just under 40° F.

Now the media took notice. Houston was swarming with hysterical, clamoring journalists. The Senate asked all private companies and public institutions to allow a short break so that their employees could pray for Almighty God's help in bringing the astronauts safely home. Pope Paul shared his hopes that the astronauts would be rescued with a crowd of 10,000 in St. Peter's Square. *Le Monde* wrote that the whole human race had been united in its concern for their fate. President Kosygin informed Nixon that every civilian and military authority in the Soviet Union was under orders to assist in a rescue, should this be necessary. England offered the services of the Royal Navy, just as thirteen other countries also put their forces at America's disposal.

In space, the astronauts were trying out some brand-new procedures developed by Houston which would make it possible for the lifeboat to recharge the mother ship's power supply prior to reentry. Here again standard procedures had to be turned on their heads. Swigert spent two hours writing down the necessary instructions on every available scrap of paper. The egg was about to save the chicken, but as yet no one could say whether the command module could be brought back to life after such a long cooling-down period.

As the earth gradually drew closer, it was possible to ease up gradually on the strict rationing, and the temperature could slowly be raised again. Water rations could also be increased slightly. After four days, it was almost time to leave the lifeboat. But first the service module had to be uncoupled from the back of the command module. For the first time, the crew saw the full extent of the damage, with the great mass of wires wafting, like so much spaghetti, through the gaping hole in the side.

A little under two hours before reentry the shaken and exhausted crew members crawled back into *Odyssey*, the Apollo module; its minimal electricity and oxygen supplies would now be adequate for the last stretch of the journey home. By this time, the three astronauts had been saving on water for so long that they were all dehydrated and slightly confused. Fred Haise had developed a severe bladder infection and was shivering with fever. The past 140 hours had left *Aquarius* with just enough power for another half day and water for approximately eight hours. That was how close to the edge they had been. They said farewell to their lifeboat, which — without ever having been designed for the job — had performed better than anyone had dared hope. Even the very tightly coupled complexity of space technology had proved to contain some leeway for improvisation and human inventiveness. And *Aquarius'* only reward would be to be burned up, soon afterward, on coming into contact with the earth's atmosphere.

Only now did the mission revert to something resembling the normal pattern. The three men were again lying side by side, with the heat shield facing toward the earth's atmosphere — just as they would have done after a perfectly executed operation. There were no words to express how much they all longed to feel the usual 5 g pressing every inch of their bodies back into the seat — a welcome sign that they were entering the earth's atmosphere.

There was considerable concern in the control room as to whether the heat shield had remained intact after the explosion in the service module. Even the tiniest hole would have the effect of a blowtorch and put an end to any hope of ever seeing the astronauts alive again.

Another concern was whether the command module would enter the earth's atmosphere at the right angle—within the narrow range of between 7.5 and 5.5 degrees—which would mean the difference between burning up in the atmosphere or bouncing off it and heading back out into space.

For over three minutes, during the radio blackout, they could only cross their fingers. For Gene Kranz and his men in Houston the seconds now felt like hours.

But everything went as planned. Suddenly, from the USS *Iwo Jima*, the scarred capsule was sighted emerging from the clouds, suspended from its three parachutes. Soon afterward, the astronauts appeared on television screens—unshaven, dazed, and strangely aged—standing on the deck surrounded by cheering crewmen.[4]

In Houston, even the most hardened of personnel wept at their control panels, while the ritual fat cigars were brought out with trembling hands. Throughout the country traffic came to a standstill. For a brief spell, no crimes were committed in Los Angeles; and in Las Vegas the one-arm bandits were stilled—a hitherto unheard-of occurrence. Word came in from news agencies in South Africa of a prominent witch doctor, Magomezulu, who had declared the whole incident to be clear proof that the United States had blasphemed against a God who did not want any more landings on the moon. But from more than one hundred countries, messages of congratulations poured in, while President Nixon gallantly picked up the astronauts' wives on his way to Hawaii. The following Sunday was officially declared a day of national thanksgiving for the safe return of the astronauts, and prayers were said for them in every church in America.

In typically wrong-footed fashion, Vice President Spiro T. Agnew tried to exploit the situation with some ill-timed taunting of the American left wing, who would—just as soon as they could once more get away with it—no doubt suggest that space research grants be poured into the nearest slum, to no good end.

The writer of an editorial in *National Observer* seemed, however, to be more in touch with national feeling when, on April 20, 1970, he wrote:

At a time when Americans were becoming bored with successful flights to the Moon, the ordeal of Apollo 13 reminded the nation of the dangers and difficulties of space. But because of courage, care and uncommon resourcefulness — because, in fact, of history's most dramatic field experiment — the astronauts made their way safely back to Earth. The first walk on the Moon was a wonder of wonders, but the return of Apollo 13 was the gladdest moment of all.[5]

But why did it happen? Again a board of inquiry was set up to investigate the accident. This time faulty welds were discovered in the oxygen tanks — which happened also to have been dropped during installation. But as to the main cause there was no doubt: Fitted inside the tank were a number of propellers designed to agitate the highly viscous, super-cooled oxygen at regular intervals. Some small heating elements had also been fitted at the outlet, to facilitate getting the oxygen out of the tanks. These warmed the chilled oxygen up to 79° F, thus enabling it to flow freely through the pipes in gas form. As a safety measure — to ensure that the prescribed temperature was never exceeded — the heating elements were connected to a thermostat valve. Naturally, this thermostat had been designed for the 28-volt system employed in the spacecraft.

But during the filling and testing of the tanks at the Kennedy Space Center, 65 volts were applied, in order to speed things up. For this reason, back in 1965, the manufacturer had been told to make alterations to all of the tanks, so that they could run at both 65 and 28 volts. The order had been passed on from North American Rockwell, the capsule manufacturers, to Beech, who made the tanks. But somewhere along the way this order had been buried under piles of paperwork and no one at NASA or at Rockwell had ever checked whether the thermostats had, in fact, been replaced. They had not, and were therefore still capable of withstanding only 28 volts.

At a test carried out at the Kennedy Space Center two weeks before the launch of Apollo 13, it became necessary to take the quite unusual step of emptying the no. 2 oxygen tank after use. When activated by a 65-volt charge, the thermostat melted instantly. At one end of the tank, the temperature rose to 1,000° F, and all of the insulation material was sweated off the wires inside the tank. No one had noticed anything amiss, so the tank was refilled for Apollo 13.

In space, it had functioned perfectly until just a few seconds before the accident occurred, when Houston decided that the propellers inside the tank should be activated, and the circuit to the exposed wires was closed:

13, we've got one more item for you, when you get a chance. We'd like you to stir up your cryo tanks.[6]

The board of inquiry's report called the cause a "procedure error" but could not pin the blame fairly and squarely on any one of the parties involved. In conclusion, the board said, exactly as if they had been reading Charles Perrow:

> The accident was not the result of a chance malfunction in a statistical sense, but rather resulted from an unusual combination of mistakes, coupled with a somewhat deficient and unforgiving design.[7]

16. After Apollo

It had all begun so well. Adlai Stevenson's belief in modern technology as a magic wand capable of granting mankind everything it desired. Kennedy, sincerely convinced that, with the help of science, we were moving toward the happiest society the world had ever known.

When, at Christmas 1968, Apollo 8 returned from orbiting the moon, the astronauts brought with them a picture of an "earthrise," in which the blue, weightless, and fragile globe of the earth was seen rising above the scarred surface of the moon. Deeply moved, Lyndon Johnson sent framed copies of this picture to every one of the world's heads of state. Even North Vietnam's Ho Chi Minh received a copy.[1]

The budding environmental movement was not slow in starting to refer to "Spaceship Earth." Its resources were common to us all and we would have to be every bit as sparing in our use of them as Haise, Lovell, and Swigert had been with theirs, if we were to survive *our* journey through space.

The tide was imperceptibly turning. NASA discovered that it was now having to fight harder for every dollar. The current was flowing outward and that same money might be badly needed elsewhere. Strangely enough, the lunar fairy tale had worked like a booster rocket for the notion that the federal authorities should finance more and more areas of American society. Eisenhower had been concerned — almost before all this began — about this trend. In 1959 he said:

> We must get the Federal Government out of every unnecessary activity. We can refuse to do things too rapidly. Humanity has existed for a long time. Suddenly we seem to have an hysterical approach, in health and welfare programs, in grants to the states, in space research. We want to cure every ill in two years, in five years, by putting in a lot of money. To my mind this is the wrong attack.[2]

Lyndon B. Johnson — who, with his vision of "The Great Society," was responsible for much of the development which Eisenhower had dreaded — had this to say:

> [Until *Sputnik*] we didn't have any Federal aid for education . . . so we started passing education bills, we made a national effort in elementary education, a national effort in higher education, where two million students were brought into our colleges. And they said, "Well, if you do that for space and send a man to the Moon, why can't we do something for grandma with medicare?" And so we passed the Medicare Act, and we passed forty other measures . . .
> And I think that's the great significance that the space program has had. I think it was the beginning of the revolution of the 60s.[3]

In this sense, it would be wrong to regard the welfare programs of the 1960s, or the Vietnam War, as factors which got in the way of NASA's space adventure. Webb had already made the point that NASA was inextricably linked to the world around it. And NASA itself, more than any other American institution, had propagated the idea that all those who hungered could be fed and all those who were cold could be clothed. NASA too had fostered the notion that high technology could be used to beat back the Communists. Actually, NASA had led the way in creating a climate in which the aerospace industry could have the expertise and the manpower standing by to meet the Pentagon's requirements without delay. So now, the fact that the Vietnam War and welfare benefits were clamoring for their share of the national budget — at the expense of NASA — only served to emphasize how all of these elements were part of the same package deal. It was simply that the space adventure had been unpacked first.[4]

But contradictions abounded. The great move to improve the quality of American education had prompted disciplinary crises and falling standards in the universities. The heavy investment in educating more engineers and technologists had created massive unemployment in those very fields. And an increased contribution to aid for Third World countries had ended, in the aftermath of the Vietnam War, in charges of imperialism and exploitation from those same countries.

During these years — when more than ever before was being invested in education — America's youth came within spitting distance of being totally anti-intellectual. No great value was attached to learning, least of all the sciences. Most people found it hard to see where technology ended and technocracy began, and both were regarded as antihumanis-

tic, imperialistic, and, quite simply, as symbols of "the system." Young
people were more fascinated by the exploration of their own inner realms
than those of outer space. They were not taken in by glib references to
the magic wand of technology. They were more likely to identify with
Thomas Pynchon's *Gravity's Rainbow*:

> Living inside the System is like riding across the country in a bus
> driven by a maniac bent on suicide . . . you catch a glimpse of his
> face, his insane, committed eyes, and you remember then, for a
> terrible few heartbeats, that of course it will end for you all in blood,
> in shock, without dignity.[5]

The people at NASA were baffled. During the last days of Webb's
directorship, he had been endeavoring—with increasingly strident zeal
—to explain to congressmen how NASA's inventions could be employed
in Vietnam, town planning, police work, firefighting, the educational
system, and the health service; and to combat airplane hijacking, drug
dealing, and pollution.[6] The key word was "spin-off." There were bonuses
to be reaped from space research. But in Congress more and more dis-
senting voices felt that it would be cheaper to tackle these problems
head-on, here on earth, rather than taking an expensive detour around
the moon and back.

Each trip to the moon with Apollo and Saturn cost the taxpayers a
billion dollars.[7] It could not go on like this. Especially since most American citizens could not have cared less about the ninety pounds' worth
of moon rock which the astronauts proudly brought back from every trip.
NASA had planned to make ten trips to the moon, but had to battle for
permission to carry out just six of these. And it became more and more
difficult to conceal the fact that Apollo was a dead end. If there was any
future in space, it lay, not in extravagant and scientifically dubious trips
to the moon, but in communications satellites and in military and civilian surveillance of Spaceship Earth. NASA was forced to come down to
earth in the most painful and literal sense.

This was easier said than done. Like all vast technological systems, NASA
had by this time built up an incredibly conservative momentum. This
organization had come into being in another age, under different conditions and with a different goal. It had grown used to being accorded
ample resources, national popularity, and political backing; to enjoying
the protection of the President and being above the wearisome battle for

resources that typified other public enterprises. The technological historian Thomas P. Hughes actually describes great technological systems as dinosaurs, whose brains and response patterns had once fitted in perfectly with their surroundings. These faculties are now out of place, but the dinosaurs are still there. They blunder about blindly, almost touchingly lost and incapable of understanding the world around them. Their strength lies in their toughness and in power that has been built up over many, many years, but their capacity for change is appallingly feeble.[8]

So NASA was slow to get the message. It made its opening bid with plans for a space station to orbit the earth. This space station was to be set up inside a Saturn rocket stage, and would serve a wealth of very earthly purposes. But when adequate funding never materialized, these plans died on the vine.

However, NASA's George Mueller did not approve of this new cautious approach or the heavy emphasis on down-to-earth objectives. In keeping with NASA's proud past, he believed that great, symbolically charged super-projects — along the lines of the Apollo program — provided the only way to win and keep the attention of the politicians. NASA ought to draw up a grand master plan, an integrated space program, and sell it as a package to the politicians and the nation.

The chance came in February 1969, five months before the lunar landing, when Nixon set up a task group and requested "definitive recommendations on the direction which the U.S. space program should take in the post-Apollo period."[9] Nixon intended to withhold all funding for this period until the recommendations were forthcoming. In its work, the task group drew to a great extent on Mueller's preliminary studies, based on internal NASA conferences as far back as December 1968.

The key words of these plans were "reusability," "commonality," and "cross-application." There were to be no more expensive one-shot rockets or million-dollar moon buggies that were simply abandoned after use. Recycling was the order of the day and it was up to NASA to lead the way. Furthermore, the same equipment had to be able to perform several functions. Specialization was a thing of the past. A space station orbiting the earth ought also to be capable of circling the moon or forming the nucleus of a lunar base. It should be possible to develop basic satellite models which could be furnished with different modules, according to their particular requirements. And finally, craft for both manned and unmanned spaceflights — unlike previous spacecraft — had to be able to share equipment.[10]

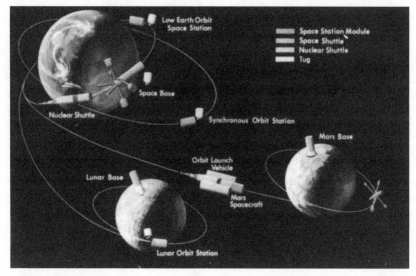

NASA POSTER SHOWING GEORGE MUELLER'S INTEGRATED SPACE VISION. (NASA)

In September 1969, just after Apollo 11, the task group issued its report, *The Post Apollo Program: Directions for the Future,* in which it said:

> The Space Task Group believes that manned exploration of the planets is the most challenging and most comprehensive of the many long range goals available to the nation at this time, with the manned exploration of Mars as the next step towards this goal.[11]

Now here was a goal that was both dramatic and dynamic, and no more unattainable than a lunar landing had seemed ten years before. So why not? Kennedy had said "Man, Moon, Decade." Now the motto would be "Man, Mars, Decade." First, a recyclable space shuttle had to be developed, to replace the expendable rockets. This space shuttle would ply back and forth, carrying the necessary modules to a low-earth-orbit space station. Later, it would form the link between earth and the space station.

The space shuttle would be supplemented by a space tug, which would remain in space and transport communications satellites further up to the desired, so-called geostationary, orbit at an altitude of 22,300 miles, where their period of revolution would correspond to that of the earth and hence they would appear to be stationary. There, they could be left to get on with their work. The tug could act as a link between several space stations.

Thus, NASA would also be able to build, module by module, toward its much-longed-for mission to Mars. NASA envisaged getting underway with the Mars mission in 1981, 1983, or 1986. A NASA poster showed departure from earth on November 12, 1981, arrival at Mars on August 9, 1982, departure from Mars in October, passing Venus on the way home and returning to earth on August 14, 1983, a little less than two years after setting out.

The finances too could be dispensed in modules. According to the figures which were submitted, the entire program would cost between $8 and $10 billion per year.[12]

To NASA's surprise and dismay, neither the politicians nor the American people evinced the slightest trace of enthusiasm. The United States had won the space race, the competitive element was gone, as was the climate of the Cold War, which had, for so long, provided NASA with its raison d'être. Nothing was left but a complicated science-fiction scheme which could be viewed with absentminded interest, before being weighed against other, and possibly more urgent, projects that might warrant funding. Just one week after the lunar landing, a Harris poll had shown that 56 percent of the population believed the Apollo project cost too much money. At NASA it had been hoped that Armstrong's achievement would open the door to even bigger budgets, but the survey found that 64 percent of the population felt that NASA was already getting too much.[13]

In Congress, Walter Mondale protested:

> I believe it would be unconscionable to embark on a project of such staggering cost when many of our citizens are malnourished, when our rivers and lakes are polluted and when our cities and rural areas are dying. . . . What are our values? What do we think is more important?[14]

During negotiations with the Office of Management and Budget, NASA had to accept having its annual budget cut by a further 10 percent, and so bid farewell to Apollo 18 and 19. Since the space station had been given such a cool reception, NASA looked into the possibility of the Europeans coming in and covering development costs. From this quarter too came only procrastination.

Nixon refused to guarantee NASA so much as a minimum budget, and the organization wriggled and squirmed. Officially, by virtue of its structure, NASA was directly accountable to the President. Under Kennedy, this had been a dream, but with Nixon it was a nightmare. This arrangement prevented NASA from openly criticizing its chief executive; instead, NASA had to let its supporters in Congress argue its case.

Even with the most optimistic time schedule, it would take at least four years from the last scheduled Saturn launch with Skylab in 1973–74 to the point where a space shuttle — constituting the first phase of the new integrated system — could resume manned spaceflight. This lull was what NASA dreaded most. Had not Webb spoken of momentum and of the necessary flying speed? Lift had to be obtained from exploits capable of capturing the interest of the voters at regular intervals. If such exploits were not served up, any pilot could tell you what would happen: the organization would stall; it would cease to fly; would go under; plummet, out of control, into the abyss below and be lost forever.

At NASA this situation was obviously regarded as catastrophic. The organization could not just turn around and start churning out what the young were now referring to as "soft technology"; nor take up the "Small is beautiful" chant. This great technological system commanded vast stores of technical equipment — installations scattered throughout the United States, all of them tailor-made for epic missions into outer space. The staff at NASA were skilled technicians and specialists in space technology; NASA's suppliers in the private sector were geared to meet its unique requirements. All of this constituted the sum total of NASA's capital. But this capital was worth something only if the system could continue doing what it did best. Much of NASA's equipment was only interesting and relevant inasmuch as it pertained to NASA, and the work of many of the organization's experts was so highly specialized that NASA was pretty much the only place where they could use their expertise. Were the rug to be pulled from underneath NASA, all of this carefully accumulated capital would become worthless and NASA's team of extremely self-assured specialists would be promptly and definitively unemployed. It went without saying, therefore, that NASA had to use every means at its disposal to fight for its own survival.[15]

Von Braun, who had spent the whole of his American life longing for the chance to lead NASA, traveled to Washington in February 1970 in a final, desperate attempt to win the ear of the Nixon administration.

But Nixon did not want to know about plans for Mars, space tugs, or atomic engines. He wanted peace in Vietnam. Von Braun still had a way with words, but he soon realized that even NASA's own people no longer believed in his visions. He became a solitary man in the corridors of NASA headquarters. A rather weird figure from the past. In June 1972, he bowed to the inevitable and left NASA, an organization he had never imagined could do without him.

For Wernher von Braun to have ended his career as head of NASA would have been the icing on the cake. But, politically, this was out of the question. He could not run away from his past. Instead he could be awarded the highest American decoration for public service, the Distinguished Federal Service Medal, which he could set alongside his *Kriegsverdienstkreuz I Klasse mit Schwertern* (Military Service Cross, First Class, with Crossed Swords), while Tom Lehrer sang, to the glee of young Americans, "Wonze tze rocket iz ap, who kares where it comes down, that's not my department, sayz Wernher von Braun."

Of the grand integrated space system, only odds and ends remained. NASA was forced to drop one module after another. Soon the organization's budget stood at just a third of what it had been in its heyday. The space station and the atomic engine were killed off in negotiations with Congress, no one said a word about the space tug, and only a real fool mentioned Mars any longer.

Even the humble space shuttle was heading down the drain when NASA's director, James Fletcher, who was now growing desperate, flew out to see Nixon at San Clemente. With him he brought the inflated prognoses for the jobs which North American Rockwell had promised to create in Nixon's home state of California, if given the green light for the development of a space shuttle.[16] Shortly before noon on January 5, 1972, Nixon stepped out into the sun for an improvised press conference, at which he announced:

> I have decided today that the United States should proceed at once with the development of an entirely new type of space transportation system designed to help transform the space frontier of the 1970s into familiar territory, easily accessible for human endeavor in the 1980s and 90s.[17]

The best part was, as Nixon told the press, that the space shuttle would put an end to astronomical expenditure on outer space and ensure that access to space would soon be "safe and routine."[18]

17. The Political Spacecraft

Never before had NASA used economy as an argument. Instead they had spoken of "human enterprise," "new knowledge for mankind," and "the inherent urge to seek fresh challenges." Now the word was that the space shuttle would save money. It would present a cheap alternative to the Saturn V, Atlas-Centaur, Titan III, and all the other expendable rockets. And it would create jobs in the beleaguered space industry. In the summer of 1971, the number of workers involved in space projects fell to 149,000—compared with 420,000 six years earlier. Back then, Lloyd Berkner of the American National Academy of Sciences estimated that every scientist in NASA created work for between five and ten engineers, each of whom went on to generate jobs for between ten and fifteen skilled workers.[1] Now these workers were being laid off—from the bottom up—and the effects were being felt in Texas, Alabama, Florida, and California.

Less mention was made of the fact that the space shuttle was now NASA's sole option in the manned spaceflight stakes. Without the space shuttle, no astronauts; and without the astronauts it would be difficult to remain a powerful, independent space organization. And so NASA desperately needed a space shuttle, to justify its own existence. Even though this shuttle no longer had anything to shuttle to.

The space shuttle was to fulfill somewhat more secular aims than Apollo. Historians divide the Great Discoveries of the sixteenth century into three phases: Discovery, Exploitation, and Colonization. As far as space was concerned, the discovery phase was over, at least for the moment; and colonization would have to wait; so now it was time for exploitation. Commercial and military considerations were now to be put before scientific curiosity. Obviously, the main motivation behind a space shuttle was still political and (in the broad sense) military—just as it was with Apollo and Mercury. The difference was that this time arguments were also made for the financial, commercial, and job creation aspects.[2] Mueller promised to bring the amount invested per pound sent in orbit down from Saturn's $1,000 to "somewhere between $20 and $50."[3] And Nixon had been promised 24,000 new jobs for California.[4]

Right from the start, the space shuttle was visualized as a manned, reusable two-stage launcher. Two vertically launched manned aircraft; the smaller one riding on the back of the other. Both aircraft would have rocket engines powered by liquid oxygen and hydrogen, just like the top stage of the Saturn rocket. These machines would also be fitted with ordinary turbofan engines, to be used for landing in the earth's atmosphere. Early plans depicted a mother ship the size of a jumbo jet, with twelve rocket engines; and the space shuttle itself as a somewhat smaller Boeing 707, with two engines.

The idea was for the mother ship to haul its burden up to a height of between twenty-five and thirty miles, while achieving a speed of up to three to four times the speed of sound. The Orbiter would then be released and carry on alone toward its predetermined orbit. The mother ship would turn around and be steered back by its crew to land safely at Cape Kennedy—where it would await the return of the space shuttle and be refueled for the next trip.

The space shuttle's predecessors included a wartime project in which Austrian Eugen Sänger put forward the idea of an intercontinental bomber driven by rocket engines. On earth, this aircraft, which would weigh 100 tons, would accelerate along a runway laid with railway tracks; and in the air it would reach the unprecedented speed of almost 3.7 miles per second. Not fast enough to go into orbit, but enough to skim the upper limits of the earth's atmosphere. During one of its dives, its bomb load could be released—over New York, for example—after which the craft could zoom onward to land among friends in the Emperor's Japan.[5]

A better-known forerunner was the American X-15 experimental aircraft from the 1960s. This airplane had collected a wealth of valuable information on the aerodynamic properties of craft flying at many times the speed of sound; on the steering of a craft out on the fringes of the earth's atmosphere by means of small altitude control thrusters and on returning to and landing in the earth's denser atmosphere. The X-15 also took off from an airplane flying at high speed—namely, the B-52 bomber. In its almost 200 flights, with the toughest test pilots at its controls, this aircraft attained a height of 66 miles and a speed of Mach 6.7, or 4,520 mph.[6]

It was test pilots like these who had regarded the astronauts with thinly disguised contempt, for having willingly submitted to being locked up in a tin can over which they had not the slightest control. It was quite another matter to fly your own plane through space and bring it back in to land at Edwards Air Force Base. Quite another matter to sit astride your own rocket engine, spurring it on — as opposed to finding yourself curled up at its tip. Then it did not matter so much that a month's wages for pioneer test pilot Chuck Yeager amounted to just $283, while the first seven Mercury astronauts, featured in *Life*, could each take home $25,000 a year over and above their basic salary.[7] The way the test pilots saw it, only now were wings and dignity being restored to space travel.

The American Air Force had also supplied raw material, in the form of the early results of tests made prior to the X-20 Dynasoar project, in which a winged craft was to be launched on a Titan rocket and — as in Sänger's project — skim over the surface of the earth's atmosphere. In outer space this craft would perform certain defense-related exercises, such as strategic reconnaissance, inspection — and, if necessary, destruction — of enemy satellites, and, possibly, intercontinental bombing raids. Finally, it would be able to land like a glider.[8] This project had been put on ice in December 1963.

The large aerospace companies were asked to submit proposals for a space shuttle that could be reused 100 times. Convair came up with the idea of a mother ship measuring 240 feet in length, helping an Orbiter 174 feet long along the first 35 miles of its route. It had a load capacity of 25 tons and its Orbiter would be able to cover the last stretch of the route taking it into its orbit — 270 nautical miles above the earth — under its own steam. Lockheed envisaged two similarly gigantic aircraft — their great size being due, not least, to the vast amount of fuel they would have to carry. North American Rockwell visualized a 263-foot-long first stage with a total wingspan of 230 feet; the second stage measured 202 feet in length and had a wingspan of 146 feet. Together, the two aircraft weighed almost 2,000 tons — much larger and more complex in scale than the X-15. The enormous first stage was supposed to accelerate its massive load to Mach 4, a speed as yet never attained by any heavy aircraft.[9] The cost of launching was estimated at $2 million.[10]

While these proposals were being submitted, NASA's share of the budget was rapidly falling to below 1 percent of federal expenditure. And there were no indications that this would have any chance of rising in the foreseeable future. The space shuttle on NASA's drawing board was budgeted at between $10 and $12 billion. When the politicians heard these figures their response was immediate and negative. NASA had better not pin its hopes on a "Cadillac space shuttle."[11] If that was what it would cost, then NASA would have to abandon those plans altogether or else go back to the drawing board and come up with something that fell within the bounds of political possibility. The Office of Management and Budget let it be known, indirectly, that these bounds of possibility lay somewhere in the region of $5 billion. That was, however, not an annual sum — as had optimistically been suggested as a minimum in 1969 — but a total cost.

NASA now found itself in the unfamiliar situation of having to cut back on what was considered to be the *right* construction. They were not to try to work toward an optimum technical standard. Rather, they had to pare the structure down to an acceptable price.

One way of doing this might be to abandon the reusability of every component in the system. It soon became clear that the operating costs were in inverse proportion to the development costs. In an attempt to lower the costs of the first phase they hit upon the idea of putting the fuel in a special, separate tank — an external adjunct to the two aircraft which would save both on weight and on space inside the shuttle. The latter could then be reduced in size but still be capable of carrying more cargo. The fuel tank, costing more than half a million dollars, could not be reused and would have to be dumped after every trip — thus bringing about a corresponding increase in the operating costs. But the development work would be cheaper and less complicated; and, as always, it was the short-term costs on which the politicians had set their sights.

Until now, the costly manned first stage had been left untouched. The total cost now amounted to $8.2 billion and the cost per launch was now given as $3.8 million. But, once again, NASA received an emphatic "no."

The first stage could no longer be kept exempt. The next thing that had to be abandoned was the idea of a manned mother ship. This would have to be supplanted by unmanned booster rockets. To begin with, they toyed with the idea of rockets powered by liquid fuel, but soon fell back on solid-fuel rockets. Since these rockets were to be the first ever to be reused — up to as many as twenty times — they would, after use, have to parachute into the ocean. No one dared think of what all the pipes and pumps and valves inside a liquid-fuel rocket would be like after a dip in the ocean. Rockets with a core of solid fuel were simpler to scrub clean and refill for the next trip.

The technology of solid rocket motors was already well known, thanks to the Minuteman missiles stored in silos all over the country. The Titan and Polaris missiles also used this kind of fuel — the advantage being that they could be launched at a moment's notice, not having to go through the long, slow process of being filled with chilled liquid fuel. But no manned spaceflight had ever employed solid rocket motors. Hitherto, when the risk was greatest, priority had always been given to the advantage of being able to "throttle down" or switch off the rocket altogether if anything went wrong — an option that was negated by solid fuel, which just fizzled away once it had been ignited.

The solid-propellant boosters needed for the first stage would be the biggest ever built — 12 feet in diameter and 149 feet long. They would provide 75 percent of the system's total force within the first two minutes and would then be jettisoned at an altitude of 28 miles, after which the Orbiter, with its own three rocket engines, would proceed alone toward its orbit.

The Orbiter itself also underwent changes along the way. As far back as January 1971, NASA had been having frantic meetings with people from the Pentagon in Williamsburg, Virginia.[12] If the Air Force was willing to back the space shuttle, then Congress would be more easily persuaded and the costs could be shared. But the U.S. Air Force had some demands of its own to make. Gone were the days when the civilian space organization could ignore the military's need for access to outer space. The Air Force had recently been pressured by Congress into giving up its own MOL (manned orbital laboratory) project, which would have been manned by their own astronauts and cost $200 million. MOL was to have been an 18-feet-long, 10-feet-wide space station, intended to compete with the Russian Salyut Station. It was to be launched on a Titan rocket and visited by Air Force astronauts in an improved version of the Gemini spacecraft. Now these ace military astronauts were being seconded, much against their will, to NASA.[13] Eisenhower's clearly defined boundaries between civilian and military space programs were becoming increasingly blurred.

The Air Force's interest lay in spy satellites circling in a low orbit — which fit in perfectly with the space shuttle. But they were particularly interested in a north-to-south orbit, which would allow the satellite to keep to the side facing the sun, while the rotation of the earth beneath the satellite conveniently brought a constant succession of different regions into its field of vision. This meant that the launches could not make full use of the earth's rotation, as it moved eastward at 1,300 feet per second; and this again meant a smaller load capacity with the given amount of fuel. But the Air Force insisted on cargo space of at least 15 by 60 feet and a load capacity in such an orbit of at least 40,000 pounds — as opposed to NASA's projected figure of 25,000 pounds.[14]

THE SHUTTLE (NASA)

NASA said yes, even while knowing that this would necessitate a significant reduction in weight.

The armed forces also wanted the space shuttle to be capable of landing on a runway somewhere between 1,100 and 1,500 miles off to one side of its ground track. It might be necessary to bring the space shuttle down to its base again after just one orbit. During the 110-minute duration of the orbit, the base would, however — due to the rotation of the earth — have made a substantial shift eastward and sideways. Hence the need for cross-range capability. Again NASA said yes.

Not surprisingly, NASA had envisaged a space shuttle with wings set more or less at right angles to the body. This would permit a suitably low landing speed — approximating to that of a standard passenger plane. They also wanted a straight and uncomplicated approach.

Now they had to do some rapid rethinking. To keep the necessary leeway on either side, the wings — and, hence, the air resistance — would have to be reduced. This meant that the space shuttle would have to have V-shaped delta wings. Such wings do not, however, hold up well at low speeds. So the landing speed had to be substantially increased and the angle of approach rendered very steep: 22 degrees, or seven times steeper than the approach of a passenger airplane. The projected air-breathing turbofan engines — which would ensure that the space shuttle could make another circuit should it come in at the wrong angle — were dropped, out of regard for weight. The same fate awaited a projected Abort Solid Rocket Motor assembly strapped to the Orbiter's tail, which would ignite if anything went wrong within the first two minutes, to lift the Orbiter free of the boosters and up to an altitude from which it could turn around and fly back for a landing. Thus the space shuttle would be the first American spacecraft with no abort system. The maximum flying altitude was also somewhat reduced — which suited the Air Force perfectly.[15]

The need for maneuverability in the earth's atmosphere made demands on the heat shield, which had to be made heavier — and this again made greater demands on the engine capacity required, if the load capacity was not to suffer.[16] The engines, which had until now been the most powerful ever to employ liquid oxygen and hydrogen — the engines from the top stage of the Saturn rocket — would no longer suffice. Lighter *and* more powerful engines would have to be developed.

Now they were just about there. The cost — it so happened — came to $5.15 billion. In its final version, the space shuttle measured 122 feet in length and had a wingspan of 78 feet. The cargo space exactly matched the armed forces' specifications. The landing speed had purposely been set so high that it only just came within the existing technological limits for undercarriage and wheels. The space shuttle would come out of its orbit on the other side of the earth at a speed of 17,500 mph, brake, go into some sharp turns once inside the atmosphere, and, if the computers had calculated the remaining power reserves correctly, approach the runway like an 80-ton glider at a speed of 330 mph, then bring up its nose at the last minute and land at a speed of between 195 and 240 mph — as opposed to a DC-9, which lands at between 135 and 155 mph. And unlike the DC-9, the space shuttle preferably had to get it right the first time, every time.

NASA hired a firm of consultants, with the imposing name of Mathematica, who calculated that every launch of this final system would cost $10.5 million.[17] If maximum use were made of its cargo capacity, the space shuttle could actually pay for itself with just 30 flights a year.[18] At this time, NASA was expecting to make one flight a week into space — once they got the hang of things; altogether, the prediction was 581 flights between 1980 and 1991.[19] So they would even make a bit of a profit. To the great satisfaction of the politicians, the space shuttle system was starting to look like a sound investment.

18. Aerospace

The space shuttle was the first aerospace craft — in the true sense of the word. Spacecraft and airplane rolled into one. It would be necessary to use all the expertise amassed by the big American concerns during the Mercury, Gemini, and Apollo projects. And it would be necessary to draw on all the lessons learned from the X-15 and subsequent supersonic aircraft. The space and aeronautics industries would both have their hands full applying this knowledge to a space shuttle.

These big companies had grown even bigger during NASA's bumper years in the 1960s. North American, McDonnell, Grumman, and Boeing pocketed over half of all the money spent by NASA in the years up to 1966. But behind these giants, as we have seen, were smaller suppliers who made a living by supplying the big boys. The food chain behind the Mercury program involved 12 main contractors and 75 subcontractors. After these came the third link in the chain — 1,500 minor companies — and, at the very bottom, the fourth link was comprised of 7,200 small businesses.[1] All with their eyes pinned on NASA, like a group of sea lions at feeding time.

During the Apollo program, the total number of companies running in the slipstream of the big firms — whether directly or indirectly dependent on the project — was closer to 80,000. And it was not only in the field of space research that this trend was evident. The enormous grants awarded to the armed forces far exceeded NASA's own generous budgets — to the extent that in 1968 defense spending amounted to half the national budget. In that year alone, the Pentagon invited tenders for contracts — for everything from shoelaces to atomic weapons — amounting to almost 26 billion dollars.[2]

At that time, 22,000 main contractors and 100,000 minor contractors were totally or partially dependent on military funding. Shipyards and aircraft factories received over half of their revenue from the armed forces. And 5,300 American cities boasted at least one major company which was dependent on the Pentagon for its livelihood.[3]

Historically speaking, this was a relatively new situation. Before World War II, the services had produced most of their own equipment, in arsenals; and huge defense budgets were acceptable only during wartime. During peacetime, the country demobilized and both the armed forces and the military budgets shrank in size. American policymakers foresaw a similar situation developing after World War II. With the weapons having spoken and created a new international scenario, the time had come for Open Doors and free enterprise, which could secure America's position as world leader by financial means.

This strategy was discontinued around 1950, with the advent of the Cold War. Everyone now had to learn to live with the unfamiliar concept of being in a permanent state of military readiness, even in "peacetime." The American forces had, at all times, to be primed for the United States's new role as peace officer of the free world.[4]

As mentioned earlier, Eisenhower found it difficult to accept this developing situation. He had been the first to speak of the dangers of the great "military-industrial complex." This combination of the two sectors would, in his opinion, become insatiable. The armed forces would always be striving for an unattainable 100 percent superiority and, for that reason alone, would never be satisfied. And industry could be expected to stop at nothing to get its share of such a profitable area. Shortly before his death the former President put it like this:

> No matter how much we spend for arms, there is no safety in arms alone. Our security is the total product of our economic, intellectual, moral and military strengths.
>
> Let me elaborate on this great truth. It happens that defense is a field in which I have had a varied experience over a lifetime, and if I have learned anything, it is that there is no way a country can satisfy the craving for absolute security—but it can easily bankrupt itself, morally and economically, in attempting to reach that illusory goal through arms alone. The military establishment, not productive of itself, necessarily must feed on the energy, productivity and brainpower of the country, and if it takes too much, our total strength declines.[5]

But even Eisenhower, during the final days of his presidency, had to yield to the military machine—albeit reluctantly. And after his retirement, the process advanced in leaps and bounds. The military-industrial complex consolidated its position of power with every contract for new intercontinental ballistic missiles, jet airplanes, aircraft carriers, submarines, and weapons systems. Not only that, but it took great pleasure in satisfying NASA's need for equipment and hardware for its space adventure.

Nowhere, along this food chain, was there any motivation to economize. Very often the systems were so advanced and so specialized that there was no competition. The supplier could set his own price, and the taxpayers had to pay up. Often, the big firms won contracts by submitting grossly underbudgeted bids. Once they had the order, they could exceed the budgets and the agreed-on deadlines with impunity, and without having to worry about competition. On one particular system, NASA had budgeted for $3.2 million, but the supplier, Rockwell, produced an invoice for $19.2 million, and NASA, as always, had to shell out.[6] On another occasion, a company shamelessly came up with a figure of $12,000 for parts which they had promised to supply for $5,000 and which later proved to be available on the open market for $2,000.[7]

By the end of the 1960s, even advocates of a strong defense could see the problem. It was generally agreed, even in these circles, that 25 percent of the defense budget was "pure fat."[8] People were speaking openly of totally superfluous "gold-plating." The press recounted grotesque tales of nuts and bolts for which NASA had had to pay Rockwell $120 but which could be bought at any hardware store for $3.28.[9] But what could you do, when such powerful forces were at work?

A simplistic, Marxist analysis would say that these forces were, of course, the big American business concerns, but the situation was more complex than that. These companies had powerful allies. Both the regional lobbies and the trade unions were — regardless of their political hue — tremendously interested in, and dependent upon, the perpetuation of this vast technological system. Often, local politicians would devote almost all their energies to dogging the footsteps of the congressional space and defense committees in hopes of taking a slice of the cake back to their own states.

James Webb always stressed that 90 percent of NASA's funding wound up in the private sector. This was a quite deliberate strategy, intended to distribute the orders as widely as possible, and thus win extensive political backing. So no one need be concerned about the vast sums of public money being invested in the space program — because they always found their way back into the capitalist market. As time went on, however, this assertion began to seem less and less plausible, as the large aerospace companies became increasingly, if not totally, dependent on public contracts for which the competition was practically nonexistent. And between 1959 and 1969, less than 10 percent of NASA's money was spent on equipment which had a fixed price and was subject to genuine competition.[10]

This trend was also reflected in monopolization. In 1966 McDonnell and Douglas merged; a year later it was the turn of North American and Rockwell.

Again, there were those who began to voice the thought that the United States had been infected by its rival in the East. In the late 1960s, Seymour Melman of Columbia University stated that the vast military-industrial complex was heading toward "a complete transformation of society [toward] the Soviet type of state capitalism." And at the time of the Vietnam War, Senator Wayne Morse of Oregon declared:

> The American people desperately need to recognize, before it is too late, that we are being run in this country today by an industrial-military complex that makes its profits out of American blood, and jeopardizes all the future generations of American boys and girls.

Walter Adams of Michigan State University described this complex as Frankenstein's monster, "threatening to control the contract state which brought him into being."[11]

But it was not only industry that had become dependent on this process. The universities too had climbed on the bandwagon, even though this cost them their much-vaunted academic freedom. And large private research laboratories such as GE and Bell Labs weren't kicking *their* heels on the sidelines either. In 1964 a full 78 percent of American research and development budgets was, for the first time, financed out of federal funds.[12]

Even arrogant, unworldly scientists picked up the methods and catchwords that would raise money for their institutions. Costly equipment was purchased, whether needed or not, so as not to lag behind anyone else. What did it matter if the same research was undertaken twice or three times, in different parts of the country, just so long as a grant could be squeezed out of the federal government? Critics maintained that the universities were prostituting themselves, that they bent over backwards in their efforts to get money. But only very few could afford to stay aloof.

In 1963, the state financed 88 percent of prestigious Caltech's budget, 66 percent of MIT's, 56 percent of Princeton's, and 25 percent of Harvard's.[13] A very large proportion of these grants came from military funds. As a result, in 1967 MIT occupied the no. 62 position on the list of main suppliers to the Pentagon — to the tune of $94.9 million — while Johns Hopkins University was at no. 73 with $71.1 million.[14] Such a state of affairs made it hard to preserve the image of free, objective, independent research. Just as disturbing was the opinion of some pundits that this would result in more expensive, but poorer-quality research and the undermining of scientific integrity.

From the very beginning, NASA's aim had been to establish a thorough system of quality control. The Apollo project would never have been a success if the organization had accepted the standards that were generally prevalent in heavy industry. Nor would the Pentagon's specifications for unmanned missiles suffice when it came to sending up astronauts. Hence NASA's policy that the space organization itself should command the highest in-house capability in any given field. Only in this way could the private contractors be kept on their toes. Because of this, the NASA inspectors were considered to be the toughest in the business, and they never had to resort to a subcontractor's own data.

It was not without some resentment that the industry had, time and again, to hear complaints from these inspectors about sloppy workmanship and procedures that left a lot to be desired. Widespread frustration was also felt over the lengthy process of obtaining written approval for the smallest thing, so that — if a problem did arise — NASA would know whom to hold responsible. Spreading the NASA spirit among the subcontractors was not without its problems. They had their own methods, and besides, they were used to getting their own way with the Pentagon.

As we have seen, both the board of inquiry's report following the fire in the Apollo capsule and the investigation of the accident on board Apollo 13 revealed serious problems in the complicated interplay between NASA and the privately owned companies.

The industry countered by wooing NASA experts with sky-high salaries, with which the space agency could not compete. One by one, the specialists deserted NASA and, the day after they had taken up their new posts, those companies could sell their services back to NASA. There was no mention now of "contractor penetration." If anything, the shoe was on the other foot. In this way several NASA centers were whittled down to a thin shell of NASA administrators around a highly educated core of privately employed former colleagues — now wearing the coveralls of Boeing, General Dynamics, and Lockheed.[15] The industry skimmed off the cream and profited from the state-educated experts without ever having had to finance their education themselves.

This was a well-known, tried and tested method. Around 1960, 1,400 retired officers with the rank of major or higher were on the payroll of large firms supplying the armed forces. General Dynamics alone employed twenty-seven retired generals and admirals.[16]

This led, during the latter half of the 1960s, to the gradual erosion of NASA's ability to criticize the work of its subcontractors. And ten years later it was seriously threatened. In the mid-1980s there were almost as many aerospace employees as NASA personnel working at the Marshall Space Flight Center.[17] A number of NASA centers were no longer able to set their own standards for the quality, cost, and duration of a given project. Instead they were at the mercy of the aerospace firms. Now and again the services of consultants from one of these firms had to be bought, to check on another. At Marshall the staff of the quality control department had been reduced, over fifteen years, from 615 to 88.[18] NASA was no longer the watchdog. Instead it was on its way to becoming a branch of one of the aerospace giants. Having thrown its own yardstick away, from now on NASA had to beg for what it needed.

The old hands at NASA did not like what they saw. William E. Lilly of the Johnson Space Center was among those who believed that NASA field centers were becoming "too cozy" with their contractors.

> I think they got real lax in the real checks and balances, the adversarial relations, really challenging and not taking their word for it. Making people prove what they said.

According to Lilly, it was highly embarrassing to have one of NASA's field centers receive an order for a particular job and then send the constructional drawings back to headquarters exactly as they had come in from the private aerospace contractors:

> . . . they wouldn't even mark off Rockwell's insignia on the goddam stuff. . . . They became dependent on the contractors. They couldn't do without them and they relied on them too closely.[19]

For the aerospace industry the space program was an attractive, if difficult, proposition. Since Apollo, NASA had been a poor customer and the armed forces had already begun easing off before the withdrawal from the Vietnam quagmire. Several of the largest companies had already initiated large-scale layoffs.

And so the tough business of lobbying had to be undertaken if one was to be in the running when the big contracts came up for grabs. Bids were due to be submitted just as Nixon's first term as President was coming to an end. Several of the biggest aerospace companies deemed it wise to make secret contributions to the campaign to reelect the President. It never did any harm to be in good standing with the White House. This fact alone is enough to prove that the presidency is not unequivocally in the pocket of the military-industrial complex. Unfortunately, when the Watergate scandal erupted, these sums came to light, revealing that Rockwell International, for example, had contributed $98,278 toward a sound investment in the future.[20]

In a farsighted strategic move, this same Rockwell International had managed to have the company's director of the space shuttle project, Dale D. Myers, installed at NASA as Associate Administrator for Manned Space Flight as early as the beginning of 1970. Myers was to be responsible for inviting bids for the space shuttle.

The first contract put out for bids concerned the development and construction of the space shuttle's three main engines. These engines were to be the most powerful rocket engines to date, fueled by oxygen and hydrogen. The conditions and specifications for bids were sent to Aerojet General, Pratt & Whitney, and the Rocketdyne division of Rockwell International. The fact that Rocketdyne had built the engines for the Saturn rocket counted in its favor, but people in the business felt that Pratt & Whitney was the best bet, since it had — in conjunction with the Air Force and NASA — already developed and tested a rocket engine whose specifications came very close to those required, a rocket that spearheaded the next generation of engines using liquid fuel under high pressure.

The commission which was to assess the various bids was appointed by Myers, and its meetings were conducted in secret — officially, in the interest of fair play, so as not to leak any design details. Not surprisingly, NASA's new director, James Fletcher, announced in July 1971 that the decision had been made: The order would go to Rockwell, through Rocketdyne. Its worth: $450 million. Pratt & Whitney complained about this decision but were told that everything had been done by the book.[21] This choice fitted in perfectly with Nixon's wishes. The President's adviser John Ehrlichman — who later found himself behind bars — has since revealed that the President had attached great importance to promoting aerospace interests in the key Republican states in the Sun Belt. And Rocketdyne, unlike Pratt & Whitney, was based in California.[22]

The next bid concerned the extremely prestigious contract for the Orbiter itself. In the running this time were Grumman Aerospace, Lockheed Missile and Space Company, and, once again, Rockwell International.

Grumman had taken the lead in developing the final space shuttle design, with the external fuel tank — a delta-wing space shuttle with two attachable boosters. Thus many people believed that this company had a good chance of landing the contract. But Grumman's plant was situated on Long Island, New York — in the heart of a Democratic East Coast which was extremely hostile to the Nixon administration. So, in actual fact, Grumman did not stand a chance. On July 26, 1972, James Fletcher made another appearance, to announce that this contract — worth $2.6 billion — had gone to Rockwell International, by way of Downey in California.[23] A jubilant Rockwell management organized a big party at a nearby restaurant.[24] But the people at Grumman were in no mood for a party and declared that they had been "screwed."[25]

Then it was the turn of the boosters, and here the favorite was United Technologies Chemical Division, manufacturers of the solid-propellant boosters most closely approximating those of the space shuttle — the 86-foot-long, 300-ton solid rocket boosters for the military rocket Titan III. This company also had experience with the compartmentation of solid rocket boosters. Moreover, it was situated in California, so there should be no problem there either.

But NASA's new chief, James Fletcher, hailed from Utah — as did the chairman of the Senate Committee on Aeronautical and Space Sciences, Senator Frank E. Moss. Moss had a decisive role to play in determining Congress's future attitude toward NASA and the space shuttle project.

Senator Moss told the press that the contract in question was worth around a billion dollars and that he was confident that Administrator Fletcher would not forget his friends back home. And Utah happened to be the home of Thiokol (later Morton Thiokol), a company which, however, had experience only with smaller solid rocket boosters, such as those for Trident.

Thiokol, United Technologies, Aerojet Solid Propulsion Company, and Lockheed Propulsion Company were all invited to submit bids. A thirteen-man commission set to work to put the various designs in order of merit, according to a number of criteria.

Aerojet received top marks for safety, because their rocket was constructed in one piece, instead of being split into sections. On the other hand, this rocket did not score well on the economy side and, when all the marks had been added up, it wound up in fourth and last position. Lockheed came first, Thiokol took second place, and United Technologies was third. James Fletcher thanked the commission and then picked Thiokol to supply the space shuttle's boosters, on the grounds of good marks for "management" and a low price.

The management result revealed nothing of how Thiokol had recently had a violent explosion at one of its plants, which had killed twenty-nine employees and left a score more seriously disabled. The federal authorities who investigated the accident cited "ineffective management" as a contributing factor.

Lockheed opted to take this patently absurd decision to a board of appeals. This body — while, in diplomatically couched terms, recognizing Fletcher's right to exercise his judgment — asked him to reconsider the matter. Which he gladly did. And once again, the winner was Thiokol — from back home.

At NASA, all this fuss about the boosters helped foster a particularly wary attitude toward this very facet of the construction; a facet so close to the boss's heart and, hence, one that enjoyed the protection of people at the very top.[26]

In August 1972, work on the space shuttle's main components had all been allocated — with Rocketdyne doing the main engines; Rockwell International, the Orbiter; Morton Thiokol, the booster rockets; and the Martin Marietta Corporation, the big external fuel tank.

Disappointed suitors were appeased in other ways. Lockheed Missiles was given responsibility for the space shuttle's heat shield; and United Technologies was allotted the job of supplying the fuel cells for the shuttle's electricity supply and of assembling the parts for the boosters which the company was not allowed to build. IBM was to supply the computers.

In addition, all of the big names — such as General Dynamics, Boeing, Honeywell, Hughes Aircraft, McDonnell Douglas, Sperry-Rand, and Westinghouse — were awarded a greater or smaller slice of the cake. The main contractors immediately set about finding thousands of subcontractors dispersed throughout the country. Rockwell farmed out work to more than 10,000 other companies, while Thiokol signed subcontracts with 8,600 smaller firms.[27] In NASA the work was allocated in such a way that the Houston center was to take care of the Orbiter and house the simulators; the Marshall Space Flight Center in Huntsville was to monitor work on the main engines, the boosters, and the fuel tank; and the Kennedy Space Center would assemble the individual components, as well as being responsible for the launches and servicing between missions.

With his work now done, Dale Myers resigned from NASA, to return to his old job with Rockwell International, which was soon going to be kept very busy—thanks, not least, to his efforts.

In December 1972, Apollo 17 returned from the sixth and final mission to the moon. A month later, on January 22, 1973, Lyndon B. Johnson died in Texas. The next day, Nixon could announce to the nation that the nightmare of Vietnam was over.

Johnson was the man who, with his drive and his fervent belief in the international significance of space research, had been the chief architect of NASA at the time of its founding in the late fifties. Sadly, Vietnam was to be the cause of his having to step down from the presidency, a broken and detested man. Nixon magnanimously decided that the Houston center would henceforth be known as the Lyndon B. Johnson Space Center.[28]

In April 1974, Dr. Kurt Debus—NASA's director of launching throughout the entire history of manned spaceflight to date—broke ground for what was to become the runway for the new space shuttle. Just as the news of his imminent departure from NASA was being made public. And so, yet another figure left the scene; one who had, with his dueling scars and his unmistakable German accent, attested to a line that ran all the way back to Peenemünde.[29] The time had come for a changing of the guard.

19. Off the Shelf

Around 1972, the idea that the space shuttle would not require any crucial new technological breakthroughs became a recurring theme at NASA press conferences and at congressional meetings. Any innovation would be confined solely to combining familiar elements in a previously untried fashion.[1] Existing stock, right off the shelf, would be used throughout, it was said. According to the simplified version to which the press was treated, it was merely a matter of converting a standard airplane, attaching rocket engines to its tail, and sticking on a heat shield. And you had a space shuttle.

Congress did get a few more details. NASA revealed that the electrical system would employ the fuel cells developed for Gemini and Apollo. The landing gear and tires would be similar to those on most modern aircraft. The cockpit would be fitted with instruments like those used in aviation. The main engines took the technology applied to the Saturn rockets a stage further. The solid rocket boosters were based on the ten-year-old technological design of the Titan rockets. And even the little reaction-control thrusters used to control the spacecraft's position were already well known from Apollo.

NASA's confidence in its abilities was clearly detectable in these assertions. Against all odds, they had succeeded in developing the Saturn V and the lunar module, and they had put men on the moon. So developing a space shuttle to fly in a low orbit around the earth ought not to present any insurmountable problems.

In March 1971, while still director of the space shuttle program at NASA, Dale Myers told a congressional select committee that it was the main engines which would entail the most significant technological innovations; that, along with the development of a new heat shield and the space shuttle's computer systems. But, he assured the committee, the necessary heat-resistant alloys were already available. The technical groundwork for an integrated electronic guidance system had also been done, although Myers did admit that its development would necessitate a certain amount of innovation. In conclusion, Myers told the select committee:

> No technology breakthroughs are required. Essentially the Shuttle will be based on 1971–72 knowledge applied to allow us to fly an operational Shuttle in 1979. Although significant innovation is required, principally to keep development costs low, the basis for the technology required is solidly on hand.[2]

This was a new departure for NASA. To openly emphasize the absence of technical breakthroughs in a new project. The reason, of course, was — as Myers let slip at the end — that the development costs were something of a hot potato. If Congress were to receive the impression that the project was going to require too much expensive and risky experimentation, then the coffer lid would be slammed shut. Many politicians felt that they had already gone too far by sanctioning the $5.15 billion.

Such attitudes were met with mild surprise by the engineers in NASA's various development departments — those in the lower echelons of NASA being well aware that the space shuttle would call for decisive breakthroughs in a multitude of technological areas. It was to abound in "never-before solutions."

Never before had there been rocket engines as powerful as those which would be required here; never before had a rocket engine filled with highly combustible liquid oxygen and hydrogen been ignited on the ground, while positioned right alongside an enormous fuel tank. And never before had it been necessary for a rocket engine to be reused, or to have a total combustion time of over seven hours. All previous rocket engines had done well if they lasted a modest number of minutes before petering out.

Never before had it been necessary to catapult a winged spacecraft up through the earth's atmosphere with such force. Never before had a manned spaceflight been accomplished by means of solid-fuel rocket boosters. And never before had such boosters had to be cleaned after use and used again.

Never before had a new spacecraft carried human beings on its maiden trip. And never, ever before had anyone tried to bring a spacecraft the size of a DC-9 back down through the earth's atmosphere. The largest object so far had been the Apollo module, measuring twelve feet in length.

And even then not down to a runway which had to be homed in on with an accuracy of a few meters from a distance of 5,000 miles by a 90-ton, computer-controlled glider that was barely capable of gliding.[3]

But remember: "No technology breakthroughs are required," the technicians told one another, shaking their heads and smiling.

The development of the main engines would prove to be one of the chief reasons for delays during work on the first space shuttle. These three engines had to be more powerful than the mightiest rocket engines to date — because they had to contribute some of the thrust which NASA had originally envisaged investing in the now too expensive manned first stage. This stage had now been replaced by two booster rockets working in conjunction with these three engines.

Each engine had to exert a thrust equal to that of one of the old Atlas rockets — 375,000 pounds-force on the ground. They were supposed to burn for approximately eight minutes on each mission, until the craft was safely in orbit. During the launch it should be possible to "throttle them down" to 65 percent of full power for short periods, so as not to put too much strain on the space shuttle. For the rest of the journey they would remain inactive, constituting nothing but dead weight. For this reason they needed to be not only powerful but also as compact and light as possible. For this reason alone they had to abandon the idea of using upgraded Saturn engines, which — while they might well be capable of providing the thrust — would involve a reduction in cargo capacity of between 13 and 18 tons, thereby effectively negating the Air Force's north-south orbit.[4]

Another reason for their having to be so powerful lay in the fact that every pound of thrust which could be transferred to the main engines could thus be subtracted from the far more expensive booster rockets — thus reducing the operating costs.

In principle there were several possible ways of increasing the thrust. First, by opting for the best fuel possible. And here, as Goddard and Tsiolkovsky had already pointed out, there was only one option: liquid oxygen and liquid hydrogen gave "more bounce for the ounce" — despite all the problems presented by these extremely cold liquids. Second, by increasing the size of the engines — but, as we have seen, in this case that was out of the question. And lastly, an attempt could be made to increase the pressure within the engine, in such a way that an unaltered weight would impart a greater speed to the gas molecules in the rocket thruster.[5]

In fact, they had little choice but to pin their hopes on this last option, and try to develop a rocket engine with a combustion pressure greater than that of any previous engine. One and a half times the size of the Saturn engines.

The fuel would be led from the fuel tanks to the engines through 12-inch-thick umbilicals, at a mass flow of 1,100 pounds per second.[6] From a pressure inside the fuel tank of between 2 and 7 atmospheres, two mighty turbopumps — one for each type of fuel — would gradually step up the pressure until it reached almost 360 atmospheres in the precombustion chambers. The gases in these precombustion chambers would drive the turbine blades, which, in turn, would supply the power for the compression, by the turbopumps, of the subsequent fuel flow. After passing through the turbine blades the gases would be led into the actual combustion chamber, where oxygen and hydrogen would be ignited at a pressure of 210 atmospheres — more powerful than any previous engine.

The combustion would produce water vapor, which would exit this high-tech steam engine at unprecedented speed, through a huge, bell-shaped rocket nozzle; the tremendous expansion of the steam and Newton's third law would take it from there.

Following a recipe devised long before by Goddard, the engine and the rocket thruster would be cooled by liquid hydrogen — a fluid which, at a temperature of −423° F, was well suited to the task — before ending up in the combustion chamber. Thus, within the space of its three tons plus, the same engine had to accommodate temperatures varying from −423° F up to 6,000° F in the combustion chamber.

It is not surprising, then, that this system, with its many feedback circuits and its interactive construction, should have presented the engineers with some very big problems once the testing got underway in 1975. Resonance led to heavy, uncontrollable vibration; the combustion chambers burned out; the turbine blades in both the oxygen and the hydrogen pumps cracked; bearings seized up and gaskets blew — all of this resulting from the extremely high speeds in the turbines, due to the very high pressure which the specifications necessitated.[7] For a long time it was difficult to see how such an engine could ever stand up to being reused fifty times under such conditions.

Normally, in the aerospace business, new aircraft were built around a previously tried and tested engine design, working from the bottom up. This time, the opposite applied. The engine had to be constructed in such a way that it could live up to the specifications which the space shuttle would have to fulfill. Building proceeded from the top down. Not only that, but the engine construction had to run parallel with that of the shuttle, under rigid time constraints. As a result of the high degree of feedback inherent in the engine construction and their tight schedule, NASA carried out all of its initial tests on complete engines rather than on individual components. In 1977, they were faced with four ruined engines. In 1978, seven serious faults arose during tests. It soon became clear that the launch they had originally hoped to stage in March 1978 was now out of the question. Launching was postponed for a year.

Again, in December 1978, two engines failed during tests, and the launch was put off until November 1979. NASA had to go cap in hand to Congress time and time again to solicit sums over and above the originally agreed figure. The number of engines would have to be doubled if there was to be enough left over for the four projected space shuttles, once the tests were completed. And within just five years, following 895 structural alterations, Rocketdyne's contract price for the engines had doubled, bringing it up to almost a billion dollars.[8]

In the spring of 1979, the scientific journalist Richard S. Lewis interviewed three test engineers at the Marshall Space Flight Center regarding problems with the main engines. If they were to speak freely, all three needed to remain anonymous. One of these engineers, whom Lewis called "Dick," said:

> We got three major problems. One. Lack of money to begin with. Two. Trying to do component development at the same time we're trying to do full engine-up testing. Three. In all other programs where you had a new component, you got it to a certain level of maturity before you put it into the system.
> We didn't do that. I'm saying I don't know why.

In confidence, however, another of the engineers, "Harry," did hazard a guess as to the reasons:

> My personal opinion is . . . they didn't start out like normal programs do. They had a time constraint, a money constraint, or something. They figured they could bring it off anyway. But I think they underestimated the difficulties. Or they overestimated the state of the art. Every engine Rocketdyne's developed has had a [rotating machinery] problem. You've got 400 pounds of machinery coming up to a speed at 40,000 revolutions per minute in four seconds. That's massive. Pressures and temperatures are higher than ever.[9]

But when the engineers had to make any comment during the guided tours which NASA arranged for the press, they pointed out, instead, how much power they had succeeded in squeezing into these compact and relatively lightweight rocket engines. The equivalent, they said with pride, of developing a powerful automobile engine the size of a fist. The hydrogen turbopump alone, which was no bigger than a modern V-8 engine, supplied 310 times as much power — or 63,286 horsepower.[10]

And always, on such occasions, the people from Rocketdyne would mention how the thirsty turbopumps could empty an Olympic-size swimming pool in just twenty-five seconds.[11]

20. Tiles

Physics textbooks teach us that energy can take various forms. It can, for example, be contained within a booster rocket's or fuel tank's supply of apparently inert chemical energy. This chemical energy can be released with great suddenness, in the form of hot gases in the exhaust, imparting both kinetic energy (speed) and potential energy (height) to the spacecraft. When the time comes for the spacecraft to return to earth, both the height and the speed must, literally, be burned away in the atmosphere through intense heat generation, to bring the kinetic energy back down to an acceptable level for touchdown on the earth's surface.

For the Mercury, Gemini, and Apollo programs, the space capsules were fitted with heat shields which were supposed to conduct the heat away from the crew while they burned away, layer after layer, on the descent through the atmosphere. This ablative heat shield had been developed through a series of painstaking experiments, in which missile nose cones were covered with various prospective materials. By the time the capsule was finally bobbing up and down on the waves of the Pacific, almost all that remained of this shield was a charred and pitted surface. But all the while it was being consumed, the heat shield was serving its purpose.

This solution would not, however, suffice for the space shuttle, because it was meant to be recycled up to 100 times. The space shuttle's ample exterior could not be fitted with a new heat shield after every mission, to replace the one that had burned off. And without the heat shield, it wouldn't work at all. Both crew and craft would assuredly be burned to a cinder on coming into contact with temperatures of up to 3,000° F as the molecules in the atmosphere were smashed against the space shuttle.

And so NASA and Lockheed Missiles joined forces to develop a totally new system, by means of which the Orbiter's aluminum surface would be covered with nearly 31,000 ceramic tiles. This aluminum exterior was akin to the hull of most standard aircraft, but the surface itself could not be subjected to temperatures of more than 350° F. The tiles were meant to ensure that this crucial limit was not exceeded on the descent through the atmosphere.

The tiles were different, depending on which section of the space shuttle was to be covered. The rudder, the anterior edges of the wings, and the nose would have to withstand the fiercest temperatures. Then came the wide expanse of the underside, which would have to stand the shock of braking at temperatures of up to 2,300° F. The lowest temperatures would be encountered on the wings, on the top sides of the auxiliary rocket engines, and above the hold, where the temperature would be 1,000° F. The "coolest" spots were not tiled, but covered with insulating felt.

Computer models had worked out the different temperature curves on the spacecraft and graded the tiles at thicknesses of between 0.5 and 5 inches, depending on the temperature they would have to withstand. The temperature curves were almost visible on the space shuttle, since the most heat-resistant tiles were black, while the white ones were designed to cope with somewhat lower temperatures.[1]

Every one of the approximately 31,000 tiles was unique and had to be graded in such a way that it not only fit but also conformed with the spacecraft's meticulously worked-out aerodynamic curves. For this reason, every single tile was stamped with a number indicating its precise position in this gigantic jigsaw puzzle.

The development of the ceramic material was a technological achievement perfected in the organization laboratories. At NASA they demonstrated the insulating qualities of the tiles by lifting them, red-hot, from an oven heated to a temperature of several thousand degrees and throwing them straight into a bucket of water. They could also be held in the bare hand while still red-hot on the inside.[2]

The tiles hardly expanded at all when subjected to heat. The same could not be said of the space shuttle's aluminum hull, where a relatively large degree of give was anticipated, due to the heat and the effects of the earth's atmosphere. So there could be no question of sticking the ceramic tiles straight onto the aluminum. The different qualities of the two materials would cause the tiles to crack, fall off, or damage the thin metal shell. This problem was solved by having a layer of insulating Nomex felt create a bridge between the tiles and the aluminum, so that any give would be absorbed by the felt.[3]

Here the problems began — because NASA had neither the money nor the time for the kind of practical experiments that had been carried out on the heat shield for Apollo. Instead — without unnecessary delays — they charged right ahead with the "bricking up" of the first space shuttle. This was in itself a complicated enough process. Each individual tile had first to be gingerly removed from its plastic bag and carefully positioned in its place in the jigsaw puzzle, with a buffer between it and the neighboring tiles. Then it was necessary to check for any potential differences in height that might cause turbulence. Only then — and once an index card had been meticulously completed for each and every tile — could the gluing process begin. The glue had first to be left to harden for sixteen hours. Then pressure had to be applied, by means of a specially developed tool, for another sixteen hours. Initially, it took one workman the whole of a forty-hour week just to finish one tile. Later on, though, the rate increased to 1.8.[4] And with one down, there were only 30,670 to go before there could be any hope of getting the shuttle into the air.

NASA assigned more men to the job, in hopes of speeding up the process, but there was a limit to how many men could work on the Orbiter at one time. NASA was pushing to send the space shuttle off on its first flight in the last quarter of 1979. That was what they had promised President Carter, and he was becoming increasingly irritated by all the delays. The space shuttle was now two years behind schedule and every extra working day was costing the taxpayers more than a million dollars.[5] But by the spring of 1979 only about 23,200 tiles had been fitted.

In March 1979, in a desperate attempt to force the pace, the people at NASA decided that the Orbiter should be flown, unfinished, from California to the Kennedy Space Center in Florida. They believed that working next to the waiting launch pad would provide the workers with an incentive. Besides, it would make the American public feel as though something was actually happening. The television news programs would have a chance to show film of the first Orbiter being wheeled out of its hangar and of the impressive coupling of the Orbiter with the huge Boeing 747 mother ship, which was to carry it on its back all the way from California. The world's biggest double-decker.

This operation ended in utter disaster. Due to worries about turbulence, just before takeoff "dummies" were stuck on, as substitutes for the missing 7,500 tiles. This way, the space shuttle would also look better on film. But even as the 747 was setting off ponderously down the runway, the dummies were starting to fly off like confetti. The aircraft carried on, and took off, but 4,800 dummies and 100 real tiles were left behind on the runway. The pilot drily requested that a man with a broom be sent out to sweep up, so that he could land again.[6]

A rush job, using stronger glue, followed, and fourteen days later a fresh attempt was made. This time the space shuttle was successfully transferred to Kennedy, where it landed on March 24, 1979. A number of reporters did, however, think that its feathers were looking a little ruffled by the time it had been gently lowered onto its own wheels.

The personnel who had been working on the tiles in California now had to be transferred to Florida, together with their families. It later came to the attention of government auditors that Rockwell had encouraged its staff to rent expensive cars and see a little of America en route. In keeping with the best aerospace tradition, the bill came to $1.8 million for this simple operation alone.[7] Some specialists, however, did not want to move, and others returned home soon afterward. NASA then recruited students to work during vacation for $6.50 an hour. About 400 were assigned to work full-time on the remaining tiles. Later their numbers were increased to 1,400.[8] NASA's reputation was not helped by the fact that it was now employing students on vacation to build high-technology equipment.

Not until September 1979, with 75 percent of the tiles in position, did NASA embark upon a series of tests in which every single tile was subjected to a suction process designed to determine whether it would stay put as it was meant to. Thousands upon thousands of tiles which had already been mounted failed this test and had to be removed. Around a third of them would have to be remounted from scratch. This process would take months and rule out any chance of a launch in 1979.

NASA had to go back to Congress to beg for more money. In the meantime, it borrowed from some of its own internal funds by delaying construction of Columbia's sister shuttles and by accepting the fact that they would have to make do with four space shuttles instead of five, as originally planned.

All in all, the summer of 1979 was not NASA's "finest hour." The main engines still kept breaking down or bursting into flames, while the tiles were having to be taken off faster than they could be stuck on. Moreover, in July, NASA had to announce that Skylab, which had been in orbit since May 1973, had now lost altitude. Any thought of launching the space shuttle in time to put it back into orbit was now abandoned. Skylab was going to hit the earth at some as yet indeterminable spot, at which time loss of human life could not be ruled out.[9]

. . .

The space shuttle was now about to be stripped to the skin in more ways than one. It was becoming more and more obvious that NASA had bitten off more than it could chew. And they were most definitely not dealing with routine tasks and existing stock. Instead, what they had was a high-technology development program feeling its way along untrodden paths and, hence, characterized by all of the delays and overspending to which such programs are always subject. NASA had no need to feel ashamed of this. The problem was, however, that for ten years they had been telling Congress and the nation something quite different.

Instead of the original estimate of $5.15 billion, the development costs of *Columbia* had now escalated to around $8.7 billion — not including the cost of the other three space shuttles.[10]

NASA issued a steady stream of new press releases, announcing the final date for *Columbia's* maiden flight. November 30, 1979, postponed until December 9, 1979 — postponed until March 30, 1980 — postponed until June/July 1980 — postponed until September/October, November 30, 1980, or at any rate, definitely, March 1981.[11]

Not until November 1980 was the last tile fitted into place and glued down tightly. On November 24, *Columbia* was wheeled out of its hangar at the Kennedy Space Center, where it had been incarcerated since March 1979, twenty months earlier. One month later, in December, the future began, at long last, to look a little brighter for the main engines, when all three engines ran for 591 seconds without a hitch. Not only that but in January 1981 they ran for 629 seconds, or 119 seconds longer than the eight minutes and thirty seconds for which they would have to function to put *Columbia* safely into orbit — three years behind schedule.[12]

21. The Flying Brickyard

Whenever a new aircraft design has to be tested, the same procedure is observed. One experienced and brave test pilot (or more) is assigned to explore the plane's possibilities. Contrary to public belief, test pilots are not reckless daredevils. So, to begin with, they always take it very cautiously when flying a new aircraft. Only later will they gradually increase the altitude and the speed and subject the aircraft to difficult maneuvers, in order to discover the limitations of that particular model. In test pilot jargon: determine the extent of the aircraft's "envelope."[1] Should they encounter any unexpected response during this test, they will instantly drop back to previously charted areas at the center of the envelope. Only after they have discovered the reason for that response will they, very tentatively, start venturing out toward the limits once more. This extreme caution notwithstanding, accidents often occur and statistics show that a test pilot with twenty years of flying under his belt runs a 25 percent risk of dying on active service.

This was precisely the line that was to be followed in trying out the space shuttle landing procedure, and it was for this express purpose that the *Enterprise* had been built. On the surface this aircraft seemed to be an exact replica of the Orbiter, but both the tiles and the engine were dummies—since this craft would never make it into outer space. The weight, however, was correct, and the *Enterprise* was equipped with all of the Orbiter's instruments and computer systems.

Over ten months in 1977, these tests etched the space shuttle program into the consciousness of the American people once and for all.

The cautious approach was adopted. First, the *Enterprise* was to be strapped to the back of the jumbo jet and run, unmanned, along the runway at progressively greater speeds. The next stage involved brief flights by the entire integrated stack—with the unmanned Orbiter still strapped to the back of its Boeing 747. Thereafter, a two-man crew could be installed in the cockpit of the *Enterprise* and the computer systems could be switched on — although still without the Orbiter being released from its mother ship.

Finally, on August 12, 1977, the two pilots, Fred Haise and Gordon Fullerton, settled themselves inside the cockpit, high above the back of the mother ship. This time it was for real. Fred Haise was one of the three astronauts who made it back from the moon on Apollo 13 by the skin of their teeth. Fullerton had joined NASA from the Air Force's MOL project.

At an altitude of just under 22,000 feet the big mother ship dipped slightly as the shuttle's fetters were loosed and, for the first time, it made the glide — just a few minutes long — back to the runway.

In October 1977, after several trial landings under varying conditions, NASA could declare this part of the program completed.[2]

This was far from being a test of the Orbiter's total "envelope." It was more like a thorough exploration of the bottom left-hand corner. NASA now knew, with some degree of certainty, how the spacecraft would behave during the last five minutes of flight, from a height of 22,000 feet. They had no realistic way of carrying out a dry run of the trip from the orbit up to these five minutes.

No wonder, then, that the space shuttle's landing procedure would strike any test pilot in his right mind as an insane exercise. When it came to the crunch, the space shuttle would come in for its first landing at a speed of 5 miles per second. From this — by means of totally untried maneuvers, and at four times the speed of the X-15s — the huge plane had to come safely down both to an appropriately low speed and to the now well-rehearsed landing routine. At no point would there be any chance of resorting to less hazardous conditions, as would normally be the case. All you could do was to hope and pray that you eventually made it safely from the high-risk situation to a point involving less risk. And you had to get it right the first time.[3]

Prior to this, it had been possible only to guess at what might happen and draw up mathematical models on the basis of data from the X-15 and Dynasoar projects. Besides carrying out wind tunnel experiments. But, on the actual day, it was not the mathematicians but astronauts John Young and Bob Crippen who would be the first test pilots ever to attempt to penetrate the "envelope" from the outside, in order — all going well — to work their way down to that little corner over which NASA now believed it had some control.

The mathematical models were fed into the computers at the Johnson Space Center, along with data from the wind tunnel and gliding experiments—thus making it possible to carry out trial runs of the reentry and landing on a simulator. And so the procedures designed to bring *Columbia* safely back to earth were established step by step. And at least in the simulator—come what may—one felt that one could get out if the situation went out of control. At any rate, the astronauts always stepped back into the daylight, dazed but alive, no matter what kind of disastrous situations they had been faced with in the electronic world.

Eventually, the criteria for the landing systems were established:

In orbit, the Orbiter traveled around the earth at a speed of 17,510 mph. At this speed, the Orbiter's centrifugal force was just kept in check by the earth's gravitational pull, thus maintaining the Orbiter's height above the earth's surface and keeping it in a stable orbit.

In weightless free fall around the earth, the astronauts did not notice their speed at all. The Orbiter encountered no wind resistance and thus—by means of its 44 reaction-control thrusters, large and small—it could veer upward or downward, stand on its head or lie at an angle. But when the time came to put the landing sequence into operation, the two large auxiliary rockets in the Orbiter's OMS (orbital maneuvering system) had to be turned to face into the line of flight. At a precisely predetermined point in the atmosphere, just about over the east coast of Africa and an hour before landing, the rocket engines would be activated for approximately two and a half minutes, thus allowing the spacecraft to lose some of its speed. This would alter the balance between the gravitational pull and the centrifugal force just enough for the earth to gain the upper hand and for the Orbiter to drop out of its orbit.

The astronauts, sitting with their backs to the line of flight, would not be able to see the rocket exhaust, and would merely feel a gentle pressure pushing them back into their seats after several days of weightlessness. The firing of the OMS would have used up whatever fuel the engines had left, and it would no longer be possible to go back into orbit in an emergency.

More than 3,000 miles west of the runway in California, the Orbiter would—twenty minutes later—enter the earth's atmosphere at an altitude of 400,000 feet. It would be traveling at twenty-two times the speed of sound, or Mach 22. All of the previous space capsules had been ballistic craft and had behaved pretty much like cannonballs careening toward their splashdown in the ocean. And the ocean was wide enough to allow some margin for error.

It was a different matter with the space shuttle. It was the first American spacecraft to be flown down, at a significantly shallower angle, for a fifth of the way around the earth. This is why the shuttle would now turn around and raise its nose 40 degrees above its line of flight, to allow the black tiles on its underbelly to take the heat.

Due to the fierce heat now generated, an ionized layer would form on the underbelly, precluding all radio communication for ten to fifteen highly crucial minutes, during which the Orbiter would be dependent on its own systems. The crew would no longer be in any doubt as to the tremendous speed at which they were traveling, as the first whisper of air resistance swelled to a deafening roar and they were pressed back into their seats. Even so, at a maximum 1.7 g, this pressure would be nothing compared to the pressures of up to 9 g which previous astronauts had had to withstand on their descent from the moon, at much steeper angles and far greater speeds. Beyond the windows, the air would be colored reddish orange by this ball of fire as it fell earthward, exchanging its altitude and kinetic energy for heat.

At a height of over 170,000 feet, the computers would pick up radio waves from the armed forces' TACAN beacon, confirming their precise bearing for the last stretch of the route to the runway.

Now the only factor influencing the space shuttle's speed and course would be a new balance among the lift from the wings, the air resistance, and the earth's gravitational pull. And the onboard computers alone were capable of controlling this balance and deploying the shuttle's power reserves in such a way that, on the one hand, enough remained to make it to the runway and, on the other hand, that they had been consumed to the point where it would be safe to land right on target. The astronauts were solely responsible for the monitoring of the automated systems.

At a height of 80,000 feet, by means of wide slalom swings through the atmosphere, the space shuttle's speed would be reduced by just under 1,700 mph. It would now go into a headlong dive, at an angle of 22 degrees and a speed of 140 mph — faster than any free-fall parachutist.[4]

The space shuttle is not really a glider. In fact, it hardly glides at all. Since it weighs approximately 90 tons, in attempting to fly horizontally it will lose at least 3 feet in altitude for every 15 feet it covers. Modern gliders, on the other hand, have a gliding ratio of somewhere around 1:60 and will cover 200 feet for every yard of altitude lost. Even a passenger aircraft of comparable size has a gliding ratio of 1:16.[5] And so, thanks to the multitude of tiles covering its underside, the Orbiter has acquired the nickname of "the flying brickyard."

· · ·

Soon afterward, while still traveling at one and a half times the speed of sound, the space shuttle would overshoot the landing area and then, shortly afterward, would bank sharply and, at a height of 38,500 feet, execute a 180 degree turn into a so-called electronic Heading Alignment Circle, which would put it right on course for the runway. At a height of 13,000 feet, while still seven and a half miles away from the runway, the shuttle's computers would pick up its microwave landing system, which would feed the electronics on board the spacecraft with data regarding its bearing and its optimum gliding angle. Its speed would now drop to 420 mph and, at this point, the pilot would choose whether to take control or to leave it to the computers to bring them in. The gliding angle would not flatten out to a more normal 1.5 degrees until just thirty seconds before landing, at a height of 1,750 feet. And just eleven seconds prior to landing, at the last possible moment, the landing gear would drop into position in the airstream, allowing the space shuttle to land at a speed of 220 mph.

The runway at Edwards Air Force Base measured three miles in length, which allowed plenty of room for slackening speed without overtaxing the brakes.

Originally, as we have seen, the plan had been to fit the Orbiter system with turbofan engines which could be activated during the last stage of the flight. These would, following a sharper deceleration to 215 mph at a height of as much 100,000 feet, have afforded the crew the possibility of making a gentle and controlled approach — and the option of pulling out and going around again if anything did not seem quite right. But this option had long since been scrapped on account of weight considerations. The runway had to be hit precisely on the first attempt. As little as 150 feet out to either side would result in certain death for the crew. In previous returns of spacecraft to earth such precision was unheard of.

This hectic and chaotic process was possible only through hitherto untried computer technology. And even that would be stretched to the breaking point. Without the computers, the space shuttle would plummet helplessly earthward like a stone. In fact, the shuttle's instruments only seemed to be the same as those of a standard airplane. As with the joystick, no direct link existed between the two pilots' rudder pedals and the Orbiter's aerodynamic surfaces. They were connected solely to the Orbiter's five computers.

Of course, when the pilot pushed the joystick to the left during "manual" flight, this move was registered instantly by the computers. They would then, however, compare this command with the input from the accelerometer, the gyroscopes, and the aerodynamic pressure gauges; just as they would check whether the order was consistent with the optimum course toward the target. Only then would the pilot (presumably) get his way, and the spacecraft bank to the left. At no point would the pilots be able to sneak past the computers to gain control of their craft, and even if they could they would have no chance of coming back alive.[6]

While working independently, the four main computers were closely bound together in one network. The fifth functioned as a totally independent backup, should all else fail with the first four. Racing against the clock, programmers fed 500,000 instructions — or twenty-five times the information required for the Saturn launch during the Apollo project — into these computers.[7] The fifth computer was programmed by a separate team, working independently of the other programmers, to ensure a genuine backup option.

During the launch these computers would pass the course directions to the thrusters in both the main engines and the booster rockets. They also calculated the precise combustion time for the OMS engines, whose job it was to regulate the orbit. And, of course, they also had to determine exactly when the rockets ought to be fired — and for how long — to begin the reentry phase.

The work of programming was rendered much more complex by the fact that, besides giving instructions, the four main computers also had to keep a constant check on each other's mental health. In actual fact, before any decision could be implemented, the computers had to put it to a vote; and they were working at a rate of fifty votes per second.

This problem was solved in a simple way. If, for instance, the computers reached the conclusion that the left inboard aileron ought to be lowered by 5 degrees, then each individual computer, working independently of the others, would send such an order to its own aerosurface servoamplifier on the aileron. The servoamplifier would check the aileron's present position before transmitting an appropriate electrical charge to one of the four servovalves, which in turn would move a piston in the desired direction. In all probability the three other pistons would do exactly the same and the airfoil would dip. Hence, the vote actually took place by means of the mechanical manipulation of the airfoil.

Now, if one of the computers were to turn "peculiar" and not send such an order, its piston would resist any movement. Collectively, the three others would, however, be strong enough to exert their will. Should the peculiar computer continue to dissent, the other three would blow the whistle on it, it would be disconnected, and a warning light would come on inside the cockpit.

The computers issue their orders as a reaction to input from more than 2,000 sensors, including the crew's control lever. Usually, the computers would receive values from four feelers at a given position and then compare the values of three, chosen at random from the four. Thereafter, they were directed to throw away the highest and the lowest values, so that they were all working from the average value. This was supposed to provide them with a common basis for their subsequent calculations.[8]

The astronauts were pleased that a winged spacecraft had at long last been produced. Not surprisingly, they were less delighted by the fact that this craft could never achieve aerodynamic stability on its own. And the worst of it was that the pilot had no control over the Orbiter. Like any programmer, he could only make suggestions to the four computers, on which they would then vote.

A good number of politicians on the congressional select committees also had their doubts about this approach, but as NASA veteran Christopher C. Kraft, head of the Johnson Space Center, told one of these committees in February 1974, there was no way around it:

> If you're going to fly this vehicle into space and get it back and reuse it, we don't see any other way to do that. You just end up with a low lift-to-drag vehicle that comes down like a rock. However we're confident that with the avionic systems that we have on board and with the judgment of the pilot on top of that, we can build a system that is flyable with proper training.[9]

When John Young, who was to be the pilot on the first trip down through the atmosphere, emerged from his first computer simulation of the descent, he looked tired and grave-faced. Rather brusquely, he informed the assembled journalists that human reaction times were far too slow to control the shuttle's meteoric descent through the atmosphere. That was something that only the computers could handle. The look on his face betrayed the fact that this was no easy admission for one of NASA's most seasoned astronauts to make.

In the autumn of 1980, President Carter asked the director of NASA, Robert A. Frosch, whether the computers might not make it possible to carry out an unmanned maiden voyage — on which *Columbia* could be put to the test, as all previous American spacecraft had been — before putting the astronauts' lives at risk. After giving this some thought, Frosch replied that, technically, there would have been nothing to hinder such an approach if it had been adopted from the start. At this late stage, however, the rewriting of all the programs would entail far too much in terms of time and money.[10]

22. Raw Thrust

As the development program began to pick up speed, it became more and more obvious to everyone involved that the main engines, the heat shield, and the computer systems all required the latest in technological advances. The overspending in these areas was evidence of that. But amid all of these trials and tribulations, comfort could always be taken from the fact that the two 149-foot-tall solid-fuel rocket boosters, which were to be strapped to the structure, gave no trouble — nor were they likely to.

In fact, these boosters were relatively uncomplicated. They did not involve any form of turbopump, and there was no maze of fuel cables forming bewildering feedback systems. They were, quite simply, two white pillars full of raw thrust. Two gigantic fireworks containing a solid fuel that had never caused problems in the past. All you had to do was ignite this fuel and, by virtue of Newton's third law, everything would move very fast — in an upward direction.

Apparently, the only special feature of this model was its size. The technology behind the reliable solid-fuel rockets had been perfected thirty years before and had reached its peak ten years earlier, with the unmanned Titan III solid-fuel rocket boosters. At that time, those had been considered huge, at a height of 86 feet and with a thrust of 1,200,000 pounds-force. The Orbiter's rockets would be 63 feet taller, would weight two and a half times as much, and each would produce a thrust 2.2 times greater than that of the Titan.[1] Together, during the first two minutes of the flight, these two rockets would do five-sixths of the work of shooting all 2,004 tons of the Orbiter system sky high.

As we have seen, the contract for the boosters had been landed by Utah's Morton Thiokol, a company experienced in producing much smaller solid-fuel rockets — and quite undaunted by this fact. They devised a system whereby the prospective rocket would consist of a number of cylinders, the steel walls of which would be less than thirteen millimeters thick and totally devoid of welding seams. After being filled with the solid fuel, these cylinders would be stacked one on top of the other, with the lower edge of the top segment sliding into a fork-shaped clevis groove in the upper edge of the segment beneath, and so on. Once assembled, the structure would be secured with tempered steel pins.

While still at the factory in Brigham City, Utah, eight of these cylindrical segments were fitted together and sealed, so that the motor now consisted of just four casting segments, each one measuring 27 feet in length and 12 feet in diameter — like four big oil drums, each weighing 136 tons. But the assembly of the motor could be taken no further in Utah, because it had to make a ten-day train journey, traveling 2,500 miles across America before arriving at Cape Kennedy. There, the stacking would be completed in the big Vehicle Assembly Building (VAB) near the launch pad.

The fuel, a rubbery mixture of aluminum powder and ammonium perchlorate, was one of the most powerful available — not counting liquid oxygen and hydrogen.

The Marshall Space Flight Center had been put in charge of supervising the building of the solid-fuel rocket boosters on NASA's behalf. All of the previous rockets with which Marshall had been involved, under von Braun, had been driven by liquid fuel. Now, however, the center had to familiarize itself with another kind of rocket technology; one which they had previously regarded with tolerant disdain. After all, no dumb brute of a solid-fuel rocket had taken Neil Armstrong to the moon.

In October 1980, George Hardy, who on behalf of the Marshall Center shouldered NASA's share of the responsibility for the boosters, took the representatives of the press on a guided tour and gave them information about solid-fuel rocket boosters. On that occasion he explained how considerable emphasis had been placed on making use of existing technology, to avoid having to carry out the kind of expensive and time-consuming experiments that had been necessary with the main engines. Basically what they had done was to scale up the Titan motor.

Hardy explained to the assembled journalists that — as the first stage of the space shuttle system — it was the solid-fuel rocket boosters' job to accelerate the Orbiter to a speed of 4,400 feet per second and to a height of 150,000 feet. Then, with their task accomplished, they would be jettisoned. By virtue of their own inertia, they would, however, carry on to a height of 220,000 feet before dropping downward in a gentle arc. At 16,000 feet, a drogue parachute would be released from the rocket's nose, to be followed at 7,000 feet by three large main parachutes. And at last, 140 miles away from the Kennedy Space Center and traveling at 60 mph, the rockets would splash into the Atlantic Ocean, where two salvage vessels would be waiting to pick them up.

Thanks to the air now trapped inside their empty cases, the rockets would float, nose-up, in the water. The salvage ships would lower a specially designed remote-controlled plug (costing $1 million) to the mouth of the rocket and pump out seawater to the point where the rocket could be tethered lengthwise to the ship's side. Thereafter, the rockets would be shipped back to the coast, taken apart, and cleaned. Then came another 2,500-mile train journey back to Utah, where they would filled with fresh fuel and sent back again. This process would be repeated as many as twenty times, until the boosters were worn out.

Hardy presented pictures of the horizontal firing of a solid booster rocket in Utah. The journalists were duly impressed by the smoke cloud billowing far up the side of a nearby mountain. Seven such firings had been carried out, said Hardy:

> There were very few and very minor changes that had to be made in the motor design as a result of the early tests. . . . one of the specific design objectives from the beginning has been to utilize the very conservative factors in safety—utilize proven, well-known processes, techniques, and materials. And that's so we build the confidence required to man-rate the solid rocket booster with such a limited number of tests.[2]

George Hardy didn't mention the element of risk involved, and none of the journalists felt the need to ask, since the system had been so thoroughly tested. It was generally accepted that the main shuttle engines presented a much greater risk. Nor did he mention that the recycling of the solid-fuel rocket boosters was turning out to be an increasingly questionable money-saving operation. More than a third of the boosters' costs had been swallowed up by the salvage systems. But no one dreamed of scrapping this solution. If Congress wanted recycling, then, of course, that is what it would get.

Obviously, despite all Hardy's assurances, some risk was involved in igniting the fuel in the hollow core of the solid-fuel rocket boosters, with the pressure inside shooting up to 30 atmospheres in just 0.6 second. This pressure had to escape downward through the booster's nozzle, but it would also explore every one of the booster's joints, searching for the tiniest hole. Such a hole would immediately release exhaust gases burning at 5,600° F into the open air, just two feet from a fuel tank containing half a million gallons of liquid explosive.

Another risk lay in the fact that the rocket could not be stopped once it had been activated. Even if all of the main engines failed just after liftoff, the astronauts would have to carry on until the boosters had been jettisoned, and if they were really lucky, the computers would guide them back to the Kennedy runway. Should one of the boosters fail or so much as lose an appreciable amount of pressure relative to its mate, the Orbiter would roll about uncontrollably and the crew would be doomed. The same thing would promptly occur if one of the two rockets failed while still on the launch pad.

On the first flights with *Columbia*, the two test pilots would be provided with ejection seats, as an extra precaution. Even so, once the rockets had been activated, they would not stand a chance. No one could survive passing through the flames of the exhaust — quite apart from the fact that the chlorine gases produced by the firing were lethal. The ejection seats could come into play only during the last phase, on the approach to the runway.

Implicit in such a risk scenario was the fact that the booster joints played such a crucial part in the construction of the solid-fuel rocket boosters. It would, of course, have been best if the booster case could have been constructed in one piece, with no joints whatsoever. As we have seen, Aerojet did make such a suggestion but their bid had been rejected for financial reasons. By now, however, the segmentation had become a vital part of the design. Thiokol had no intention of building the rockets at Cape Kennedy. It was up to them to create jobs at home in Utah. And 27-foot-long drums were as much as the railroad could carry.

There were two kinds of joints. First, there were the joints that connected the aft segments while still at the factory, prior to the train journey. These so-called factory joints had been mastered; they presented no problem. The crucial joints were the so-called field joints used to assemble the "oil drums" once they reached the Kennedy Space Center.

For these, Thiokol's engineers had improved upon a system used on the Titan III. Two synthetic rubber rings, known as O-rings, were seated in machined slots on the inner flange of the clevis groove. The O-rings measured 146 inches in diameter and were just 0.280 inch thick. Each one was molded in one piece to span the entire circumference of the booster. The Titan III had only had one O-ring, but in this case the number had been doubled because the cargo would be living, breathing astronauts. These two rubber rings were designed to ensure absolutely no leakage between the individual booster segments.

The journalists were given the impression that both NASA and Thiokol knew what they were dealing with, and that they had built fail-safe mechanisms into the solid rocket boosters. This same impression was received by those responsible for approving the Orbiter's various subsystems. On September 15, 1980, the solid-fuel rocket boosters too were certified for flight and were able to take their rightful place alongside the other main components of the Orbiter.[3]

23. Creative Bookkeeping

Around the end of 1980 and the beginning of 1981, the last pieces of the puzzle were falling into place. All the different components had been assembled to form what was probably the most complex piece of machinery ever built by man. One might reasonably have expected NASA to be pleased and proud. Once again the space organization, in collaboration with the foremost American aerospace companies, had solved a complex technological problem. And solved it, this time, despite a political climate totally unlike that of the Kennedy years and under quite different economic circumstances.

But the mood at NASA was far from euphoric. Getting this far had taken a terrible toll on the old NASA spirit. Personnel were overwrought and exhausted. Management was feeling the strain and, with fresh delays looming on the horizon, they hunted around arbitrarily and unreasonably for a scapegoat. Throughout the 1970s this same management team had had to answer to three Presidents—Richard Nixon, Gerald Ford, and Jimmy Carter. All three had so little genuine interest in space questions that NASA's Administrator was often referred to some minor government official. And all three had far more important things on which to spend public funds than NASA's space shuttles.

When, in the late summer of 1980, Aaron Cohen, the manager in charge of work on the Orbiter module at the Johnson Space Center, had to write an article on the project, he could not help but remark:

> It is incumbent upon us as a nation to expand our frontiers into the unique space environment to help solve some of the major problems on earth. But in recent years, our hopes and aspirations have not been matched by a correspondingly high commitment of resources. Indeed, real federal outlays for space have steadily declined since the Apollo era. In GFY 1979, only 0.8 percent of all federal outlays is committed to space.

Aaron Cohen has learned his lesson well. The costs of space exploration have to find their justification in problems here on earth. And yet, even though he is a public servant, he cannot resist criticizing the government's indifference to its own space agency. Toward the end of the article he takes a deep breath and resorts to a touch of good old space-age rhetoric — an art practiced with much more aplomb by Jim Webb and von Braun:

> Over the next two decades, an acceleration of private and public support will be essential to our taking full advantage of the capabilities of the space transportation system. But the benefits will far exceed the costs . . . The Space Shuttle will herald a new era, with the promise of answers to problems and needs of mankind. We cannot afford to miss the opportunity.[1]

The Orbiter has to be sold on the strength of its practical applications and, to be viable, the project as a whole has, as it were, to stand on three legs. Three areas should, and must, lend support: the military, the private sector, and science.

Science was, however, not particularly interested in the Orbiter. Instead, many scientists regarded it as a rival for limited grants. At best, it would drain the space budget of funds that might be spent on the much more interesting unmanned probes, which could reach far out into the solar system and send home pictures far more interesting than the bird's-eye views of the earth offered by the Orbiter. But the staff in NASA's own scientific departments felt that they were constantly under pressure to develop projects which would lend themselves to the Orbiter and, hence, legitimize the scientific angle.

The private sector, on the other hand, showed itself more willing to participate, if NASA's rates for satellite launches could compete with those of unmanned systems. And if NASA could meet its deadlines for getting the satellites into position.

For the first time, NASA was going to try its hand at trading on the free market. At one business conference after another, well-groomed NASA employees projected tinted transparencies onto large screens — pictures demonstrating how the Orbiter could do everything that commerce could ever dream of. Reference was made again and again to the Mathematica prognosis of 1972 — now rapidly becoming outdated. In this, NASA guaranteed cargo rates of between $100 and $160 dollars per pound. Not so much was said about the fact that this prognosis presupposed a full hold on every trip *and* 60 flights per year, with all four space shuttles in operation at all times.[2]

At the middle management level in NASA personnel were feeling very unhappy about this prognosis. Over the years it had become something of a bad joke. Nor, by this time, did anyone really want to be reminded of the letter which James Fletcher had written to the U.S. Comptroller General in 1972. In this letter Fletcher said:

> The Shuttle will provide quick and routine access to space and eliminate the constraints imposed by the present mode of space operations, which is characterized by high risk, long lead time and complex systems.[3]

Or, for that matter, another Fletcher quote from the same year:

> By the end of this decade the nation will have the means of getting men and equipment to and from space routinely, on a moment's notice, if necessary, and at a small fraction of today's cost.[4]

Everyone now knew that such economic assurances were totally unrealistic. And yet neither James Fletcher nor the senior management would revoke them because, for them to stand any chance of pushing anything through Congress or the White House, the Orbiter would have to be sent up often enough to recoup its own development costs.

But commerce could still be offered the desired, competitive rate per pound — all they had to do was make a few adjustments to the number of flights on the computer models.[5]

According to the Mathematica models, a saving of $13.4 billion dollars would be made on 580 Orbiter flights (1972) as opposed to missions employing unmanned booster rockets. Of this, $5.1 billion would be saved on expenses. The remaining $8.3 billion came from the fact that the satellites required no special protective shield when carried into orbit in the sheltered surroundings of the Orbiter's hold. After that it was simply a matter of deducting the $8.04 billion in development costs of the four Orbiters from the saving of $13.4 billion. *Et voilà:* The Orbiter system could expect to make a profit of $5.36 billion in its first twelve years.

Such manipulation of figures was already being heavily contested by serious critics in the early 1970s. In April 1972, physicist Ralph Lapp told a Senate committee that NASA was tampering with the figures. The prognosis referred to a cost per flight of $10.5 million. But if the total expenditure for research and development of the Orbiter system were taken into account one arrived, instead, at a figure of 514 flights — costing $25.5 million per flight. Lapp also maintained that, in the Mathematica model itself, it was pointed out that, on average, only 7.9 percent of the cargo capacity (or 5,000 pounds) would be utilized. Thus prompting Lapp to tell the Senate committee:

> If we divide $25.5 million by 5,000 pounds, we get a unit price of $5,100 per pound placed in orbit. This I maintain is the true price since it is based on a full accounting of new appropriated dollars for the space transportation system. . . . Shuttle proponents may argue that their price is based only on operating costs, but that isn't the way General Motors figures it when I buy a car . . .[6]

Today, John Naugle, a member of NASA's scientific department since the Apollo years, views these attempts to conceal the true costs as the one factor which contributed most to the undoing of what James Webb had created in the 1960s. By making top personnel believe in the myth of the cost-conscious space shuttle, James Fletcher did a lot of damage.

> [The shuttle] would be cheaper than the expendable launch vehicles. It would be better than all the expendable launch vehicles. It would replace all the expendable launch vehicles. . . . Well there was a feeling that we were on the razor's edge. That if we said the wrong thing, or anything like that, the shuttle would be killed.[7]

There wasn't much of the old NASA spirit left. These days management wanted to see only data that supported the official NASA line — and everyone had better toe that same line. Whereas in NASA's formative years the basic assumption had been that data was data and that differences of opinion had to be settled by means of testing and verification, nowadays it seemed to be more and more the case that data could pretty much be tailored to fit. Suddenly it had become necessary to differentiate between convenient and inconvenient facts. Of course, this would come as no surprise to those who had experience with other bureaucratic cultures, but NASA had always prided itself on *not* being just any other bureaucratic culture. NASA had been a technological corporation, with little experience with petty political compromises or the tactics of tampering with the facts. To some people, NASA was no longer NASA.[8]

At the senior management level in NASA it was increasingly felt that the merest hint of internal strife would be viewed as a serious sign of weakness by a government on the lookout for the slightest chink in their armor. A sense of insecurity and of tactical procrastination filtered its way downward, spreading throughout the organization. This was how John Naugle described it in May 1986:

> Up until that era there, I never worried about saying what I felt. I always felt my bosses . . . while they might not agree with me, they might slap me down, they might quarrel with me, but they were not going to throw me out just because I brought them bad news. And somewhere between the time Fletcher came on board [March 1971] and the time he left [May 1977], I no longer felt that way.[9]

Fletcher's predecessor as head of NASA, Thomas O. Paine, is quite convinced today that NASA should never have agreed to develop the Orbiter system for anything less than the $10 to $15 billion figure with which they had originally been operating. Once NASA had agreed to the $8 billion, they were stuck with it. As that sum shriveled up and contracted under the strain of escalating development costs, the politicians kept referring to what James Webb had said right at the start: that Apollo would cost $20 billion. And $20 billion it *was*. But Webb had been smart enough to take the first estimates and double them.[10] This time around it was a very different story, as Paine explains:

> The road to hell is notoriously made small step by small step. The devil has always a very tiny bargain. . . . You wanted the shuttle for 8 billion and you got 5.2, well, gee, you are not going to quit over that. My God, you got the little thing and it's very exciting and now let's go . . . and then you get that last-minute call from the White House and the President wanted to tell you that we're gonna have to take another 75 million out of NASA . . . and then next year another 120 million. It goes on and on.[11]

NASA, like every other public enterprise, was finding itself perpetually in the spotlight — a situation which prompted the organization to throw overboard everything on the Orbiter that could possibly be dispensed with. That was how the Orbiter came to lose the system that would have shot the cabin free of the stack in the event of anything going wrong. In 1971, $292 million had been budgeted for this system, but it was axed a year later by James Fletcher, who believed NASA could no longer afford it.

General James Smart, who was acting as liaison between NASA and the Pentagon at the time, has since commented on such decisions:

> NASA didn't do it because it was stupid. They did it because they were forced to do so . . . We can never be perfectly safe and it's always a human decision as [to] just how much is enough. But in this case how much was enough was not determined by the scientists and engineers but by the politicians.[12]

The Orbiter's designers were slow to realize that the most obvious customer for a space shuttle capable of carrying heavy loads into a low orbit around the earth was the military. The time had come for this civilian space agency — which had always taken pains to keep the Pentagon at arm's length — to bite the bullet and come to some sort of arrangement. Since the mid-1950s one of the underlying principles of the American space program had been that military projects had to be developed secretly, under cover, while civilian projects had to be given as much publicity as possible. But obviously, from a technical, administrative, and financial point of view, it was impossible to keep these two sectors totally separate. Thus, all the civilian space missions availed themselves of the armed forces' bases, tracking and communication systems, and so on. And it was only natural for the space capsules to be picked up by Navy vessels. Nevertheless, for political reasons, the division between the two was obstinately insisted upon.[13]

This division also manifested itself in the allocation of grants. Every year since 1961, NASA had been awarded more funding for its space programs than the armed forces. In the mid-1960s, up to three times as much. During those years, when all the talk was of NASA, the American public had little or no knowledge of the grants awarded to military space programs. This covert status was the direct result of an order issued in March 1962 prohibiting the American press from referring to or reporting on military launches from Cape Canaveral or Vandenberg Air Force Base in California.[14]

So it was bound to come as a surprise to many Americans that money had not only been poured into NASA, and that experts estimated that — over the first twenty years of the space age, up to 1979 — the United States had spent a combined total of more than $90 billion on both military and civilian space programs.[15]

Not until sometime around 1980 did the armed forces, after twenty years as the underdog, find their grants overtaking those of a crippled NASA.

The military applications of outer space were no longer such a taboo subject as they had once been. For the first time, in October 1978, President Carter had admitted publicly that the United States employed spy satellites. He may have been referring to the Keyhole satellites, launched into space as far back as 1976, which had for quite some time been sending digital pictures back to earth; pictures which, rumor had it, were so sharp that you could read the license plates on parked cars.[16]

NASA could no longer afford to put on airs. Instead, as we have seen, they were already courting the Pentagon when the time came to determine the specifications for the Orbiter. The generals too were treated to a tinted slide show — this time presenting attractive military applications of the Orbiter:

> (a) rapid recovery and replacement of faulty or failed spacecraft essential to national security; (b) examination of unidentified and suspicious orbiting objects; (c) capture, disablement or destruction of unfriendly spacecraft; (d) rapid examination of crucial situations developing on earth or in space whenever such events are observable from an orbiting spacecraft; and (e) rescue or relief of stranded or ill astronauts.[17]

And again, during the last stages of the development work, when money was becoming very tight, the civilian space agency had to crawl back to the Pentagon.

It was at this point, in the summer of 1979, that the American public first learned, through congressional hearings, of the military's plans for the space shuttle. Dr. William J. Perry of the Defense Department assured the politicians that the military envisaged staging 113 military launches over the ensuing ten years. Toward the end of this period the armed forces would themselves finance and carry out fifteen launches a year. The Space Transportation System was going to take over from all unmanned expendable rockets.

Even back in the mid-1970s, when Keyhole 13, the military satellite, was developed, the idea was for it to be serviced by a prospective space shuttle. Keyhole had to be able to refuel in space and be furnished with fresh photographic equipment and other sensors as required. The launch of the first satellite of this type was scheduled to take place no later than 1981.[18]

But since NASA no longer felt able to keep its promise of putting the necessary 40,000 pounds into a north-south orbit, the military would themselves cover the costs of developing more powerful Titan strap-on boosters for the Orbiter, to take the place of the Thiokol model.

Despite the criticism implicit in this move, NASA was happy to have the badly needed support of the military, coming as it did at a highly critical juncture. And it took some of the sting out of the armed forces' announcement that they were going to set up their own control center in Colorado Springs, because they did not trust the security at the NASA base.

NASA was also pleased that, at long last, they had a President who took a lively interest in space questions. The newly elected President Reagan and his advisers had, so it was said, big plans for outer space.

24. Fresh Energy

On December 29, 1980, *Columbia* was wheeled slowly and majestically out of the vast VAB hangar, on the same "crawler" that had transferred the Saturn V rockets to the launch pad ten years before. Not that this was in any way meant as a sentimental acknowledgment of past ventures; again, it was merely an attempt to cut corners wherever possible.

During the first months of 1981, all of *Columbia*'s systems were subjected to a final run-through on the launch pad. And in March the all clear was given for the launch which everyone at NASA had been looking forward to since Apollo. The long trek through the desert had come to an end and America was about to regain the lead in manned spaceflight.

The launch was set for April 10, 1981. But days before this, journalists were already pouring in once more from all over the world — just like the good old days. Some older reporters grew wistful at the sight of equipment that had survived from the Cape Canaveral days. Back then, it had been spanking new; now it was marked by the ravages of time and lengthy sojourns in the salty air of the flat isthmus. And behind the premises which now housed the Orbiter engineers lay the deserted ruins of some of the old buildings. For want of other victims, the younger journalists resorted to the old ploy of interviewing their senior colleagues on memorable moments from the 1960s.

The two astronauts did not, however, manage to take off on April 10. Just nine minutes before liftoff, the backup computer refused to communicate with the four others — having received its signals 0.04 second after them. It would take the programmers a full day to correct this problem, so the astronauts were taken out and a start was made on emptying the fuel tanks while almost three-quarters of a million disappointed spectators made their way back to their cars and motels. The date for a fresh attempt was set for Palm Sunday, April 12, 1981.

. . .

For once, everything went like clockwork for NASA. The astronauts, John Young and Robert Crippen, shot heavenward in a ball of flame just twenty years to the day after stunned Americans had learned that Yuri Gagarin had become the first man in space.

John Young was NASA's most professional astronaut, with experience on Gemini 3, Gemini 10, Apollo 10, and, finally, Apollo 16—on which he had landed on the lunar surface. *Columbia*'s maiden voyage would be his fifth trip into space. Crippen was an experienced test pilot, but he had never been into space before.

Young and Crippen were now the first astronauts to experience how it felt to lie on one's back alongside two big solid-fuel rocket boosters; to hear them roaring and thundering; to feel themselves being pushed back into their seats and, through their side windows, see the steel girders of the tower vanish in the blink of an eye, while they headed onward, upward, at greater and greater speed. They felt the Orbiter slowly turning around, just as it was supposed to do, so that they were now climbing at more of an angle, feet first. Veteran John Young passed his initiation test with flying colors, registering a pulse that stayed steady at 85 throughout the launch, while Crippen, the rookie, gave himself away with a pulse of 135.[1]

After just over two minutes in the air, the pressure against their seat backs subsided, along with much of the din, when the two boosters were jettisoned right on schedule and headed off to either side in two graceful arcs. Just before this, the crew had been advised that—in the event of any problem with one of the main engines—they would still be able to make it to the emergency runway in Rota in Spain under their own steam. But the main engines worked perfectly until switched off, as scheduled, after eight minutes and thirty-four seconds. Shortly afterward, Young jettisoned the big fuel tank, which, having now fulfilled its purpose, disappeared in the direction of its eventual splashdown point in the Indian Ocean.

Just after the firing of the two OMS engines, sited above the main engines, *Columbia* became the heaviest spacecraft ever to go into orbit—more than twice as heavy as Apollo, including the service and lunar modules.[2] It was orbiting 870,000 feet above the earth; from this height large towns and airports were discernible with the naked eye. The astronauts could now unfasten the seat belts which had kept them strapped down during the launch, and start enjoying the spaciousness of the Orbiter—so unlike the capsules.

Just like other astronauts before them, in the course of one ninety-minute orbit, Young and Crippen passed from day to night and back to day again. During the first twenty-four hours they witnessed sixteen sunrises and sixteen sunsets, one every forty-five minutes, and each one more beautiful than the one before. The same sight which had, in the infancy of the space age, so enthralled Scott Carpenter on Mercury 7 that he forgot to push his buttons.

As if watching a film that has been speeded up, the two astronauts were treated to sunrises and sunsets at a rate seventeen times faster than normal — moving, as they were, seventeen times faster than the earth could rotate. Since their sleeping patterns obviously could not be governed by this continual shift from light to darkness, these were synchronized with Houston time.[3]

Back on the ground, it seemed almost too good to be true for everything to be going so smoothly. So their worst fears were confirmed when Crippen informed Houston that, looking through the window, he could see some gaps in the facing where tiles were missing from the two humps of the OMS system on the Orbiter's top side. It looked as though the worst had, after all, happened. The tiles had not been able to withstand the pressure during the launch, and now the crew would be incinerated on their descent through the atmosphere.

The engineers asked Young and Crippen to train their cameras on the areas in question, to allow them to assess the damage. What they saw did not give them too much cause for alarm. Around fifteen tiles had fallen off the topside, where the temperature would not be so intense. It would have been much worse if any of the black tiles on the underside had been missing. But there was no way the crew could check the underside. This particular mission was not equipped for EVAs (extra vehicular activities). Instead — in deepest secrecy — NASA asked the astronauts to turn the Orbiter around, so that the U.S. Army could train the cameras of a Keyhole satellite on the heat shield.[4] These pictures, which have never been made public, set the engineers' minds at rest. The Orbiter was going to make it.

Two days and thirty-five orbits later, the time came to turn the Orbiter's stern into the line of flight and — following a brief burst from the rockets in that direction — head for home. They were now about to discover whether it was, in fact, possible to reach the bottom left-hand corner of the envelope and whether a lump of metal the size of a standard aircraft could be flown down through the atmosphere at twenty-five times the speed of sound.

An hour after the space shuttle had left its orbit, the crowd of spectators near the runway at Edwards Air Force Base in California heard two sonic booms as the space shuttle flew in across the runway high above their heads. It heeled over onto its side and, with staggering precision, followed the invisible electronic cylinder through 180 degrees to come zooming back toward the runway, where it brought its wheels down onto the landing strip at a speed of 225 mph. Moments later it lowered its nose wheel. *Columbia* then ran on for a full minute in this odd, slightly stooped position, before finally coming to rest in the middle of the runway.

It looked battered and sweaty, its hull still searingly hot after its descent. But after running over a few checklists, Young and Crippen were able to leave *Columbia* by the hatch they had been helped through two days earlier in Florida. The date was April 14, 1981, and the space shuttle had proved that it could cover a distance of one million miles and return safely to earth.

The response was immediate and ecstatic. Americans could once again feel proud of their astronauts and of their NASA. The civilian space organization had produced the world's most complex piece of machinery and it had worked, the first time.

Just as President Nixon—without having lifted a finger—had been able to reap the harvest of Kennedy's Apollo project, when Armstrong set foot on the moon in 1969, so the newly elected President Reagan was able to take the credit for the space shuttle. He wasted no time in turning it into a symbol of the new resolve which was to prevail in the United States in the 1980s.

Amid all this national rejoicing, only NASA's engineers and designers, and those of their contractors, fully realized just what an achievement this had been and how much could have gone wrong.

25. Just Routine

In 1862, during the Civil War, Congress passed the Pacific Railroad Act. The railroad was about to span the entire continent from the Atlantic to the Pacific. The Union Pacific Railroad would start laying track in a westerly direction from Omaha, while the Central Pacific would work eastward from Sacramento. The remuneration received by each of these two big companies would be proportionate to the amount of track each could lay. Work on the railroad was well underway, and the Civil War had at long last come to an end, when one historian wrote that this project proved how nations could not only be divided but also be united by the work on an engineer's drawing board.[1]

True to the best American tradition, the work had been designed as a race between the two railroad companies. Thousands of laborers set to work at either end. As the project was reaching completion, the Central Pacific boasted that it could lay almost a mile of track per day; the Union Pacific countered by laying six miles per day. Then the Central Pacific had to exceed this by covering ten miles in one day. The track may not always have been quite level, but it was down. And it could always be given a good going-over later, when there was more time.

Finally, in 1869, the two teams could see one another and on May 21 of that year — at Promontory Point, north of the Great Salt Lake in Utah — the two company presidents raised a silver hammer to jointly drive home a gold spike in the tie linking the two stretches of track. Within seconds, the telegraph lines running alongside the track had spread the news across the United States.[2]

Naturally, President Reagan could not resist referring to this glorious moment when, on Independence Day 1982, he welcomed *Columbia* home from its fourth flight into space. Hundreds of thousands of people,

Stars and Stripes everywhere, military bands, and a President who said that with this, the last of the four test flights, the "gold spike" had been driven home and this new transportation system for projects in outer space could be said to be operational.

While the President was speaking — with the *Enterprise*, the "guinea pig" shuttle behind him — *Challenger*, the brand-new space shuttle, flew over the heads of the crowd on the back of its transport plane.[3]

In his speech, Reagan said:

> The United States Space Transportation System is the primary space launch system for both national security and civil government missions . . . The shuttle's first priority is to become fully operational and cost-effective in providing routine access to space.[4]

NASA leaders applauded enthusiastically, even though their own press office had, of course, written this speech. The lengthy development phase was over and the Orbiter could now get on with performing all the scientific, commercial, and military tasks set for it, and with recouping its own costs. The two ejection seats were removed. With crews of as many as eight astronauts there would no longer be room for them; nor should they be necessary in a space shuttle which was no longer a prototype but an effective tool. It was now up to the system to show what it could do on a workaday basis.

Later in 1982, NASA published its plan of action for the near future. There was no longer any talk of 500 to 600 flights per decade. Now they were aiming at approximately one flight every two weeks from 1988. As stages en route to this eventual goal, twelve Orbiters would be launched in 1984, fourteen in 1985, and seventeen in 1986 and 1987.[5]

As with any other kind of ferry service, the key to achieving such ambitious goals lay in reducing the time spent in "dock," thus allowing this "ship" to get back out to sea and earn its keep. In aviation too, aircraft were designed for the minimum turnaround time at the airport; up in the air is where the money is earned. For the first time ever in the American space program, guaranteeing the optimum turnaround efficiency was of vital concern to NASA.

Between *Columbia*'s maiden voyage and its second launch in November 1981, seven disappointing months had elapsed. Another four months were to pass before the third launch was staged, but after that the Orbiter spent just three months in harbor prior to its fourth launch. These sojourns on earth were taken up by service overhauls, major and minor repairs and adjustments, and fitting out for the next mission. Things were moving in the right direction, but they were still a long way from NASA's eventual goal: a turnaround time of 160 hours (just short of a week).[6]

But NASA was bound by its own promise to deliver "routine access to space." And it would have to operate with all the regularity of a scheduled aircraft. Space travel no longer needed to be exciting. Quite at odds with its own tradition, NASA could look forward to a time when the space shuttle's achievements would simply be taken for granted.

Not yet, of course. Thanks to the space shuttle, the press had taken a new interest in space exploration. And NASA was more than happy to receive a little attention. Glossy magazines and news weeklies featured splendid color spreads of photographs from outer space. Occasionally a centerfold would be given over to shots of large and easily recognizable areas of the earth's surface, or to astronauts floating weightlessly around their cabin. In the magazine articles one could read how an astronaut could write and read, lost in his own thoughts, while the whole Orbiter quietly shifted position around him until, suddenly, when he looked up from his floating book, the ceiling had become the floor.

Readers also heard how the theories of Newton and Tsiolkovsky worked in practice. A weightless astronaut could only move by pushing off from walls or ceilings; no amount of flailing and kicking would propel him forward. This also meant that an astronaut floating just a handbreadth away from the Orbiter in space could not be saved unless he was wearing a lifeline or unless some force was imparted either to him or to the Orbiter, making their paths collide. On his own, he would never be able to kick or "swim" the few inches to safety.

One also read with wonder how liquids behaved in a weightless state. For instance, they just would not leave a glass. You could not pour them down your throat. So the astronauts had to use straws; but if you were sucking through a straw in weightlessness and you let go of the straw, the liquid would continue to squirt out through the straw, simply because a force had once been impressed upon it to make it move in that direction. So NASA had to develop special straws fitted with stopcocks.

As for the culinary side of things, the Orbiter had its own little kitchen containing more than a hundred different types of food and twenty different soft drinks. The food was arranged on molded trays like those used on passenger planes. No wine or liquor was carried on board. Originally, back in 1973–74, NASA had planned to allow a small supply of wine for the lengthy stay on Skylab. But the Women's Christian Temperance Union objected to this suggestion. NASA, always very sensitive to public reaction, immediately dropped the idea, with the result that the Orbiter too was a dry area.

The word "toilet" did not figure in NASA's dictionary; instead they spoke of the Orbiter's "waste disposal system." But no matter what you called it, it was a huge improvement — and greatly appreciated by veteran astronauts such as John Young. Previously, when the astronauts had to appear in public and make speeches on NASA's behalf it was a safe bet that someone would eventually ask, hemming and hawing and sniggering, about the waste disposal problem in space. After a while, the astronauts grew used to this and would explain matter-of-factly about the tubes with their rubber sheaths and the rather more undignified diaper arrangements whereby astronauts on the Apollo missions even had to store their excrement in special containers for subsequent analysis on earth. After all, in some areas they were still "aeromedical test subjects."

But NASA's engineers had been working hard to turn going to the toilet into yet another routine reminiscent of conditions on earth. They had developed a lavatory in which a powerful current of air supplanted the nonexistent gravity and swept solid material down through a spinning drum called a "slinger" to be dried in a vacuum. They were no longer interested in subsequent analysis. The only problem was that the astronaut had to sit very firmly on the toilet seat, and that could be difficult in a weightless state. Experiments were carried out with different types of seat belts, as well as straps to fasten over the legs. Urine was still dealt with by means of a special unit attached to a tube. Here, however, the engineers came up against a delicate anatomical dilemma. Because, for the first time in the history of the space program, the Orbiter's facilities would also have to serve as a women's toilet. But they managed to come up with a special unisex answer to this problem.[7] Not that going to the toilet became totally routine. The astronauts' training manual allocated two hours for reading the toilet handbook and one hour for toilet training in a weightless state (it still intrigues me, how they do that) — then you were on your own.[8]

The accompanying laboratory animals were not given quite the same training in waste disposal. Which led to the astronauts on the seventeenth flight finding themselves in extremely unsavory circumstances, with the excrement from the monkey cages floating all over the cabin. Incensed by this, the Orbiter's commander, Robert Overmyer, informed Houston that he'd always said those cages were no good and that it was no fun having ape shit flying around the cockpit.[9]

. . .

But, obviously, all these articles on the astronauts' doings in their much-vaunted weightlessness were just incidental. The Orbiter had not been created for astronauts to lark around playing ball and turning comical somersaults for the amusement of American television viewers. Primarily, the astronauts were there to work.

Their actual tasks consisted of:

Launching, hooking up to, and refueling of satellites
Repair of satellites in space
Observation of earth
Making charts by means of cameras and radar
Medical experiments
Experimenting with industrial production in weightlessness
Astronomical observations[10]

Step by step, the newspapers were able to report how the individual tasks were accomplished as scheduled.

The Orbiter itself proved to be an exceptionally stable platform for various instruments designed to survey the earth. Because of its great bulk, cameras and radar equipment were not subjected to anywhere near as much vibration as they would have been had the satellites, which were so much smaller, been used as platforms. The Orbiter per se constituted the world's most advanced spy satellite to date; more flexible and sophisticated than the CIA's Keyhole 12.[11] Understandably, such capabilities were not given prominent press coverage. Instead, the media latched on to the fact that the Orbiter's surveys had already led to the revision of several geographical charts.

Extravehicular activities (EVAs) also became possible once more. On the sixth mission, astronauts stepped out through the airlock to practice maneuvers in space on the end of long lifelines. On the tenth mission they tried out the "flying armchair" (the MMU, manned maneuvering unit), which had cost $15 million. Astronaut Bruce McCandless II seated himself in this electronic chair and put over 300 feet between himself and the Orbiter by means of small bursts of nitrogen gas from twenty-four small thrusters positioned at various points around the chair. The entire maneuver was carried out without a lifeline, thus making Bruce McCandless the first human satellite. For this reason the chair was also fitted with a beacon and radio facilities which should enable his colleagues on the Orbiter to find him, should they become separated in the boundless expanse of outer space.[12]

In 1983 *Columbia* was joined by *Challenger*, a year later by *Discovery*, and finally, in 1985, by *Atlantis*. The program was now up to full strength, with four operational Orbiters.

There were too many shuttle astronauts for all of them to become national heroes. But, of course, the entry of American women into space did not pass unnoticed. Sally Ride was the first, in 1983, but close on her heels came Judith A. Resnik, Kathryn D. Sullivan, Anna L. Fisher, Margaret R. Seddon, and Mary L. Cleave. The women's movement was more than satisfied with this turn of events and a wax model of Sally Ride was put on display at the National Air and Space Museum in Washington.

Also, the age limit for astronauts was raised. After all, the Orbiter was not nearly so physically taxing as earlier space travel systems. The record was set by a fifty-eight-year-old astronaut in July 1985.

Nor did the astronauts now have to be American. The Germans and the Dutch sent up a team, after financing the Spacelab module—to be carried in the Orbiter's hold—through the European Space Agency (ESA). And after West Germany had coughed up $175 million, NASA permitted certain sections of this mission (61-A) to be directed from a control room in Oberpfaffenhofen.[13] The Germans waxed enthusiastic about the *Deutschland I* mission, which had put their country back in touch with rocket technology, albeit on a somewhat more peaceful basis than last time.

For the first time, VIPs could—with a little luck—be allocated a seat on board the shuttle. Thus, in June 1985, the Saudi Arabian prince Sultan Salman Abdul Aziz al-Saud was taken along as "Payload Specialist," after NASA received a valuable order for an Arabian satellite system. At just twenty-eight years of age he was one of the youngest astronauts so far. The control center gladly supplied the coordinates to help him face Mecca at the appointed times.

Room could also be found, at short notice, for powerful politicians who were in a position to influence NASA's future budgets. Senator Jake Garn and Congressman William Nelson were both taken up, even though the astronauts, who had been training for years, found it hard to conceal their resentment. As always, however, NASA considered happy politicians to be more important than discontented astronauts. Though the astronauts may have been cheered up slightly by the news that Senator Garn had spent two full days of his trip into space suffering from space sickness, feeling nauseous and vomiting. Old hands particularly relished the fact that the senator had volunteered — as part of a biomedical experiment — to be fitted beforehand with microphones which would pick up the sounds of his digestive system.[14]

Even the big commercial concerns could buy space on board — although, to their annoyance, they were not permitted advertising space on the sides of the Orbiter. When Coca-Cola made a big song and dance about developing a special can for use in outer space, Pepsi immediately booked space on the same flight. A former Reagan adviser, Michael Deaver, who was now a lobbyist for Coca-Cola, was furious with the NASA chiefs. He more than hinted that he still had very close links with the White House. This cola war was resolved by the astronauts promising to taste the Coke before the Pepsi. As it happens, they thought both brands tasted strange under those conditions.[15]

NASA did not hesitate to exploit the PR possibilities of the space shuttle. Experiments by high school students were taken on board in special containers, which anyone could buy their way into for a reasonable sum. One flight carried thousands and thousands of first-day covers; another, commemorative coins. The Orbiter was also fitted, at great expense, with cameras designed to shoot film in the special IMAX format, thus making it possible for audiences in planetariums all over the world to experience the awe-inspiring sensation of flying through the heavens with the space shuttle.

After twenty "operative" launches, the Orbiter system had sent up twenty-five commercial satellites for the United States and a number of other countries; and it had launched the European Spacelab. Two satellites had been repaired in space, and two had been salvaged. The system had proved that it could carry out work that no unmanned system could ever have done. And in this particular field, the United States led the world. No other country had a "machine" like the American space shuttle.

26 ■ Backstage

Behind the scenes, a lot of very hard work was being done to maintain the image of easygoing routine. Twelve flights had been promised for 1984. There had been five. According to the 1982 schedule, in 1985 fourteen Orbiters should have been launched, but only eight left the launch pad. It was still a far cry from the twenty-four flights projected for 1988.[1]

There was just no way around it for NASA. Budgets had not increased at all for the space shuttle. Reagan celebrated *Columbia's* first launch by cutting $604 million off NASA's allocation. In 1981–82, for the first time in almost twenty years, the armed forces were given more than NASA to spend on space projects. And in every year thereafter, NASA's share shrank proportionately. It was not enough that the space shuttle was a technological success. It also had to prove that it could pull its weight financially, meet the promised launch deadlines, and keep its customers happy. Only in this way could NASA justify the need for a civilian space agency.

But on closer inspection, even in technological terms the space shuttle was only a qualified success. On its first flight, a number of tiles fell off. Before the second mission, the countdown ground to a halt just thirty-one seconds before liftoff, and when the Orbiter did at last get off the ground one of the fuel cells failed, cutting the mission short by a full three days. The third launch included television cameras and radio equipment that did not work and breakdowns in the toilet system. On the fourth mission, both boosters were lost when the parachutes failed to open on descent. During the flight the steering was affected by steam emanating from tiles that had been standing too long in the rain on the launch pad.

The astronauts on the fifth flight were plagued by nausea and vomiting and had to abandon a scheduled EVA when it was found that two small plastic pins were missing from space suits costing $2 million apiece; and without these pins the suits would not be airtight. But the certificates completed by the contractors clearly stated that the pins had been affixed. Once the shuttle had landed, further checks were made on the suits. In one of them, metal shavings were found to be blocking an exhaust-vent in the suit's oxygen supply system. If the suit had been used it would have swelled up like a balloon until it exploded, when it could well have blown a hole in the side of the Orbiter. The contractor received a severe reprimand.[2]

Both on this flight and on several others, the braking system failed during landing due to the high landing speed. The brakes overheated and the wheels locked. And things really started to go wrong when attempts were made to move the landings from California to the new runway at the Kennedy Space Center, to save precious turnaround time. Often weather conditions were unreliable, and in April 1985 the crosswinds were extremely high. But once a space shuttle has left its orbit, there is no way back. So the shuttle landed anyway. In order to keep the nose wheel on the center line the main wheels on the windward side had to brake extra hard, which led to the potentially disastrous explosion of a tire.

Several other flights had their landing site changed at the last minute because of problems with the weather in Florida. And on every occasion this led to further delays—with the Orbiter first having to be transported back to Kennedy.

The main engines too were fragile. In August 1984, the computers shut down the engines just three seconds before the solid rocket boosters were to be ignited. This abrupt stoppage caused a fire in the hydrogen gases in one of the thrusters—though this was soon extinguished. Other missions were called off just thirty-one and fourteen seconds respectively before liftoff. In July 1985, things went badly amiss when the computers disengaged one main engine just four minutes and fifty-five seconds after takeoff. Soon afterward, the computers reported that another engine was about to shut down. With only one engine, the Orbiter would not stand a chance of getting into orbit and would have to attempt an emergency landing on Crete. The pilot, Gordon Fullerton, told the computer to override complaints from the engine and—only just—made it safely into space after all by letting the OMS engines burn longer than originally planned.

From the state of the engines after each mission, it was obvious that they would never stand up to fifty flights—far from it. On several occasions, the ground crew had to replace cracked rotor blades, gas pipes, bearings, and even complete turbopumps which had not been able to withstand the tremendous pressure.

The tiles too were an ever-present thorn in NASA's side. Time-consuming replacement of sections of tiling, large and small, was normal after most flights. Even after flights on the back of the mother ship, loose tiles were discovered. On other occasions, they had to be replaced following damage arising from collisions with cranes or other machinery.

Generally speaking, the computers constituted one of the Orbiter's most reliable components, but in November 1983, two of the four main computers broke down just before commencement of the landing procedure. Shortly afterward, three of the inertial navigation systems also failed. Attempts to restart the computer were successful, but it cut out again after landing.

When the engineers boarded *Columbia* after landing, they discovered that the computer breakdowns were only part of the story. A small fire was also blazing in the motors which drive the space shuttle's hydraulics. This had been caused by leaking gaskets in the fuel hoses.[3]

None of these flaws ended in disaster and—depending on how you look at it—you can either be disturbed by the fact that there were so many or feel grateful that in each instance the space shuttle's reserve and fail-safe systems were able to intervene and prevent a catastrophe. On the other hand, it is obvious that all of these flaws are a reflection of a system that is nowhere near to achieving a routine, operational level.

These failures are evidence, rather, of a system which is still in the development stage. It resembles the first phase of any high-technology research and development program, and in this it is no different from the first years of the Mercury project in the early 1960s. But then the problems had been anticipated and were eventually solved. Now, every single breakdown was regarded as an embarrassing exception, to be explained away and then corrected, under wraps, as quickly as possible—so as not to damage the space shuttle's image as a standard piece of technological equipment.

Bill Lilly of NASA told how, during the development years, he had discussed this very point with the director of NASA at the time, Robert A. Frosch:

> [Frosch] thought they really intended to have an operational kind of thing. He also did not believe that it was going to be operational for a long, long time. I know I used to tell him as far as being

operational, Frosch, that son of a bitch [the shuttle] is going to be R&D [research and development] on out into the year 2000 . . . He said, "Hell yes!" He agreed. That didn't mean you couldn't do more and more. But in my mind, the thing is such a complicated business that it takes the complete dedication of people at all times . . .[4]

And yet, quite at variance with this view, the Orbiter was hustled through four test flights — not six, as originally planned — before being declared to be in good working order. This did not, of course, have any effect on the failure rate but it made a crucial difference in the atmosphere in which the shuttle's technicians had to work.

As the backlog of delays gradually increased — along with the pressure to achieve a faster and faster turnaround — seventy-five-hour or even eighty-hour workweeks were not unusual.[5] A number of reports, which came to light only after the *Challenger* disaster, spoke of workers who were downright groggy from overwork. So groggy were they that the workplace became infected by an air of fatalism and it became increasingly difficult to uphold safety routines to the letter. Not surprisingly, the technicians took shortcuts to save time. In the twenty-seven days prior to the fatal *Challenger* flight, not one key technician at Kennedy had a day off.[6]

Previously, the policy at NASA had been that anyone who freely admitted a mistake was let off the hook. But now firings and transfers had become an everyday fact of life when employees reported slipups. And this, of course, led to workers covering up their mistakes. This did not apply only to the technicians on the shop floor; contractors were not eager to admit to mistakes made by their companies either. The rot even spread as far as the NASA field centers. Each separate organization reasoned that neither headquarters nor the other centers need be told any more than was strictly necessary. They were seen more as rivals than as colleagues. So, for as long you possibly could, you kept your problems to yourself.[7]

At one point, NASA calculated that every minute of flight by the space shuttle needed three man-years of preparation. The lion's share of this work lay with the 14,000 workers at the Kennedy Space Center, some of whom were employed by NASA while others were on the payrolls of the main contractors.[8] Tensions ran high, both when all the stops were being pulled out to get a job done and when one group of technicians had to hang around for hours, seething with impatience and unable to get on with the next job until another group completed some minor task.

The situation became even more complicated when several Orbiters had to be made ready simultaneously. This was a totally new situation at the Kennedy Space Center, where they had been in the habit of concentrating on just one mission at a time. Even though separate teams were assigned to each Orbiter, key workers found themselves having to be everywhere at once. Thus, during the month prior to the *Challenger* flight, preparations were being made simultaneously for three separate missions.

Hanging over the workers' heads was NASA's declared goal: to make any given Orbiter ready for its next space flight in just 160 hours. This was still a totally unrealistic goal. The technicians did their utmost but, before the *Challenger* disaster, they had never succeeded in bringing the turnaround time down below a disappointing 1,240 hours, or fully fifty-one days. And as if that weren't enough, the bosses—with their big satellite customers on the line—kept telling the people on the shop floor that they weren't pulling their weight.

Things were not improved by the fact that NASA employees were now working under a President who had conducted an election campaign against everything that might conceivably come under the heading of "Washington." Under Reagan, particularly, NASA was regarded as an inefficient public mastodon which would function much better if it were privatized. This attitude was a bit hard to swallow for people who had dedicated their lives to NASA. In retrospect, one now says:

> I don't know how others work, but I know the people I worked with. Shit, we used to work seven days a week for years. . . . You never thought about it. . . . But to be put in the same class as the Post Office Department or Agriculture or a bunch of clerks . . .[9]

In 1983, in the hope of improving efficiency and in line with Reagan's ideas about privatizing public bodies, NASA hired Lockheed to take care of preparing the space shuttles for flight. Lockheed had no prior experience with the Orbiter systems, but it would be their job to act as co-ordinators between Rockwell and the people at Thiokol and Martin Marietta and, at the same time, maintain contact with personnel from at least three NASA field centers. Seasoned NASA technicians were profoundly shocked to see people with the Lockheed logo on their coveralls moving in to work on the Orbiter with unauthorized tools. Instead of promoting better coordination, this new echelon of workers only added to the confusion.[10] And the situation was definitely not improved when

NASA sought advice and expertise from the airline companies—who ought to know something about timetables and routine procedures. The former Apollo astronaut Frank Borman was appalled:

> There's just no way that I can understand in God's green earth that an airline could undertake, with its normal procedures, the operation of a space shuttle. When NASA came out with the request for a proposal from airlines to run the shuttle like it was a [Boeing] 727, I called them and told them they were crazy. The shuttle is an experimental vehicle, and will remain an experimental, highly sophisticated vehicle on the edge of technology.[11]

Very soon, in an effort to reduce the turnaround time, components were being borrowed from one Orbiter to use on another—a practice known as cannibalization. Of course, they ought to have waited until they could afford a decent stock of spare parts, but both time and money were in short supply. Meanwhile, if they wanted to get anywhere, they would have to borrow. Because of this cannibalization, NASA could keep only two of its four Orbiters in the air.[12] The risks grew with every component borrowed, since this involved tinkering with the innards of vitally important systems. The chances of something being damaged were also increased, but, then, they had managed to get an Orbiter off the ground, whereas it would otherwise have had to languish.

During these years, the American press, which had, since the Watergate revelations, been much given to gloating over its own exposés, did not delve too deeply into these matters. In most cases, they had absolutely no idea that any problems existed. NASA's press office was smart and slick. Journalists were given exciting guided tours and came away with the impression that they had seen all there was to see. But they were not present when an exhausted engineer discharged eight tons of liquid oxygen from a fuel tank four minutes before liftoff. At that point, the engineer had been on duty for eleven hours straight after doing two twelve-hour night shifts.[13] The engineers also missed seeing that a sensor from one of the fuel hoses had snapped off and been sucked into the Orbiter, where it proceeded to block a valve. This fault was discovered at the last minute, purely by chance. If that Orbiter had been launched, the engine could have exploded.[14]

Most space journalists had, in fact, put their names forward to be the first journalist into space on a mission scheduled for the autumn of

1986 — just after the first teacher. There was no need to put this unique opportunity at risk by writing untimely articles or making unfair judgments.

But no journalist had been privy to the project's best-kept secret. Nor, at the end of 1985, were the people at NASA headquarters any the wiser. No one had told the astronauts, the journalists, or the senior administrators at headquarters that the solid-fuel rocket booster joints had proved — on closer inspection — to have failed, to a greater or lesser degree, on twelve out of twenty-four flights; that as much as three and a half years earlier — before Columbia's first flight — the design had been deemed problematic; and that, even so, no one had made any serious move to correct this problem.[15]

In the Interests
of National Security

The U.S. Air Force was somewhat better informed than the press as to
the true state of affairs at Kennedy. In a confidential report written in
October 1985, the armed forces' observers painted a gloomy picture of
the situation surrounding NASA's pride and joy. The spare parts systems
were one big mess. No proper documentation was made of what repairs
had been carried out and the requisite service inspections were simply
skipped if they were not seen to be absolutely critical.[1]

Air Force experts considered the solid-fuel rocket boosters to be the
system's Achilles' heel and — according to an article published later in
The New York Times — estimated the risk of failure at 1:35. If their calculations were correct, the American space shuttle was one of the most
dangerous technological systems ever built.[2]

NASA's Administrator, James Beggs, knew that the people from the
Air Force were looking over his shoulder and was aware that they were
not well disposed toward his organization. In fact, it might even be that
it suited them rather well for NASA to have its problems. And so, as far
back as March 1985, Beggs had issued a directive to all NASA employees
telling them that now, more than ever, it was important to live up to the
set schedule. Around the same time he told a journalist from *Aviation
Week and Space Technology*:

> The next eighteen months are very critical for the shuttle. If we are
> going to prove our mettle and demonstrate our capability, we have
> got to fly out that manifest.[3]

If they could not pull this off, the space shuttle's opponents would —
still according to Beggs — have a good excuse for saying that the American space shuttle was not to be trusted.

.　.　.

This was exactly what the Air Force had already said, in June of the previous year, when asking Congress for hundreds of millions of dollars to spend on powerful expendable rockets. In justifying this request, the armed forces explained that it was in the interests of national security that they be able to send up reconnaissance satellites at short notice. And, they said, the space shuttle was obviously not up to the job.

NASA was shocked by this attack. They knew only too well that one of the most telling arguments for NASA's existence would be invalidated should the armed forces withdraw their support. The agency approached the President, who solved the problem with a compromise whereby the armed forces were granted some rockets in return for promising to make use of the space shuttle for at least eight missions a year from 1988 onward.[4]

But at NASA a pattern was slowly beginning to emerge. It was obvious that the generals in the armed forces had grown much more assertive under Reagan. It was also obvious that they were now giving vent to many years of pent-up spleen over civilian space flights. This fitted neatly with the fact that, in March 1981, the Air Force had succeeded in having the physicist Hans Mark appointed as NASA's Deputy Administrator, under the ineffectual James Beggs. Mark had previously been Secretary of the Air Force.

Hans Mark emigrated from Germany as a teenager and had been taught in the United States by the "father of the hydrogen bomb," Edward Teller, a man who had been ostracized by virtually everyone in the world of physics after his virulent attack on his colleague J. Robert Oppenheimer, the popular head of the wartime Los Alamos laboratory. Oppenheimer's sole crime had been to believe that the development of the hydrogen bomb would initiate an insane and pointless nuclear race. To Teller, such a view was an unmistakable sign that he harbored a soft spot for the Communists. And wasn't there something about Oppenheimer himself having been a Communist back in the 1930s?

Since then, an embittered Teller had retreated to devote himself to developing ever more powerful atomic weapons. At no time had he ever made a secret of the fact that he did not trust the Russians' motives during the so-called détente. So the sight of American and Russian astronauts shaking hands in space in July 1975 was bound to get his back up. In President Reagan, the now aging Teller had at last found a President who would listen to him. He could have given Reagan's "Evil Empire" speech himself.

Soon after Reagan's election victory, another of Teller's pupils, George A. Keyworth — who had worked with atomic weapons at Los Alamos in the 1970s—was appointed as the President's scientific adviser. Together, Hans Mark and George Keyworth made a powerful team, with neither making any bones about the fact that their enthusiasm for civilian NASA went only so far.

Hans Mark later stated that he had had to fight hard to keep the Pentagon interested in the space shuttle. Many generals viewed manned space flights as unnecessarily expensive. Surely modern automated electronic systems could do the job cheaper and more effectively. But if it was unavoidable, then the space shuttle ought to be under the complete and unrestricted control of the consumer (i.e., the military). It was intolerable that military personnel had to operate through a civilian agency.

If they were not happy with NASA's terms, civilian customers could, for example, switch to the French unmanned Ariane project, which had been able to stage satellite launches in the mid-1980s at prices far below the Americans', in spite of all NASA's promises.[5] But of course this option was not open to the military. American military engineering would have to remain exclusively in the hands of Americans, and they were therefore referred to the newly developed space shuttle system.

Even before the shuttle's first flight, the Pentagon had demanded the right to remove any civilian cargo from the Orbiter whenever they saw fit, and take over that space themselves—should it be deemed necessary in the interests of national defense. After crossing swords with NASA a few times, the armed forces had been granted this "assured access to space" as requested—a move which did not make it any easier for NASA to sign contracts with private customers. Even so, the military were still not happy with a civilian space shuttle.[6]

George Keyworth was one of those who believed that the Orbiter would be far too expensive to run. After all, it could not put a simple satellite into orbit without lugging along tons of equipment—the sole purpose of which was to ensure that an otherwise superfluous crew could breathe and be brought home safely. He did not believe in NASA's arguments for the indispensability of the astronauts. In a later interview he put it like this:

> Of all the organizations that I have dealt with . . . I have only seen one that lied. It was NASA. From the top to the bottom they lie . . . The reason they lie, of course, is because they are wrapped up in a higher calling. In their eyes they are white lies. They tell lies in order to do what has to be done. Because in the end the result will be for the betterment of the public. So they are not lying from evil. But, nevertheless, they are lying.[7]

Meanwhile, Edward Teller was telling an intrigued President of the possibilities of waging war in space. Star Wars. Satellites could keep the whole of the Soviet Union under surveillance and would be able to shoot down Russian missiles with laser weapons the minute they took to the air. Russian satellites could be fought with special attack satellites. This way, the United States would be protected by an electronic umbrella. Ronald Reagan thought this sounded wonderful. And he was happy to appoint George Keyworth on the strength of Teller's hearty recommendation.

It did not worry him that this would be the most expensive weapons system to date, at an as yet unknown cost of between $200 and $1,000 billion.[8] He asked defense experts to get onto it right away. The Orbiter too was to be involved in laser experiments.

Former colleagues of Teller's among the world's leading atomic physicists were, however, extremely perturbed by this prospect. Hans Bethe, who had also been involved in the development of the first atomic bomb during the war years, adamantly opposed such notions. Again and again he stressed the point that discussions on arms reduction and political agreements offered the only hope of reducing the nuclear threat. Anything else was just a technological illusion which could only have a destabilizing effect. If, that is, it was at all possible to make these systems function — which Bethe and many other physicists seriously doubted. In any case, it would never be possible to discover whether they really could halt all incoming missiles until it was too late.

> We need to understand the other fellow and negotiate and try to come to some agreement about the common danger. That is what has been forgotten. The solution can only be political. It would be terribly comfortable for the President and the Secretary of Defense if there was a technical solution. But there isn't any.[9]

According to its critics, the Star Wars program would only lock both superpowers into an endless spiral of escalating defense expenditures engendered by a new arms race. But this did not worry the hawks on the American political scene. They were in no doubt as to which of the two superpowers would win such an economic steeplechase.

When, in August 1983, the Russians suggested the drawing up of a treaty forbidding the militarization of space, a treaty which would preclude any military application of the Orbiter system, the Reagan administration gave this idea the cold shoulder. They were not going to cut themselves off from any likely opportunities in outer space — which looked set to become the most important theater of operations in the future.[10]

In April 1983, amid this atmosphere of renewed interest in space exploration, James Beggs and Hans Mark succeeded — to Keyworth's annoyance — in reviving the project for an American space station to be serviced by the Orbiter. Reagan could see the sense in the argument that the Russians, with their Salyut Station, had more experience than the Americans in prolonged stays in outer space. So the United States would have to invest $8 billion in this area — though, fortunately, the lion's share of this sum would not fall due until after Reagan was out of office.

Reagan could see the military possibilities, but he was also fascinated, in a superficial fashion, by the idea of a space station possibly constituting the first step on the road to Mars. So Beggs really played up this very remote possibility in his presentation to the President, even though he himself had no faith in it. All in all, Administrator Beggs did not think much of his President.

> He was almost technically ignorant. Not quite, but almost. He grasps a few of the broader concepts, but when you start talking to him in any kind of detail about the broader aspects of the program, his eyes glaze over.[11]

It took a while for the public at large to become aware of the game being played out between the Pentagon and NASA. Reagan seemed so convincing, with his soft voice and tender eyes, talking about this umbrella that was to be spread over the heads of American children, so that they could sleep safe and sound in their little beds. Nevertheless, an element of concern did begin to creep in at the thought of the snow-white space shuttle being sent up to perform obscure and mysterious tasks on extremely murky military missions.

To the Reagan administration's great annoyance, the backlash was given its clearest expression in Congress by a Republican, Harold Hollenbeck — a leading light on the House of Representatives' Space Committee. Throughout his political career Hollenbeck had supported NASA. Now he was becoming more and more disturbed by what he saw being played out behind the scenes in the Senate. What angered him most was the way in which, on the one hand, the Pentagon never missed an opportunity to run down the space shuttle system, while at the same time they were obviously hoping that one day it would be all theirs. For

Hollenbeck, the last straw came when Keyworth secured for the Pentagon funding for expendable booster rockets that were sure to undermine the economy of the space shuttle system.

In the aftermath of the *Challenger* disaster, this grant was interpreted by George Keyworth as a cynical masterstroke:

> We don't talk about these things publicly, but . . . we perceived that someday, someone could very well get killed on the shuttle . . . And we knew that despite all the critical military assets that we needed to put up, we were not going to send up a shuttle after an accident until we've analyzed it twenty-five times over. We had estimated a minimum of a two-year delay.[12]

In August 1982, Hollenbeck — who had, by this time, had enough — took the initiative in setting up a hearing on the government's plans for NASA. At this hearing he gave a speech in which he said:

> NASA with its rich history, its great accomplishments has been turned into a situation-comedy father. Like its TV counterpart, it is snickered at because its wallet is bled by ungrateful children, but its quiet good nature seems endless. The tragedy is the American people are not aware of the politicizing and militarizing of the civilian space agency. The Congress and the press have failed to do their jobs. How depressing it is to know that America's space policy for the 1990s is classified "Top Secret." . . . The arrogance of this administration, of the military, and of some of the contractors towards a civilian agency that accomplished so much does not go unnoticed . . . I speak for a strong civilian space program not simply to annoy those that now run NASA, but because it was a civilian-run team that put us on the moon, that did it in the budget, and made damn few excuses. I, for one, do not want a gold-plated space program that is part of some Star Wars Pentagon . . . I can only hope the next generation of Americans will not look back upon those of us here today as the leaders who sat in silence as America turned a noble endeavor into an interstellar war machine.

It was no accident that this speech contained references to the original space policy outlined by Democrat John F. Kennedy in what was fast becoming the dim and distant past. As Hollenbeck saw it, from having been an out-and-out civilian agency, NASA was on its way to becoming merely a subcontractor for the Pentagon. He went on:

We went into space as a new frontier and now we drag the hate and bitterness of earth into the heavens as if it is the right of man to make war everyplace. We need a good and brave leadership who will stand up and say, "Enough!" The greed machine of contractors, revolving-door jobs, the endless excuses to build more military hardware now is being applied to virgin territory. What we are talking about is not national security. What we are talking about is big bucks. Big bucks that can be spent without the glare of the press or congressional oversight because the government can invoke the magical phrase "national security." The space policy of the Reagan administration is a national tragedy. . . . The goodwill that was space is gone. The NASA that set sail on this new sea is a romantic memory.[13]

At NASA, Harold Hollenbeck's comments were viewed by most of the somewhat older employees as a pretty accurate expression of their own thoughts on the matter. In the open-plan offices of headquarters, however, he was inevitably regarded as a man who could not come to terms with the new order. John F. Kennedy was gone forever and, without the Pentagon, NASA had no future.

At NASA headquarters much greater concern was evinced over the latest figures on the space shuttle finances. The latest Orbiters had cost, not $675 million apiece as projected in 1972, but $1.47 billion (all figures converted to the 1986 dollar rate). The fuel tank, which was supposed to cost a little over $4.5 million a time, instead came to $28 million — all of which would vanish, hissing, into the Indian Ocean after every trip. The cost of a launch was to have been $28 million. Instead it worked out at $280 million — and up. And that all-important cost per pound sent into orbit was nothing like the promised $100 to $270. Instead, as Ralph Lapp had already predicted, it came to $5,264 per pound.[14]

28. TISP

It was NASA's James Beggs who came up with the idea of TISP. Within the organization itself there was some resistance to the project, and the astronaut corps were particularly unhappy about it. In fact, this idea came very close to threatening the astronauts' holy of holies: the right stuff — the reason why astronauts had originally been chosen from among the test pilots. They already belonged to a profession in which a twenty-year career carried a 25 percent risk of dying in service.[1] They knew the risks, they accepted them, and every one of them was convinced that he, at least, would come through in one piece.

Later, they had been joined by astronauts from other fields — veteran fighter pilots, Army personnel with an engineering background, and, eventually, scientists. But the pecking order was clearly defined: from the test pilots at the top all the way down to the scientists at the bottom.

The space shuttle brought about changes in this setup. In a crew of seven or eight not everyone needed to be a superman or be made of the finest stuff. Older, and younger, astronauts joined the ranks, as did women; and engineers were selected from the big clients in the satellite industry. Places were even reserved, as we have seen, for influential politicians — albeit trained military pilots.

But TISP was a radically new departure. A perfectly ordinary American citizen was to join a flight into space. As John Young remarked, this could be anyone at all, just so long as the person concerned was alive and breathing.[2] The big plus for NASA with this project was that it seemed to indicate that the space shuttle was as safe as a passenger airplane, and it was guaranteed to rekindle the dwindling interest of the press and the public. Such considerations were far too important to be thwarted by prima donna astronauts — even if some of them *had* been waiting nineteen years for the chance to use what they had learned in training.[3]

The original idea had been to choose a journalist from among the host of media people who had covered space matters for their newspapers and television networks. That way they would be getting not only an ordinary civilian but also an individual who could communicate his impressions in such a way as to maximize the effect of the project.

The White House immediately agreed to the idea in principle, but they also had their own priorities. The election campaign to secure Ronald Reagan his second term in office was already underway in 1984. Again and again, throughout the country, Reagan's opponent, Walter Mondale, spoke of Reagan's failure to produce results in the field of education. Mondale's people drew up a special report card in which Reagan received an F in every subject except sports and drama. They persistently depicted him as being anything but intellectual; a President who was well on the way to ruining the American educational system.

The opinion polls showed that this message was getting through to the voters. The President's campaign strategists were growing anxious and felt that he had to make a dramatic gesture toward education. It was decided that Reagan would make a major speech on the education question at Washington's Jefferson Junior High School on August 27. In their rush, they forgot that this fell in the middle of summer vacation; nevertheless, they managed to drum up a decent enough number of students from the surrounding area to make the whole thing look convincing.

The President clearly relished the dramatic buildup to the sensational conclusion of his speech.

> Until now we hadn't decided who the first citizen passenger would be. But today I am directing NASA to begin the search in all of our elementary and secondary schools and to choose as the first citizen passenger in the history of our space program one of America's finest — a teacher.
>
> When the shuttle lifts off all of America will be reminded of the crucial role that teachers and education play in the life of our nation. I can't think of a better lesson for our children and our country.[4]

The idea had borne fruit: TISP (Teacher-in-Space Program). In the space of two months, NASA received 11,000 applications from teachers in every sector of the American educational system. From these, 114 semifinalists were selected — to undergo the most thorough screening and assessment. Because, of course, the question of who this perfectly ordinary, random citizen would be was not altogether irrelevant. On July 1, 1985, NASA published the names of the ten finalists. And, at long last, on July 19, the choice was made: the lucky winner was thirty-seven-year-old Sharon Christa McAuliffe.[5]

Christa McAuliffe was a high school teacher from Concord, New Hampshire, who taught history, law, and economics. The selection committee had been impressed by the way she referred, in her application, to the inspiration she had drawn, as a child, from John F. Kennedy's words on sending a man to the moon.

> I remember the excitement in my home when the first satellites were launched. My parents were amazed and I was caught up with their wonder. In school my classes would gather around the TV and try to follow the rocket as it seemed to jump all over the screen. John Kennedy inspired me with his words about placing a man on the moon, and I still remember a cloudy, rainy night driving through Pennsylvania and hearing the news that the astronauts had landed safely.

She also introduced a feminist angle in her application:

> As a woman I have been envious of those men who could participate in the space program and who were encouraged to excel in the areas of math and science. I felt that women had indeed been left outside of one of the most exciting careers available. When Sally Ride and other women began to train as astronauts, I could look among my students and see ahead of them an ever-increasing list of opportunities . . .
>
> I watched the Space Age being born and I would like to participate.[6]

Christa McAuliffe was, however, fully aware that she would not be an astronaut. She would be more of an observer among the astronauts, a layman in a high-tech world which only very few could fathom. But she might be able to act as an interpreter between this world and ordinary Americans. This aspect too was covered in her application:

> Much information about the social history of the United States has been found in diaries, travel accounts and personal letters. This social history of the common people, joined with our military, political and economic history, gives my students an awareness of what the whole society was doing at a particular time in history. Just as the pioneer travelers of the Conestoga wagon days kept personal journals, I, as a pioneer space traveler, would do the same . . .

> My perceptions as a nonastronaut would help complete and humanize the technology of the Space Age. Future historians would use my eyewitness accounts to help in their studies of the impact of the Space Age on the general population.[7]

She compared herself to the women of the pioneering days, faithfully writing to the loved ones they had left back East, as they trundled across the wilds of America in covered wagons. She too, in her modern-day covered wagon, would keep her journal, as she traveled the last frontiers remaining at the end of the twentieth century. And perhaps, in 1986, children would be as inspired by her voyage into space as she had been by President Kennedy.[8]

At any rate, from the moment she was selected McAuliffe became public property. Reporters swarmed in from all over the United States to portray this ordinary citizen's perfectly ordinary life with her lawyer husband, Steve, nine-year-old Scott, and seven-year-old Caroline. In no time she was being viewed as a source of inspiration, not just for American children but — perhaps chiefly — for American women who, like McAuliffe, manage to combine a career with running a family, having a husband and children. Just as John Glenn had once been — though for very different reasons — Christa McAuliffe became an American idol. She did it all. She was good at her job, loved and supported her husband and children, did volunteer work, and taught Sunday school. And now she had shown the courage to go careering off into the heavens in the company of genuine astronauts.

What Christa had to say, on anything under the sun, was of interest. Very early on she told the reporters how much she admired John F. Kennedy and the whole Kennedy clan. This was too much for Reagan's White House. Michael Deaver of the President's staff rang James Beggs and asked him to make sure that McAuliffe was quite clear about what one could and — most importantly — could not say.[9] She confined herself to more diplomatic utterances thereafter.

McAuliffe's hometown, Concord, declared a Christa McAuliffe day.[10] Just as the space shuttle had risen like a phoenix from the ashes of a decade marked by Watergate and Vietnam, Christa was regarded as a symbol of a new feeling of confidence and patriotic optimism in 1980s America. She was released from her teaching and embarked upon a nationwide round of public engagements which was interrupted only by the necessary astronaut training. According to her, this consisted of learning which buttons she must, on no account, push.

The people at NASA were delighted by her naturalness and her complete lack of prima donna qualities. She was every bit as down-to-earth and straightforward as one could wish for. In her, NASA's vast, impersonal organization had found a sympathetic public face, one with which people could identify. And, unlike John Glenn, McAuliffe accepted the great circus that had grown up around her as being part of the deal. John Glenn had been wonderful, unforgettable. But he had also seemed unnervingly far out of reach in his superhuman perfection. Christa McAuliffe was wonderful simply because she was not out of reach. She looked like the girl next door. And if *she* could do it, then why not try to aim higher yourself? Even her husband, Steve, was on the verge of becoming a role model, having had no problem with taking care of the kids for the seven months that his wife had to spend away from her family.

Christa McAuliffe wanted her spaceflight to show that ordinary people had their part to play in great historical events. She hoped that her pupils would get the message and be seized by enthusiasm for American history, because they themselves would have a hand in it. History was not something beyond them, or above them. They *were* history and could determine which course it should take.[11] In saying this, she was, of course, instinctively playing to perfection the role earmarked for her by NASA and the White House.

NASA announced that McAuliffe would be traveling on mission 51-L in January 1986. In September of that year, a journalist would be sent into space and possibly, later still, a representative of workers in the American labor unions.[12]

NASA's press office was asked to prepare a draft for the part of the President's State of the Union speech that dealt with space. The speech would be given while 51-L was out there, so it seemed only natural to make use of the McAuliffe effect in this instance too. The press office therefore suggested that the President introduce that section with a reference to her:

> Tonight, while I am speaking to you, a young elementary [sic] school teacher from Concord, New Hampshire, is taking us all on the ultimate field trip as she orbits the Earth as the first citizen passenger on the Space Shuttle.
>
> Christa McAuliffe's journey is a prelude to the journeys of other Americans and our friends around the world who will be living and working together in a permanently manned Space Station in the mid-1990s, bringing a rich return of scientific, technical and economic benefits to mankind.[13]

NASA hoped that the President's advisers would accept this draft, as it reaffirmed the government's undertaking to support a civilian space agency which might otherwise find itself hard pressed. The President's fondness for folksy, sentimental symbolism was well known. So they were pretty certain that he would not be able to resist mentioning the person behind the term TISP—mother, housewife, and teacher, Christa Mc-Auliffe. In so doing, he would be able to bolster his own reputation regarding education and a little of the gloss was bound to rub off on NASA.

Meanwhile, that teacher was preparing for two fifteen-minute lessons, to be broadcast from space, in which schoolchildren throughout the United States would, first, learn about day-to-day life in a space shuttle and then see Newton's laws, which they knew about from their physics lessons, put into practice.

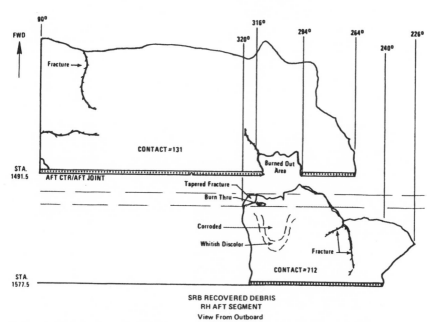

SRB RECOVERED DEBRIS
RH AFT SEGMENT
View From Outboard

THE CRUCIAL PIECES OF WRECKAGE FROM *CHALLENGER'S* RIGHT BOOSTER. (LEWIS, 1988)

29. One Day to Go

January 1986, and the waiting was at an end. At long last, Christa McAuliffe and the other crew members of the mission code-named 51-L would be on their way. The *Challenger* flight had originally been scheduled for December 23, 1985, but had been put off several times due to delays in the *Columbia* launch 61-C.

This, the last launch before *Challenger*, had been subjected to widespread ridicule in the press, where it was finally dubbed "Mission Impossible." The situation had been exacerbated by the fact that the crew on this flight included Congressman Bill Nelson, who had stated publicly that he believed it to be God's will that he travel in space. On the face of it, though, it didn't much look like it. Mission 61-C was postponed again and again, seven times in all — three of these with the crew already installed in the shuttle cabin.[1] On one occasion, the launch was interrupted just thirty-one seconds before the computers were to activate the main engines; another time a halt was called only fourteen seconds before liftoff. Every time, more disappointment and frustration, and a crew who had to be brought down stiff-legged after a long and uncomfortable wait. Bill Nelson's constituents were cursing too. He had personally invited thirty-seven busloads to the launch he had arranged with his God.[2]

Every time, the fuel tank had to be emptied — only to be slowly refilled at a later time. Then the countdown — 2,000 pages of it, contained in four thick ring binders — could begin all over again.

Not surprisingly, these postponements led to frayed tempers at the Kennedy Space Center. It was no joke for NASA to be made a laughingstock on the television news evening after evening. Obviously, all of these postponements were well founded. Safety limits of one sort or another would have been exceeded and the whole point of these procedures was to ensure against any risk of losing the spacecraft and the crew due to substandard launch conditions.

Even so, the press could not have cared less about the technical reasons. The point was that even after six attempts NASA, always so effectual in the past, could not get *Columbia* off the ground.

At NASA, there was nothing they would have liked better and all the different teams were coming under greater and greater pressure. *Columbia* simply had to go up, in order to come back down and be ready to take off again, as scheduled, on March 6, on what had been christened the Astro mission — America's contribution to the study of Halley's comet. The comet was not about to wait for NASA and it would be another seventy-six years before they had another chance for a close look at it.

In order to carry out this mission *Columbia* would have to leave on January 10 at the latest, and attempts *were* made at a launch on both the ninth and the tenth — both times unsuccessfully. It was January 12 before, at long last, they pulled it off. And only then did personnel at Kennedy have the capacity to concentrate properly on *Challenger*.

Mission 51-L was now scheduled for January 23, but almost immediately it ran into delays caused by difficulties in bringing *Columbia* back down. To save time, arrangements had been made for a landing on the special space shuttle runway at the Kennedy Space Center field. And *Columbia*'s mission had been cut short by one day — that too should provide them with some breathing space. But in Florida, which was experiencing the wettest winter in living memory, weather conditions were not favorable. Instead of redirecting the space shuttle to the more stable climate of California, it was decided to leave *Columbia* in orbit for an extra day. But the next day, conditions in Florida were no better. So *Columbia*'s computers were given orders to calculate for a descent to Edwards in California instead.

Landing in California would involve an extra week's "production time," since the shuttle would first have to be transported back to base. So NASA's senior management gave orders for yet another day's wait in outer space. Not until the next day, January 18, was any hope of a landing at Kennedy reluctantly abandoned and *Columbia* was allowed to land where the sun had been shining all the time. And so another week was notched up on the debit side of the balance sheet.[3]

Under such circumstances, there was no possibility of launching *Challenger* just five days later. Especially not when vital parts would have to be borrowed from *Columbia*. *Columbia* had no sooner touched down than technicians were climbing on board to remove certain all-important sensors and one of the five computers. This done, one of NASA's training jets promptly took off for Florida with the parts.[4] Of course, the general public was told nothing of all this. The cannibal approach was not suited to engendering good press coverage. Nevertheless, it was necessary if this hectic launch schedule was to be met. *Challenger* had, therefore, also borrowed several vital components from *Discovery* and *Atlantis*.

It was essential that *Challenger* should take off, so that the launch pad could be made ready for *Columbia*, which would need to use it just six weeks later. Naturally, the personnel and management at the Kennedy Space Center could not help but see things differently from the astronauts, who had been waiting for so long—although, of course, they shared their impatience. But whereas for the astronauts the launch represented the culmination of a long, tough training program and, by that token, an end in itself, for the people at Kennedy it was just one piece in an increasingly complex jigsaw puzzle.

Challenger also had to fit in with a prospective launch schedule that was growing ever tighter. In the spring of 1986, NASA planned to send two exploratory missions to the sun and Jupiter. On May 15, *Challenger* was to take off on the Ulysses mission—carrying, for the first time, an extremely powerful Centaur rocket stage which would, after being jettisoned from the Orbiter, propel a satellite toward Jupiter. This satellite would utilize Jupiter's gravitational pull to hurl it farther out, into a new trajectory which would put it on course for the sun. Once there, two and a half years later, this satellite would give scientists their first chance of studying the solar poles.[5]

Doubts had been expressed about carrying the powerful Centaur rocket in the Orbiter's hold. NASA would much rather have an unmanned space tug undertaking such tasks. Then the satellite could have been transferred, in space, to the tug—which would be in orbit already—and it could have taken it from there. But the space tug had never gotten beyond the drawing-board stage. So the Orbiter would have to carry the huge rocket stage itself—which meant that, for the first time ever, a space shuttle would be flying with liquid fuel in its hold. Actually it would be carrying an enormous bomb. All the previous rocket stages used in the launching of communications satellites from the Orbiter had been much smaller and been powered by solid fuel, which was much safer.

NASA engineers had been opposed to the Centaur stage for safety reasons, but were overruled by their superiors. Centaur's contractor, General Dynamics, had adopted exceptionally aggressive tactics to ensure approval of this project. In the past, the company had supplied this rocket stage for the old Atlas rocket, which was now to be phased out, in line with the lower priority being placed on expendable rockets. If their investment was not to go down the drain, they needed to be on that space shuttle. So they hired James Beggs as a consultant. His appointment, soon afterward, as Administrator of NASA made doing business with that organization a whole lot easier.[6]

Five days after the Ulysses mission, on May 20, *Atlantis* would take off from a neighboring launch pad on the Galileo mission. The purpose of this mission too was to put a satellite into orbit around Jupiter by means of a Centaur stage. This would be the first time that two Orbiters had been prepared for flight and launched almost simultaneously. Everyone involved foresaw tremendous strain being put on key personnel at Kennedy. But they had no choice—Jupiter wouldn't wait for NASA either. The "launch window" amounted to just a few days when the position of the earth relative to Jupiter would make it possible to reach the planet in 600 days. If either or both Orbiters were to miss this crucial time frame, it would be another fourteen months before they could take another crack at it.[7]

Within NASA, criticism had also been leveled at the Ulysses and Galileo missions because the Orbiters would be carrying plutonium— in Ulysses' case, 23.1 pounds; in Galileo's, 46.2 pounds. The plutonium was to provide the satellite with atomic energy on its long journey to the sun and Jupiter. It needed no great stretch of the imagination to work out the risk of an explosion on the launch pad. Nevertheless, NASA's management decided that it was worth taking this risk. Both missions would give NASA's poor standing in scientific circles a badly needed boost.

It is highly unlikely that Christa McAuliffe had anything to do with these tortuous deliberations, but there is no doubt that she was affected by them, albeit indirectly. At the Kennedy Space Center the "production line" was under pressure and no exception could be made for *Challenger*, even if a schoolteacher *was* to be on board. The space shuttle had to be launched as soon as possible, to allow *Columbia* to take over its launch pad, ready for the Astro mission. As soon as *Challenger* returned to earth at the beginning of February it would undergo the extensive alterations necessary for it to accommodate the Centaur stage and dispatch the Ulysses mission on May 15. The embarrassing delays incurred by 61-C were already threatening to affect later missions. It was now absolutely vital that *Challenger* not put more strain on the schedule.

NASA would have preferred to launch *Challenger* on Sunday, January 26, the day on which Vice President Bush had to fly from Washington to Honduras for the inauguration of a new president. It would be most opportune if George Bush could make a brief stopover in Florida and give the 51-L launch extra media exposure. But the Vice President's timetable could not be disrupted if there was the slightest risk that the launch might not take place. One thing NASA did not need was yet another embarrassing delay — this time witnessed by Bush.

The meteorologists were predicting a cold front moving in from the northwest. The first wave of wind, rain, and low cloud would reach Cape Kennedy early on Sunday morning. And the Orbiter tiles would not be able to withstand a launch in the rain. Even the gentlest raindrops were transformed into projectiles at those speeds. This front was backed up by a ridge of high pressure bringing an exceptionally chill wind from the polar regions. On Tuesday, the temperature in Tampa was expected to drop down to 14° F. Such temperatures would contravene the official launch criteria, which prohibited launches at anything under 31° F.[8]

So Sunday was out, because of the rain; Tuesday, because of the cold. Bush was therefore advised that he would have to miss out on the launch and it was put off until Monday, January 27.

The weather did not, however, turn out according to the meteorologists' predictions. The front came to a halt north of Florida and the weather on Sunday was beautiful. Ideal for a successful launch. But by then it was too late for a change of plan. An air of suppressed indignation prevailed among the *Challenger* crew, because regard for the Vice President's comfort and convenience and for NASA's reputation had prevented them from taking advantage of this chance.

Meanwhile, the cold weather was building up behind the stationary cold front. Much faster than expected. The weather on Monday would be colder than anticipated. New forecasts projected a temperature of 40° F at launch time — a full 15° F colder than previously forecast.

On Sunday afternoon, belatedly, the rain fell. During the night the weather cleared, as the temperature dropped and the fuel tanks were filled.[9]

When McAuliffe and the other members of the crew were awakened on the morning of Monday, January 27, the temperature was hovering just above freezing, but weather conditions looked promising. Spirits were high, because at long last it looked as though the tension was about to

be relieved. Even though they were extremely rushed, the NASA personnel attending the astronauts were efficient and cheerful. But they could
not entirely deny their technical training and love of accuracy. And so
the crew were awakened right on schedule at 5:07 a.m. EST — not a
minute earlier, not a minute later.[10] After breakfast, they went through
the ritual of parading out in single file for the photographers, and by
7:57 they were all strapped down tight, on their backs, in their seats.

This mission should not present any big problems. *Challenger* was to
launch a large navigation and communications satellite, TDRS-B, which
would enable NASA to contact its Orbiters without using one of the
numerous earth stations. Fully developed, this system would consist of
three satellites. One had already been launched by *Challenger* in 1983.

The hold also contained another satellite, known as Spartan-Halley.
This would be hoisted out of the hold to execute twenty-two orbits while
making photographic and spectrographic analyses of the comet. It would
then be hauled in again and brought back to earth.

Finally, Greg Jarvis was to carry out a series of experiments on the
behavior of liquids in weightlessness for the Hughes Aircraft Company.
Small, transparent flasks containing various liquids were to be spun at
varying speeds while being filmed on video.[11] These experiments would
be of value in developing future techniques for satellite refueling in outer
space.

And then, of course, there were Christa McAuliffe's much-publicized
space lessons.

Only now did those characteristic hitches that had also plagued 61-C
start to crop up. A warning light showed that the hatch had not been
shut properly behind the crew. It certainly seemed to be shut properly
but the light refused to confirm this. If the warning light was right, the
hatch could spring open during the launch, causing a lethal decompression of the pressurized cabin. On the other hand, they had come up
against failures in the warning light microswitches before. After some
discussion between the control centers and the engineers in the tower a
method was devised whereby McNair could verify from the inside
whether the latching pins were properly secured. He reported that they
were and so it was decided to ignore the warning light.

With this out of the way, they could proceed to releasing the special
three-legged auxiliary handle — the "milk stool" external hatch handle —
which had been used to close the hatch, so that the technicians would
not need to come into direct contact with the tiles. Even a tiny scratch

from a fingernail had, in the past, necessitated the replacement of a tile. Hence the development of this special tool which could be screwed to the aluminum underneath through holes in the tiles — thus providing a secure door handle while the astronauts were being installed. Once the hatch was safely closed — as it was now, at long last — the handle had to be unscrewed and the three holes sealed with little tile plugs.

All went well with two of the legs, but at 9:10 frustrated technicians had to report that the third bolt just spun around without catching and could not be pulled out of the hull. The thread was obviously stripped. Thus the auxiliary handle could not be removed and a launch costing somewhere between $200 and $300 million was brought to a hopeless standstill by the ruined thread on a bolt costing a few dollars. The technicians requested permission to drill the head off the captive fastener with a battery-driven drill. Ordinary electric drills were too dangerous to use so close to so much explosive. After some consultation, permission was given, but it took almost forty-five minutes for the Black & Decker to reach the tower.

The battery for the drill had been lying in a toolbox outside in the cold and could barely turn the drill bit. It took another thirty minutes for nine more batteries to be sent up. Eight of these were also flat; only one worked, but it was not powerful enough to drill out the fastener. The technicians then requested permission to use standard AC instead. Management — now growing exasperated — waived the prohibition on using electricity, but the drill bit was not long enough and the machine could not get a decent purchase on the alloy of the screw. Finally, permission was requested to use a hacksaw to cut through the leg and the screw. Another wait. NASA managers were not happy about having a hacksaw anywhere near the brittle tiles but there did not seem to be any other way. The crew had now been lying on their backs in an exceedingly awkward position for over three hours, while the technicians struggled with the door and dealt with certain other unresolved problems. But this problem was solved by the hacksaw, and at 11:30 — after wrestling with the hatch handle for almost two and a half hours — relieved technicians were able to give the all clear.[12]

While these problems were being overcome, the wind had risen. Wind in itself was not a major stumbling block, but the wind direction was not in their favor. One of the most important links in the space shuttle emergency procedures involved an option known as RTLS (return to launch site). Should a problem arise with the main engines, the shuttle ought to be able to turn around and be brought back in to land on the Kennedy Space Center runway. The computers were furnished with a program whereby the shuttle freed itself from its fuel tank, turned around, and glided back to base. This landing had never been put to the test, but NASA was sure it could be done. However, should problems arise during the first two minutes of flight, while the boosters were still operating, there was no escape.

A RTLS landing was dependent, however, on the crosswinds over the runway staying below 23 feet per second. And at this point the wind speed across the runway was being gauged at between 35 and 43 feet per second. Under these conditions *Challenger* would have no chance of landing safely. And so, at 12:35 p.m., the launch was canceled. By that time, McAuliffe and the others had been lying on their backs, unable to move, for almost four hours. The technicians could now get ready to empty the fuel tank once more. The postponement of the launch — this "scrub" — would inevitably put more pressure on a team already working under great strain. Not only that, but it would add another $300,000 to the debit side of the budget.

Christa McAuliffe's parents had been waiting patiently throughout those same four hours. Her father found it hard to conceal a touch of annoyance, and told those around him that he, for one, could have got hold of that hacksaw a lot faster than NASA.[13] Other spectators wondered how any foreman could countenance a workman turning up for work with a flat battery. Meanwhile, the astronauts' spouses tried to explain to their disappointed children that the space shuttle would probably take off tomorrow.

That evening, Christa McAuliffe spoke on the telephone to a woman friend who asked her how it felt to lie there, waiting, inside the space shuttle. The teacher replied:

> Go borrow a motorcycle helmet. Lie on the floor with your legs up on the bed. Lie there for five hours. You can't read, you can't have anything loose around you. You're strapped down real tightly with oxygen lines and wires coming out of your suit. The only people who have anything to do are the pilot and the copilot.[14]

The weather forecast for Tuesday, January 28, said that the wind would die down and the temperature would drop as that icy ridge of high pressure gradually advanced. The meteorologists were predicting eleven hours of frosty weather, with temperatures dropping to 18° F just before dawn. After that the mercury would slowly start to rise again. According to the forecasts, however, the temperature would not reach the prescribed 31° F until somewhere around noon. "It's going to be close," the Director of Shuttle Operations, Bob Sieck, told *The New York Times*.[15]

30. Liftoff

They don't have much use for antifreeze in Florida, nor for draining water pipes — simply because they don't get much in the way of frost. So NASA hadn't drained the miles of water piping on its launch pads either. Regulations stipulated that in the highly unlikely event of freezing conditions, the entire water system had to be drained. But the regulations were of no help in this situation. Draining the water pipes would also eliminate any hope of a launch on Tuesday. NASA couldn't just close down for the winter right in the thick of things.

It was therefore decided that all of the faucets and showers around the pad, and on all the work platforms on the tower, be left to run at a trickle, to prevent the water in the pipes from freezing. Thousands of liters of antifreeze were poured into the big troughs underneath the two solid-fuel rocket boosters, to keep them free of ice.[1] The purpose of these troughs — along with the veritable deluge of water in which the whole platform was drenched — was to moderate the blast when the main rocket engines were ignited.

While the *Challenger* crew were settling down for the night, a host of technicians was preparing to refill the fuel tanks with liquid fuel, the temperature of which made the icy wind on the launch pad seem like a warm summer breeze. But the cold also had an effect on the refueling process. At the top of the nose cone on the big fuel tank was a vent, designed to permit liquid oxygen to escape as vapor when the topmost layers began to boil. The small amount of oxygen lost would never be missed and, in any case, the fuel lines kept topping up until just before liftoff. But in order to function properly, this vent must never be allowed to become too cold. Even on the hottest summer day it would become chilled due to the temperature of the fuel. The NASA manual stipulated a minimum temperature of 45° F and stressed that there should be "no visible ice buildup on the nosecap fairing exit area."

To keep within this limit, the vent was fitted with electric heating elements. But the heating elements were not equal to this unexpected cold snap, even at the highest setting. The previous day, the technicians on the launch pad had reported that the 45° F limit had been exceeded. Mission control had given the matter some thought and then issued a formal waiver, extending the limit to 28° F. With this document signed and sealed, the formalities had been observed and the preparations could proceed.

Late on Tuesday night this limit, too, was exceeded, when the temperature dropped to 24° F. The ice-cold liquid oxygen was on the verge of winning the battle against the heating elements. The gases in the vent were now down to 12° F. Once more the technicians reported back and requested that an official decision be made. The temperature was now 33° F below the limits specified in the manual. A new waiver would have to be issued if the rules were to be observed. Alternatively, the refueling would have to be called off and the launch postponed. In a tape transcript published later, NASA managers can be heard making their evaluation.

THOMAS: *Okay. The only outstanding item we have right now is the one on the waiver on the cone temps.*
LAMBERTH: *Okay. It looks like we probably could say about 10 degrees and be okay on that one.*
THOMAS: *Okay. We'll use 10 degrees then.*

The waiver was issued. The nose cone vent was now sanctioned down to 10° F. It could hardly get any colder, so the launch would still fall within the set limits. Although, admittedly — owing to "production deadlines" — these had gradually been tailored to fit the circumstances.[2]

In the middle of the night, while the crew were sleeping, a special ice team inspected the platform and the tower. Under normal circumstances this team were also supposed to report any buildup of ice from the cooling down of the fuel tank exterior. It would do the Orbiter tiles no good to come into contact with chunks of ice from the tank; and the rocket engines could be damaged if they happened to take in any ice with the powerful suction occasioned by their ignition.

This time the ice team really had their work cut out for them. The sight that met them when they climbed out of their pickup truck prompted one of them to mutter something about *Dr. Zhivago* into his walkie-talkie. The drains under several of the fire hydrants on the tower had frozen over, with the result that water had washed across the various platforms and steel stairways and poured down over the girders, to solidify into frozen waterfalls and stalactite formations of icicles. The same applied to the showers situated at strategic spots so that technicians who came into contact with toxic fuel substances need waste no time in rinsing their hands, faces, and eyes. Both the west and the north face of the tower were completely covered in ice, from the topmost tap, at 235 feet, downward.[3] The ice team found some icicles which were as much as eighteen inches in length.

The crew access arm which the astronauts were to use, if they had to get out fast, was covered with a layer of ice several inches thick. Seven panic-stricken astronauts were supposed to run across this glassy surface to reach a number of specially designed baskets which, if the system worked, would then slide down steel cables to deliver them at top speed to a special armored personnel carrier which would carry them to the safety of a nearby bunker.[4]

The leader of the ice team, Charles Stevenson, reported what he had seen. He also photographed the ice at selected spots, to enable the mission management team to evaluate the situation later, and help them to make better arrangements another time. But it was not up to him. Again it was the mission management team, in the persons of Gene Thomas and Horace Lamberth, who had to make the decisions.

LAMBERTH:	*. . . Charlie's worried about it, Gene — the acoustics re-leasing it and it being free when the Orbiter comes by.*
THOMAS:	*Boy, he's really stretching it.*
LAMBERTH:	*Oh no, I don't know whether that's stretching it too much or not.*
THOMAS:	*Well, I mean if we can ignore it, we need to feel comfortable about it . . . We need to all know if we don't get back into tanking as soon as possible we could possibly blow it just for that.*
LAMBERTH:	*Yeah, we understand, Gene.*[5]

Several times during the night, Launch Director Thomas's subordinates reminded the technicians that a tight schedule was on the line here. Thomas was stressed out and impatient. He was continually having to commit himself, making crucial decisions. The night before, the television news had been poking fun at NASA, showing pictures of the technicians struggling with the hacksaw in the tower. The anchorman,

Peter Jennings, opened his program with a bit of sarcasm: "Once again a flawless liftoff proved to be too much of a challenge for *Challenger*."

By now, the problems with the ice—together with a technical fault in one of the liquid oxygen storage tanks—had extended the deadline by almost two hours. Although one hour could be won back by curtailing a scheduled pause in the countdown.

The ice team were asked to check on the situation again just over an hour before the crew were due to board the Orbiter. The news, this time, was even worse than in the middle of the night. Using a fishing net, the team had hauled large ice floes out of the troughs under the Orbiter. Mission management decided to postpone the launch for another hour, to give the ice on the tower time to melt. There would be no launch, as planned, at 9:38 a.m.; with any luck it would be staged at 11:38. The ice team were to make one final inspection twenty minutes before liftoff.[6]

Because of the delays incurred during the night, the crew were roused an hour later than planned on the morning of Tuesday, January 28, 1986. But by then they were all already up. This gave them a bit more time for a last call to their spouses and for going over their personal checklists.

Again they waved to the representatives of the press as they walked out to the NASA bus in the chill morning air. No one could say for sure whether they would get away this time. Everyone remembered how often Bill Nelson and the other astronauts had to go through the smiling and waving routine on the preceding mission. Each time, only to have to clamber down from the tower again and drive back for another night at Cape Kennedy. Press photographers remembered just how often they had had to scrap their pictures; and yet, every time, they too got caught up in the mood of the moment and snapped away with renewed enthusiasm. Astronauts were still something special.

At 8:03, the crew arrived at the tower and were met by the men who were to help them into their seats in the upended Orbiter. As a matter of course, the mission commander, forty-seven-year-old Francis "Dick" Scobee, was handed in first, to the front left-hand seat—which is always reserved, in any aircraft, for the captain. This mission marked the pinnacle of Scobee's career. He had started out as an airman in the Air Force, had trained as an aircraft mechanic, and then fought hard to become a pilot. After more than 6,000 flying hours in forty-five different types of aircraft, as well as a tour in Vietnam, he had become an astronaut. And now he was about to make his second flight into space—this time as skipper.

Then it was the turn of the copilot, forty-year-old Michael Smith, to be eased in next to Scobee. This was Smith's first time, and he was clearly overawed by the situation. The last thing he wanted was to make a mistake. All those hours of training in the simulator could not be allowed to go to waste. Like the others, he was settled in by engineer Manley "Sonny" Carter, who had crawled into the Orbiter with the crew. Besides helping them into their seats Carter also had to make sure that the radio link to the astronauts' lightweight helmets was working properly. Astronauts no longer wore space suits during launches, but the helmets had been retained to protect their heads from the vibrations caused by the fierce acceleration, and as a means of providing them with oxygen in an emergency.

SMITH:	*Wow! Boy! The sun feels good this morning.*
CARTER:	*You should have been here at two a.m.*
SCOBEE:	*Ice skating on the MLP [mobile launch platform]? You guys up here working?*
CARTER:	*It's a lot of fun.*[7]

A couple of days earlier, Mike Smith had told a journalist how fed up he was with people asking whether the space shuttle wasn't dangerous. No, it wasn't, he said. The space shuttle system was just great; it was a program of which the whole country ought to be proud.[8]

After Mike Smith, Sonny Carter helped Ellison Onizuka to climb aboard. He too would be sitting on the top deck, like the two pilots — though in a seat directly behind them. One of his tasks would be to film Halley's comet during the flight. Onizuka had been born in Hawaii, the grandchild of Japanese immigrants. This group of Americans had waited a long time for NASA to give them their chance.

CARTER:	*You know, I think these visors are cold from being outside and are fogging up. Here's the pocket checklist.*
ONIZUKA:	*Okay, thank you. See you later on. Kind of cold, this morning.*
SMITH:	*Up here, Ellison, the sun's shining in. At least we've got the crew arranged right for people who like the warm and the cool.*[9]

Then it was time to bring in the female members of the crew, Judith Resnik and Christa McAuliffe. Resnik was a bona fide astronaut, trained as an electrical engineer. She had been into space before and knew what to expect. The tension made her overly cheerful and perky. Before clambering up beside the two pilots and Onizuka she nudged McAuliffe — her senior by a year — who would be seated on the lower deck, and said, "Next time I see you, we'll be in space."[10]

ONIZUKA:	*My nose is freezing.*
CARTER:	*Good morning, Judy.*
RESNIK:	*Cowabunga!*
CARTER:	*Heyyy!*
SCOBEE:	*Loud and clear, there, Judy.*[11]

Christa McAuliffe was neither perky nor especially cheerful. She waited her turn calmly and rather pensively, and was glad of some help from Sonny Carter. It was also obvious that he took extra trouble with her. This ordinary American citizen was tucked in and fussed over, for all the world as though Carter were her mother and she was a three-year-old kid again.

CARTER:	*Okay, real good. Put your other arm through here and I'll hold it for you. Okay. Talk to the OTC on that button.*
MCAULIFFE:	*OTC, PS-1.*
OTC:	*Loud and clear.*
MCAULIFFE:	*Good morning — I hope so too.*
CARTER:	*Now, just put your visor down on the right.*
MCAULIFFE:	*It's down.*
CARTER:	*Now tighten your helmet a little bit on the back. Make sure it's snug but not too tight, though, and then push this button right here and tell the LTD com check.*
MCAULIFFE:	*LTD, PS-1. Com check.*
JENICEK:	*Have you loud and clear.*
MCAULIFFE:	*Good morning.*
CARTER:	*Okay. Now raise your visor with a little push with your right hand. Right hand here. It'll fog a little; those things are cold, and we're a little warm. Now we'll put that beauty [emergency air pack] where it feels most comfortable for you, about there [beside the seat]. Feel where that is now. Okay? Okay you're ready. Doin' good. Watch your arm there, Christa. There you go. Christa, while I hook up Greg [Jarvis], you're gonna lose com for a while. Good morning, Greg.*[12]

Greg Jarvis was installed in the seat next to McAuliffe. Neither of these two would be able to look out during the launch.

The middle deck had only one small window, in the hatch through which they had entered, and this window was now behind them. Jarvis was a satellite expert from Hughes Aircraft. It was he who would be carrying out the experiments with liquids in weightlessness. He had originally been scheduled to go up with *Discovery* in April 1985, but his place had been taken by Jake Garn, the politician, at the last minute. Amazingly, the same thing had happened to him again, when he was ousted from *Columbia*'s New Year's flight by Bill Nelson. Evidently, politicians ranked higher as passengers than Payload Specialist 2 Greg Jarvis. PS-2 was now strapped into a recumbent position beside PS-1. Unlike his crewmate, he was quite at home with the technical jargon on board and was treated accordingly.

JARVIS: *OTC, PS-2 radio check.*
OTC: *Copy.*

Astronaut Ronald McNair was installed just inside the hatchway. Carter instructed him to keep an eye on the hatch locking mechanism again. There had been no time to replace the faulty warning light switch. McNair was black and had often been held up as an example of how African-Americans could make it to the top, if they really tried. Like Scobee, Resnik, and Onizuka, McNair had had a shot at flying with the space shuttle before. He ran through his routines quietly and professionally. Friends had heard him say that pretty soon he was going to put the astronaut life behind him and find something else to do that would give him more time with his two small children.[13]

Sonny Carter waved to the three on the middle deck and edged his way out. This time, the hatch closed as it was supposed to and the hatch handle unscrewed without a hitch. It was now 8:36 and everything was still going according to plan.

For the next three hours, the crew passed the time with small talk. They had hoped to be given the signal for liftoff at 10:38 but the countdown was delayed for an extra hour because of the ice. There was not a great deal they could do. Most of the hustle and bustle was taking place outside the Orbiter — at the Kennedy Space Center and in the control center at Houston, which would assume control of the Orbiter once it was free of the tower. Computers at both these places were checking thousands upon thousands of different parameters—all of which had to be acceptable if the Orbiter was to take off on time. Right now, the crew's main job was just to be there, in the cabin. Only as the time for liftoff approached would the professional astronauts again have to be on their toes. During this phase, the two civilians lying amidships would be relegated to the role of spectators.

Everything the astronauts said to one another was transmitted to mission control and taped. They knew this, and were used to it after all their hours in the simulator. But they would be tipped off when the moment came, just before liftoff, when the American public would also be able to listen in.

Judith Resnik was particularly chatty and high-spirited. When Smith complained about being in such an awkward position, saying he already felt as though he had been lying there for four hours, as they had done the day before, Resnik piped up:

RESNIK: *I feel like I'm past it. My butt is dead already . . . Okay,*
               Ellison, get out of there.
               [Laughter]
ONIZUKA: *That's too low.*
               [Laughter]
SMITH: *Crew [garble] . . . crew gynecologist.*
               [Laughter][14]

Variations on this theme recurred several times. When Judith Resnik again returned to the subject of her numb rear end, Jarvis suggested that he and Onizuka could arrange for a bit of massage. Resnik said she was disappointed in her neighbor Onizuka—who did not look as though the idea appealed to him.

Other than that, the main topic of conversation was the weather. At one point it looked as though it was snowing, when some hoarfrost blew off the fuel tank and was swept past the cockpit windows. Onizuka also remarked on the dripping faucets in the tower.

ONIZUKA: *Where are they getting all that water?*
RESNIK: *Your tax dollars.*
SCOBEE: *Yeah, that's probably special grade water.*

Resnik went on talking, with Ellison Onizuka putting in a word here and there. Scobee's interjections struck a slightly more authoritative note. After all, he was the commander and clearly respected as such. The copilot, Mike Smith, gradually relaxed and joined in the conversation. Greg Jarvis was also delivering a running commentary from between decks. Ron McNair lay quietly looking out of his little round window.

The tape did not pick up much from Christa McAuliffe for almost two hours. She was apparently collecting her thoughts for the terrifying experience that was now just minutes away. Shortly before the flight she had taken out a life insurance policy for a million dollars and she had not been trained, as the early astronauts had been, in desensitizing.

Just once, she seems to have felt that she had better say something:

MCAULIFFE: *It'll be cold out there today.*

Her mother and father were in the spectators' stand, well wrapped against the cold and patiently tolerant of the press photographers moving in closer and closer, to capture their expressions as their daughter soared heavenward in a cloud of smoke and steam.

In the administration building, the astronauts' immediate families, around twenty adults and children, left the televisions and made their way up onto the roof, to watch the launch from there. Because of the cold, they had waited until they were sure that the countdown would proceed. The children ranged in age from Scobee's twenty-one-year-old son, Richard, who would soon be graduating from the Air Force Academy, to McNair's eighteen-month-old daughter, Joy. McAuliffe's children — nine-year-old Scott and six-year-old Caroline — were there with their father. Scott had brought the whole of his third-grade class to Cape Kennedy, but his schoolmates were out in the spectators' stand. He had given his mother his favorite cuddly toy, Fleegle the frog, to take with her on board.[15]

On the launch pad, the ice team were now finished their final inspection. Again, large chunks of ice had been fished out of the troughs and a layer of ice cleared from the launch pad itself. Over their walkie-talkies, Stevenson and his team reported the continued presence of a lot of ice on the shady side of the stack, although this was starting to melt in the sunshine. Mission management thanked him for this information and asked him and his team to clear the area.[16]

The doors into the "firing room" were closed. Twenty minutes to go and armed security guards would now prevent anyone from distracting the decision makers and the engineers. Although by now the computers were taking over more and more of the decisions. The mission management team too were rapidly being relegated to spectator status.

The radio lines crackled with technical jargon on activating the APUs (auxiliary power units), fuel line pressure, and the imminent ignition of SSMEs 1, 2, and 3. Just before liftoff—in just fifteen seconds—300,000 gallons of water was shed across the launch pad from pipes so huge that there was no way they could freeze up.

T minus 7 seconds. Immediately thereafter, all three main engines were activated. The computers measured their performance. They could still be stopped. Because of the stack's asymmetrical construction, a force of more than a million pounds tipped the fuel tank over to one side; this pushed the pillars of the solid-fuel rocket boosters three feet off center —though they were still held in the iron grip of the mighty detonator bolts at their foot. The joints between the various cylindrical segments gave a little under the pressure.

Not until these giant rockets were also ignited at 11:38:00:010 a.m. EST did the bolts release their grip and the spacecraft shot into the air through a flurry of ice raining down over the iron girders of the tower. The spectators cheered. For a moment the crew forgot their strict professional code and they too cheered. Any astronaut who—like Scobee, Resnik, and McNair—had once experienced that tremendous acceleration was never done talking about it and never stopped longing to experience it just once more. And now here they were again, at long last. As a rookie, Christa McAuliffe is bound to have experienced it differently.

All at once, the deeply arched boosters were no longer bolted down at the bottom. The energy was therefore released in a predetermined rebound, and both the nose and tail sections juddered and sprang around the midsection. A certain amount of flexibility was allowed in the booster joints for this oscillation which continued until the rocket was totally clear of the tower. The columns of smoke stood out beautifully against the ice-blue Florida sky. It was the coldest launch in the history of American manned spaceflight.

Flight control was officially passed to Houston. But Cape Kennedy was, of course, still keeping track. In Dakar, in Senegal, the lights were lit on the emergency landing strip. If the main engines should break down, the Orbiter would be able to land there after approximately eighteen minutes of flight. Wind conditions at the space center were perfect. An RTLS landing should not present any problems either, if anything went awry early on in the flight.

The computers were churning out a continuous stream of data. Some sensors were reporting in several times per second; others at long intervals. On the control center VDUs things were looking good. All the vital readings were A-OK.

Like a big whale, the Orbiter rolled lazily onto its back, now clearly heading east. The engine output decreased exactly as it was supposed to do. The solid fuel in the boosters was also molded in such a way that their thrust would decrease correspondingly at this point. Here was where the Orbiter encountered the greatest atmospheric effect — with its speed increasing but the atmosphere still not thin enough to be discounted. There was quite a bit of turbulence at this level, but the computers were voting on what to do about it at lightning speed — scores of votes per second — and passing corrected steering coordinates to the five thrusters, in order to stay on course. The engines were now back up to full power.

Everyone was starting to look forward to the moment — less than a minute away — when the two solid-fuel rocket boosters would be jettisoned, having served their purpose. Then, before any VDU had a chance to register anything abnormal, the long-range cameras caught it — a distant flash of light. The displays on the big screens were reading T+73 seconds.

Over the radio link came a sudden "Uh-oh!" from the copilot. After that, no downlink, nothing but hushed static in the ether.

After seventy-three seconds, *Challenger* exploded in front of husbands and wives, children and grandparents. Many were still giving an optimistic thumbs-up as they started to scream. The press photographers took pictures of Christa McAuliffe's parents as the fact of what had happened slowly dawned on them, and excitement gave way to incomprehension. "The craft has exploded," a NASA official explained softly to McAuliffe's parents.[17] "The craft has exploded," repeated Christa's mother, as if in a daze; incapable of accepting what she was now gradually grasping. Her husband's face turned ashen. Pinned to his chest he had a big badge bearing a smiling portrait of his daughter, Christa, who was now gone, along with *Challenger*.

Only on the radar screens could *Challenger* still be seen, but what had been one object was now several.

The stray boosters were exploded in midair by a radio signal from the Air Force range safety officer. One of the parachutes suddenly opened, quite unprompted, and drifted slowly down over the sea, secured to its rocket nose cone. For a brief instant hope blossomed. Maybe the crew had been saved by some abort system after all; but no abort system existed to cover this situation, and, hence, no escape.

Others hoped against hope that Scobee's voice would come over the radio, apologizing for the break in radio contact; and that the space shuttle would, soon afterward, come zooming in across Kennedy and execute one of those breathtaking 180 degree turns before making a perfect emergency landing on its runway. But, of course, that did not happen either. There was no hope of any RTLS. No return to launch site.

The Cape Kennedy range safety officer issued an official statement confirming the explosion. In the office buildings of the space center, employees and secretaries were now openly weeping.

The stunned relatives were assembled in the astronauts' building, which the crew had left just over three hours earlier. The bereaved were served coffee and doughnuts as they wept and tried to make sense of it all.

Late that evening, a plane arrived carrying Vice President George Bush and Senators Jake Garn and John Glenn. On this occasion too, Glenn was the most fortunate in his choice of words. He explained how the astronauts felt that death could be an acceptable price to pay for exploring new frontiers. The first seven astronauts had often wondered how many of them would survive. John Glenn also described how he had felt pain akin to theirs when three astronauts to whom he was very close burned to death in their Apollo capsule nineteen years before. Among them, one of the first seven, Gus Grissom. During the Apollo 13 space trials too, he had learned to accept that space would claim its victims. It was just a wonder that it had not happened earlier. Space travel was a risky business but the astronauts knew the risks and they accepted them.[18]

By virtue of his charisma and the weight lent by his own experiences in space, he gave the bereaved, if not comfort, then at least a breathing space in which to come to terms with what had happened. Their own instincts had probably told them that this was a highly dangerous exercise, but everything they had heard had led them to believe that the space shuttle flew as regularly and as reliably as a passenger airplane. This fact alone had to be sufficient proof—for anyone who bothered to give it a moment's thought—that the system must be 100 percent sound.

31. The Morning After

The following day, with every newspaper filled with reports of the disaster, the speculation began. The President had risen above all the different theories in his televised tribute the previous evening. He had concentrated simply on according the tragedy some symbolic meaning to unite the nation in its grief.

> The future doesn't belong to the fainthearted; it belongs to the brave. The *Challenger* crew were pulling us into the future, and we will continue to follow them.[1]

The disaster led to a sharp increase in national interest in space exploration. Even though everyone had been presented with unmistakable proof of the hazards involved, a *Newsweek* survey showed that 55 percent of the population believed civilian exploration of space was important, regardless of the risk. On the question whether time constraints could have been responsible for a reduction in NASA's safety standards, 48 percent gave a definite "no" — although 44 percent believed the opposite. A large majority — 67 percent — wanted to keep manned missions, and 76 percent were in favor of unaltered or increased grants to NASA, as compared with 69 percent two years previously.[2]

Jay Shaeffer, a high school teacher from Los Angeles, had been one of the other finalists for TISP. The morning after, a number of his students had come over and put their arms around him. It could have been him. Yet even he was not really daunted by what had happened. He told *Time* magazine: "I would go today, right now, I wouldn't even go home to change." He had been heartened by his students' expressions of sympathy, regarding them as a positive sign. He understood why they were so deeply shocked by the disaster. To the students, he explained, "A teacher in space becomes their teacher. Do you know an astronaut? Everyone knows a teacher."[3]

During those first shocked days, NASA was overwhelmed by the nation's almost unqualified sympathy. The nation's heart went out to the stunned engineers and managers interviewed on television as to the possible causes of the disaster. All NASA employees were offered special counseling. Again and again NASA personnel reiterated that, for the moment, they could give no explanation for the disaster. Jesse Moore, Associate Administrator for Space Flight, who had ultimate responsibility for the launch, said that all the relevant data would have to be assembled and subjected to careful analysis before he could make any comment.[4] But what he could state, categorically, was that there had been no sign of anything out of the ordinary prior to the explosion itself. No alarm bells had sounded inside the Orbiter before that flash in which the crew disappeared. At least they could take comfort from the thought that it had all been over in an instant.

The technicians stated that nothing could have been done, either in the cockpit or in the control center, to avert the disaster, because it had occurred before the solid-fuel rocket boosters had burnt out. Only fifty seconds later and the astronauts might have had some chance of disengaging themselves from the fuel tank and making a landing at the Kennedy base. A little later, and they would have had the possibility of an emergency landing in Dakar, in Africa. And finally, if the malfunctions had held off for a while longer, there would have been the chance of an emergency landing — after just under one orbit — in California.

Briefly worded press releases from NASA could not prevent the media from indulging in speculation as to the possible causes. And it was not very hard to find sources in aerospace circles and among former NASA employees who were able to point out the failings in the space shuttle system.

One prevalent theory had it that one of the boosters must have burned through its steel casing at a weak spot, perhaps due to an absence of insulation. According to some sources, NASA had shaved a fraction of an inch off the solid rocket boosters' steel casings to save weight.[5] If the booster casing had been broached, flames would have hit the fuel tank with the force of a blast from a blowtorch. This would have resulted, almost immediately, in an explosion as inevitable as that witnessed by television viewers across the country.

Another theory suggested that a radio signal might accidentally have triggered off explosives on the fuel tanks, causing the spacecraft to explode. It was not exactly common knowledge, but all rockets from Cape Kennedy — whether manned or unmanned — were wired with explosives so that they could be blown up by remote control if they should happen to pose a threat to residential areas. Not even the urging of the astronauts themselves could persuade the military safety officers to make an exception in the space shuttle's case. And it was these very explosives, triggered by a radio signal, which had disposed of the two runaway boosters after the explosion — thereby possibly destroying vital evidence.

But what if a radio signal had been transmitted by mistake? Or, even worse, if some sick individual had gotten hold of the secret frequency and pressed a button? Even sabotage on the part of the enemy could not be ruled out.

The Kennedy Space Center dismissed this theory as highly improbable. The frequencies were kept secret and changed regularly. Besides, the signal was in a code that would be almost impossible to crack.[6]

Other sources singled out the main engines as being the Orbiter's Achilles' heel. Time and again, cracked turbine blades had been discovered after use. If such a blade broke off completely during flight it would slice through the fuel lines, the pumps, and the fuel tank's thin wall with the force of a piece of shrapnel. An explosion would be inevitable. Just as it would if one of the hefty turbopumps in the engines exploded, as had previously happened during the development phase.

It was also conceivable that a crack might have occurred in one of the fuel lines when they were subjected to that tremendous atmospheric effect immediately prior to the explosion.[7]

Amid great misgivings and a fair amount of beating about the bush, the one theory that is always aired after a major systems malfunction was trotted out once again. It might — with all due respect — be a case of "operator error." After all, the space shuttle's systems were very complex. What if the pilot was a bit premature in pressing the button which was to disengage the Orbiter from the fuel tank seven minutes later? In that case, the detonator bolts would have gone off, the fuel lines would have been cut and — it went without saying — ignited by these small explosions. Ergo: The explosion took place because the astronauts themselves made a mistake.

An offshoot of this same theory said that personnel at Kennedy had happened to damage one of the attachment struts between the Orbiter and the fuel tank.

Experts rejected both theories as being highly unlikely. There were no indications that anything of the sort had happened. And the accident on the launch pad had been a minor one, involving just the equipment on the tower. But some people insisted. Everyone knew that even astronauts could slip up under pressure. And instances of shoddy workmanship were not unknown even at Kennedy.[8]

Toward the end of the week, however, suspicions were beginning to center on the solid rocket boosters. NASA released some film footage showing flames flickering between the right-hand booster and the fuel tank in the seconds before the explosion. These pictures were supplemented by data which, on closer inspection, showed a sudden drop in pressure in the same booster ten seconds before *Challenger's* demise. It had lost 4 percent of its thrust and the left-hand thruster had gimballed to compensate and maintain course. The drop in pressure might have been due to a leak, allowing the gases to escape from somewhere other than through the slender thruster at its foot. This leak might, for instance, have occurred in one of the solid rocket booster joints if a joint seal had been seriously compromised.

This theory too was rejected by some experts, who maintained that any rocket booster would somersault wildly if it had more than one "hole" in it. But after the explosion, these boosters had continued to fly perfectly. That alone was enough to rule out this theory. Others said that the booster's own guidance system could easily have coped with losing, through one extra hole, the equivalent of less than 4 percent of its total thrust.[9]

Many observers found it shocking that they were only now learning just how much could have gone wrong. More than 700 individual components in the Orbiter system were designated Criticality 1 — that is, if that particular component failed to function, the Orbiter and its crew were as good as lost.[10] In 700 instances no backup system existed. Now, one could understand if the wings, for example, and the nose wheel were not covered by the fail-safe measures — but 700 parts? That was a bit steep. "They wouldn't let an experimental fighter plane fly with so many single-point failure nodes," one old hand in the aerospace business told *Newsweek*.[11]

The spotlight was also turned on the project's finances. *Newsweek* noted that NASA had quoted a figure of only $38 million for a full hold to attract customers when, in actual fact, the cost of a launch in 1985 was estimated at between $250 and $300 million. And in 1986, at the insistence of Congress, this figure would be increased to $71 million. Even so, they were still a long way from the point where the finances would come anywhere near to tallying. In an interview with *Newsweek*, Roger Knoll, an economist from Stanford University, said:

> It was a commercial flop—it cost too much and it performed too poorly . . . If NASA had been a bottom-line, rapacious capitalist company, it would have pulled the plug. If it had been a sensible government body, they would have built one [shuttle] for experimental purposes, and reallocated the rest of the money for expendable rockets.[12]

From other quarters too came words of warning against building a new shuttle to replace *Challenger* without giving the matter serious thought. Perhaps the space shuttle was a piece of obsolete, early 1970s machinery which ought to be replaced by other, quite different and more up-to-date devices.

32. Commissions

NASA promptly embarked upon the task of collecting any and all material which might shed some light on the cause of the accident. Every data base was sealed up. Film from all the cameras dotted around the base was brought in, and journalists, press agencies, and television networks were obliged to hand over anything of relevance to the investigation. The smallest scrap of paper containing notes on the launch was confiscated.

Whereas after a plane crash a search has to be made for the aircraft's "black box," in this case they already had the flight recorder at the space center, even though the Orbiter had not, as yet, been located. The Orbiter had been sending information back to base at the rate of several thousand reports per second. Altogether, NASA had access to millions of readings taken during those fatal seventy-three seconds. The clue to the disaster must lie hidden somewhere in this vast store of data.

During a launch, the operators at the VDUs in the control room took note only of data directly relevant to the operation which could, if necessary, be influenced. So, in fact, their data represented only a very small sample of all the data available. And in this sample, as we have seen, no one had noticed anything out of the ordinary. Now, rather than making do with the salient contours, the picture as a whole would need to be thrown into relief.

Off the east coast of Florida, the biggest marine salvage operation in history got underway. Ships from NASA, the Coast Guard, and the Navy worked alongside a number of chartered salvage vessels to comb the ocean floor for evidence and for the crew.

Analyzing this wealth of material would be a job for the most highly skilled specialists at NASA and the aerospace companies' joint command. And they would need time to reflect and to set up simulations of possible accident sequences or scenarios. Jesse Moore, the Associate Administrator for Space Flight, immediately set up a commission to take charge of the numerous investigations. This commission consisted of directors from the main NASA centers — Johnson in Houston, Marshall in Huntsville, and Kennedy in Florida — along with senior personnel from headquarters in Washington. Robert Crippen, who had made four trips with the shuttle, and Robert Overmyer, with two trips to his credit, were also assigned to this commission.[1]

True to form, the press was not content to hang around waiting for the results of any long-drawn-out investigation. Suddenly, a more inquisitorial undertone was creeping into NASA press conferences. On the very day of the accident, Jesse Moore was asked whether there had been "any unusual weather conditions aloft, or any unusual weather conditions during the launch."

MOORE: *None that I recall. We did put up some weather balloons this morning. We did look at load conditions as we normally do, and winds aloft looked good. We didn't have any excedences as far as load indicators are concerned, to my knowledge, and we thought everything was in good shape for a launch this morning.*[2]

Moore did not mention that never before had a launch been staged at such low temperatures.

Journalists also wanted to know whether he had felt pressured to launch *Challenger.*

MOORE: *There was absolutely no pressure to get this particular launch off. All of the people involved in the program to my knowledge felt that* Challenger *was quite ready to go. And I made the decision along with the recommendations from the teams supporting me that we launch.*[3]

The next day, several journalists heard a rumor that North American Rockwell, the chief Orbiter contractor, had advised against the launch because of the icy conditions on the launch pad. Apparently they had been particularly concerned that icicles might smash against the Orbiter hull and damage vital tiles. NASA's Administrator, William Graham, who was very new to the post, passed the microphone to Jesse Moore.

MOORE: *There were a series of technical meetings yesterday morning about the ice on the launch pad. The ice team went out and did an inspection early in the morning. And then came back and reported. And the technical people did sit down—all the NASA people involved as well as the contract people involved—and did feel that the conditions at the launch pad were acceptable for launch and basically recommended that, you know, we launch.*[4]

This answer did not satisfy the journalists. Had Rockwell been ignored, misunderstood, or overruled? Or had the company not actually been concerned at all? The NASA representatives on the podium moved on to the next question.

Later that afternoon, the press had a meeting with Jay Greene, who had been Flight Director at Mission Control in Houston during the launch. Greene was to have taken over control of *Challenger* from the people at Kennedy once the Orbiter was safely free of the tower. (There had never been any practical justification for decreeing such a division of labor; it was more of a political legacy from that wily Texan President, Lyndon B. Johnson.)

The ladies and gentlemen of the press were now starting to pose questions worthy of the lawyers on television courtroom dramas.

You said earlier that the weather was good, it was acceptable, but it was pretty cold. When does the cold play a factor in launching the shuttle and what specific concerns do you have about the cold? I understand that there is some concern about ice forming on the [external] tank that could possibly nick the tiles. What are the other concerns, and what specifically do you look for to make sure there will not be a problem in the cold?[5]

Jay Greene said that this question would have to be answered at a later date by the people in Florida. He had only been involved in that section of the flight from the tower upward. And in this case, "the people in Florida" meant Jesse Moore, who had not been especially forthcoming.

It gradually became apparent that the people from NASA were playing a dual role. In a sense, the agency was investigating itself. Official statements were extremely defensive, and after a while some journalists started talking of a cover-up.

They were not, however, dependent solely on official statements. Unofficial contacts could be much more outspoken, as long as they could remain anonymous. Word got out that, among NASA engineers, one of the field joints on the right-hand solid rocket booster was gradually beginning to emerge as the chief suspect. Careful studies of photographs taken by remote-control cameras just after liftoff revealed a black cloud of smoke billowing out of the field joint, betraying the fact that it was not airtight. Puffs of smoke continued to issue from the joint for as long as the camera was able to follow its course. Eventually, however, it became impossible to distinguish between it and the exhaust. On Friday, a hard-pressed NASA released these pictures to the media. As yet, however, they did not wish to comment upon them.

That same day, Nancy and Ronald Reagan attended a memorial service at the Johnson Space Center in Houston. The President and the First Lady sat between June Scobee and Jane Smith, the widows of the two *Challenger* pilots. The other astronauts' spouses and children were also present, together with other relatives. Altogether, 10,000 people — whether directly or indirectly involved in the space shuttle program — took part in the service.

Reagan knew that the American people had been united by *Challenger*, and he was not content to leave the investigation of the disaster to NASA alone. He was even less keen to have Congress getting involved in the affair. And so, just six days after the disaster, on February 3, 1986, he made it known that he intended to appoint his own commission of inquiry, which would be answerable to the President alone. The commission's instructions were succinct and clear-cut:

> The Commission shall:
> - review the circumstances surrounding the accident to establish the probable cause or causes of the accident,
> - develop recommendations for corrective or other actions, based on the Commission's findings and determinations.
>
> The Commission shall submit its final report to the President and the Administrator of the National Aeronautics & Space Administration within 120 days of the order of the President.[6]

Seventy-two-year-old William P. Rogers, a former Secretary of State and Attorney General, was appointed to chair the commission. At a hastily summoned press conference he denied all speculation that this latest development implied a repudiation of NASA. Certainly, the NASA board of inquiry would now have a different role to play. Its task now would be to assist the presidential commission. But, for his own part, he had full confidence in NASA and expected to be working closely with members of the agency. The commission would initiate its own independent hearings and make a number of recommendations, but this should not involve any conflict of interest between themselves and NASA. Both parties were united in their desire for the speedy clarification of the causes and course of the disaster.[7]

> We are not going to conduct this investigation in a manner which would be unfairly critical of NASA, because we think — I certainly think — NASA has done an excellent job, and I think the American people do.[8]

The other members of the commission were selected by the President and his advisers, with careful regard for the space shuttle's various "interest groups" and for the publicity angle.

Reagan appointed Neil Armstrong, the first man on the moon, as vice-chairman and also assigned the first American female astronaut, Sally Ride, to the commission. Joining them would be the greatest test pilot of them all, the absolute epitome of the right stuff, Chuck Yeager.

From the world of aerospace came Robert W. Rummel, formerly vice president of TWA. Boeing's airline division supplied Joseph Sutter, another vice president. Hughes Aircraft, for whom Greg Jarvis had worked, was represented by the director of the company's space communications division, Albert D. Wheelon. Another member, David C. Acheson, also hailed from the communications satellite business.

From the academic world came the physicist Arthur B. C. Walker from Stanford; Eugene E. Covert, professor of aeronautics at MIT; and the American Nobel Prize winner in theoretical physics, Richard P. Feynman from the California Institute of Technology (Caltech).

The press was represented by the former editor of the trade periodical *Aviation Week and Space Technology*, Robert B. Hotz.

From the Pentagon came Major General Donald J. Kutyna, who had been in charge of work on the Orbiter system's prospective military installation at Vandenberg Air Force Base in California.

Finally, representing the Reagan administration, there was Alton G. Keel, a government official whose job it would be to lend administrative support to the commission's chairman.[9]

. . .

Thus, when the commission was sworn in on February 6, 1986, every base seemed to have been covered. The press was there, as well as aerospace, the Pentagon, and science. And the mythical dimension had also been taken into account. Most of the commission's members boasted degrees in the sciences. Not so with the chairman, however. On the other hand, he was a shrewd lawyer, and a former Attorney General to boot.

Not everyone in NASA felt happy about the turn events had taken. Ostensiy what had been set up was a board of inquiry, but it had also been invested with such judicial authority and power that it was beginning to look more like a *court* of inquiry.[10]

33. Richard P. Feynman

Commissioner Richard P. Feynman was no ordinary professor of theoretical physics. Of course, the Nobel Prize awarded to him in 1965 would in itself have made him an obvious choice for the commission; but, if anything, his main qualification for this task was his total lack of respect for any form of "pomp, convention, quackery and hypocrisy."[1]

At the time when he was invited to join the commission, Feynman was sixty-eight years old. Like most other men of science, he had never taken much interest in the space shuttle. He had not come across much about the advances it was meant to bring in its wake in the pages of his scientific journals. He was far more interested in NASA's unmanned space probes, which had sent back such stunning photographs — those shots of Mars, for example.

Among his colleagues he was known as a wonderful lecturer who had revolutionized the teaching of science with the publication of *The Feynman Lectures on Physics*, a collection of his lectures given between 1961 and 1963 to first- and second-year students. This was an exercise that fascinated Feynman, because it entailed making extremely complicated subject matter clear and comprehensible to young students. Familiar topics had to be given a new twist, be viewed from unfamiliar angles. And this he succeeded in doing, although in the long run his lectures possibly had a greater impact on his teaching colleagues within the field of physics than on those at whom they were primarily aimed.

One of the guiding principles behind Feynman's teaching methods was his way of looking at the world as though it were a game, or a rebus — where all of one's abilities had to be brought to bear in order to arrive at the correct solution. All natural phenomena were based on laws, known and unknown. His students had to use those laws which were already known to predict new phenomena or new "behavior patterns." At other times, they had to work their way back from specific phenomena or patterns to the unknown laws which had to lie behind them. Later, they would of course come upon cases where the rules did not apply, and would expound new laws to accommodate these fresh observations.[2]

The whole idea of this method was that the rules should not be memorized from some textbook, but that they should be established from scratch in a give-and-take between teacher and students. The students were not to take anything for granted just because the teacher said it was so. He might very well be wrong. He might even be trying to trick them. They had to think for themselves, come up with counterarguments, or, in the end, give in because all of a sudden they realized just how simple and elegant the explanation was.

Typically, Feynman himself did not write his famous lectures. He did not write textbooks. He taught. Often his only prop would be a single sheet of paper covered with scribbled notes. Everything else, circumstances would have to dictate. Feynman's lectures were committed to paper by his colleagues at Caltech, with whom he co-edited them.

As a young man, immediately after taking his Ph.D., Richard Feynman had joined the group responsible for building the first atomic bomb in Los Alamos. One of those on whom he made an impression, while there, was the visiting professor Niels Bohr. Bohr took a special liking to him because he was not weighed down by awe of the famous Danish physicist. Feynman's way of asking questions, contradicting, offering criticism — quite without fear — made him the perfect whetstone on which to hone Bohr's reasoning. He was a very American physicist, representing a new generation that had never visited those shrines to physics in Cambridge, Göttingen, or Copenhagen, and that could not have cared less — since it ought to be one's arguments that counted.

Of course, not all of Feynman's colleagues were so delighted by his manner. Over the years, a wealth of Feynman anecdotes had been spawned, all depicting his knack of cutting through so much scientific waffle — humorously, simply, and lucidly — to get to the point. Feynman himself was a great fan of these anecdotes, not all of which proved, on closer examination, to conform absolutely to the truth. Often, his observations had cost him no small amount of painstaking, orthodox physics work at home in his study. It just wasn't supposed to look that way. Instead, he was fond of presenting himself as an unsophisticated, naive soul, winning his victories solely by dint of insight, humor, and common sense.[3]

To Feynman, the area in front of the blackboard was a stage on which he could present his one-man show — with no holds barred. Any and all tricks were permissible, even cheap ones, if they led to greater understanding. His students were his verbal sparring partners, but they were also an audience for whom he loved to perform. As the years went by, he became more and more convinced that the unorthodox approach was always the best. More skeptical colleagues sometimes felt that he occasionally took wide detours in the name of unorthodoxy, when traditional, albeit more prosaic, equations could have provided a sensible shortcut to the goal.

Richard P. Feynman was appointed to the presidential commission on the recommendation of NASA's Administrator, William Graham. Graham himself had once been one of those enthralled students in the auditorium. And William P. Rogers accepted his recommendation — it could do no harm to have a Nobel Prize winner on the commission. Little did he know what he was letting himself in for.

Feynman himself was not exactly overjoyed about his appointment. He knew that he was living on borrowed time, after several major operations for cancer, and he was engaged in an exciting project to develop a totally new type of computer. It did, however, comfort him to know that the commission had been given only 120 days in which to produce its findings; and he was intrigued by the fact that the disaster involved a mystery that had to be solved if all the various phenomena were eventually to be explained through the exposition of simple, underlying causes. Hence, the commission's work could be regarded as a textbook example of a scientific experiment. Hypotheses would have to be advanced, and then exploded or confirmed, until the solution stood out, crystal clear.

Immediately after his appointment, Feynman took himself off to NASA's Jet Propulsion Laboratory (JPL) to learn something about the space shuttle system. Since the laboratory was run by Caltech, this would amount to an informal chat with colleagues. Feynman was given an excellent and rapid briefing on all of the system's vital components.

He was told all about the solid rocket boosters, the joints between the individual segments, and the seals whose job it was to keep that terrific combustion enclosed within the rocket casing. One of the engineers described how several instances of charred seals had been discovered when used boosters had been fished out of the Atlantic.

He learned about the pressure inside the booster, about the solid fuel and its production, its composition, the percentages of the various constituents, etc.

And finally he was told about the main engines and the countless problems with the powerful turbopumps.

Feynman was in his element, surrounded by professionals. The various specialists involved with the space shuttle system spoke his language; they did not take offense when he interrupted them and leapfrogged over several stages in a train of thought; and, above all, they were quite at home with scientific dialogue, as a tool of the trade.

It is evident that Richard Feynman was going to have trouble dealing with lawyer, and commission chairman, William Rogers. Throughout his years of schooling and study, Rogers had been taught that it was rude to interrupt. In the courtroom and at congressional hearings, one testimony or plea was heard at a time. Then there was a chance to cross-examine witnesses, who, finally, after consulting with their lawyers, were given the opportunity to comment upon the record of their testimony as taken down by the court stenographer.

This method had evolved over centuries of legal practice. It was elaborate and lengthy, but it had the indisputable advantage that every single testimony was documented and had been commented upon from several sides before being filed. Contradictory testimonies were eventually confronted with one another, and finally, one reached — if not the truth — then at least what the court deemed to be the most likely explanation of events.

Feynman had absolutely no patience with this kind of approach. He had enough of a job sitting still on a podium, in a leather chair, simply listening to speeches which presented him — methodically and at some length — with facts with which he might already be fully conversant. He wanted to get straight to the heart of the matter and was prepared to work around the clock if necessary, as he had been in the habit of doing at Los Alamos in his youth and later while working alongside kindred spirits at Caltech. You couldn't work with one eye on the clock when inspiration was upon you; there would be time for sleep later, once you ran out of steam.

It was a case of science versus law; the laboratory versus the courtroom. Both methods were right, according to their perspectives and the traditions of their professions, but they found it hard to live in harmony.

At the first, informal sitting of the commission, William Rogers asked each member how much time he or she could spare for the work of the commission. Most replied that they had rearranged their schedules. Feynman said that he could devote himself to the job 100 percent — from then on!

But this had only been an informal get-together. They would not get down to work until the following day, February 6. All of the commissioners were picked up in limousines and driven to the swearing-in ceremony. After that, they had to be issued with identity cards and sign a number of documents. Only after all that would they hear the first testimonies from top NASA personnel — supporting their statements with slides.

Already, this early in the game, Feynman was irritated by the way everything was accorded an acronym. Instead of talking about the high-pressure fuel turbo pump, they called it the HPFTP. The main engines hid behind the term SSME (space shuttle main engine). The fuel tank was ET (external tank). Not only that but all diagrams included bullets, oversized periods which came before

- all
  - salient
    - subheadings.

This style of presentation was, of course, calculated to emphasize NASA's technical proficiency, but it did not prevent senior executives from caving in very fast when questioned in detail by the trained scientists on the commission. Again and again they had to state that they would submit additional material on this or that question at a later date. Rogers fully understood their inability to supply answers off the cuff, but it exasperated Feynman.[4]

Worse was to come, when the session proved to be of relatively brief duration — and there was the limousine, ready to drive him back to his hotel. Work was already over for that day.

> The main thing I learned at that meeting was how inefficient a public enquiry is; most of the time, other people are asking questions you already know the answer to — or are not interested in — and you get so fogged out that you're hardly listening when important points are being passed over.
>
> What a contrast to JPL, where I had been filled with all sorts of information very fast.[5]

Feynman tried to draft an appropriate work schedule to save all of the commissioners from having to sit through every testimony. Again, quite in keeping with standard scientific working methods. Not surprisingly, Rogers did not particularly like the idea.

During one recess, the Pentagon representative on the commission, Major General Kutyna, told the other members how the armed forces had carried out an inquiry after the explosion of a Titan rocket. Feynman's attention was immediately caught. The approach adopted there was akin to what he himself had in mind — except for being even more refined, technically speaking. Several other commissioners also wanted to embark on such an inquiry. For Feynman, it was now simply a question of deciding who should do what.

At this point, Rogers returned. He made it plain to Kutyna that the commission would never be given access to the amount of technical information that had been made available to the armed forces after the Titan accident. The vice-chairman, Neil Armstrong, backed Rogers up on this. The commission should certainly not expect to be able to carry out that kind of technical investigation.

Feynman was furious. Partly because he disagreed totally with Rogers — NASA was in possession of many times more technical data from the wrecked space shuttle than the armed forces had had from the Titan rocket — and partly because he himself was *only* interested in carrying out a technical investigation. If that was not the intention, then he was wasting his time in Washington.[6]

Rogers informed the commission that he had organized a trip to the Kennedy Space Center for the following week. At Kennedy, the commissioners would be given a guided tour and an on-the-spot briefing. Feynman doubted whether there was any point in such a tour. It might all too easily turn out to be nothing but window dressing and lecture after lecture — a waste of time. Besides, the trip was not scheduled for another five days. What was he supposed to do in the meantime?

> I went over to Mr. Rogers and said, "We're going to Florida next Thursday. That means we've got nothing to do for five days; what'll I do for five days?"
>
> "Well what would you have done if you hadn't been on the commission?"
>
> "I was going to go to Boston to consult, but I canceled it in order to work 100 percent."
>
> "Well why don't you go to Boston for the five days?"
>
> I couldn't take that, I thought, "I'm dead already! The goddam thing isn't working right." I went back to my hotel, devastated.[7]

Feynman rang NASA chief Bill Graham, the former student responsible for his being appointed to the commission in the first place. He told him that he would go crazy if he wasn't allowed to *do* something soon. He wanted to talk to some of the NASA technicians and engineers — at the Johnson Space Center in Houston, for example. He wanted to try to get back to the feeling he had experienced in his discussions with the people at JPL. Do what he was best at, rather than die of frustration.

Graham saw his problem and promised to arrange something. But he would have to check with Rogers. The chairman was not happy about individual commissioners sniffing around on their own. And certainly not Richard P. Feynman.

Graham then suggested that Feynman could meet NASA technicians at his office in NASA headquarters in Washington. This seemed like a workable compromise, but still the chairman opposed it.

> Then Mr. Rogers calls me; he's against Graham's compromise. "We're all going to Florida next Thursday," he says.
>
> I say, "If the idea is that we sit and listen to briefings, it won't work with me. I can work much more efficiently if I talk to the engineers directly."
>
> "We have to proceed in an orderly manner."
>
> "We've had several meetings by now, but we still haven't been assigned anything to do!"
>
> Rogers says, "Well, do you want me to bother all the other commissioners and call a special meeting for Monday, so we can make such assignments?"
>
> "Well, yes!" I figured our job was to work, and we *should* be bothered — you know what I mean?[8]

A hard-pressed Rogers finally agreed, reluctantly, that Feynman could talk to his technicians in the interim. Later, he was to regret this decision, but by then it was too late. Feynman had already spoken to them.

34. Discord

On Saturday, February 8, Richard Feynman went his own way and spent the day with experts from NASA who briefed him on the various components of the space shuttle system. Particular mention was made to him of the field joints in the solid rocket boosters, on which their suspicions were gradually centering.

Here, for the first time, he learned that, even before the first flight of the space shuttle system, it was known that the joints were not functioning as they were supposed to. The edges of the two booster segments were not squeezed against the two synthetic rubber seal rings by the rocket thrust as expected. Instead the gap between the edges widened as the pressure increased, because the booster casing was thinner than the area around the joint. The O-rings were, in fact, meant not just to act as a stationary gasket; they were also supposed to seal a crack which suddenly widened during the joint rotation.

The contractor, Morton Thiokol, had experimented with thicker rubber rings and various kinds of shim, but still the joint was never totally airtight. On several flights, hot gas had seeped through one or both rubber seals and scorched through sections of them. The Marshall Space Flight Center, which was responsible, on NASA's behalf, for the solid rocket boosters, had complained to Thiokol about the design on several occasions. A number of different adjustments had been suggested; but all of them would have put a stop to any flights until the joints had been redesigned—an idea which did not appeal either to Thiokol or to Marshall. Consequently, Feynman learned, no action had been taken.[1]

The next day, Sunday, February 9, William Rogers received a rude awakening on opening his *New York Times*. The newspaper had got hold of two memoranda from a budget analyst at NASA named Richard C. Cook. One of the memos had been written by Cook to his boss as far back as July 23, 1985, six months prior to the *Challenger* disaster. In this, he advised his boss that he had spoken to engineers at NASA who had

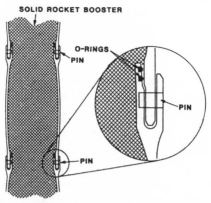

SOLID ROCKET BOOSTER

O-RINGS
PIN

PIN

PIN

Schematic presentation of a solid rocket booster under pressure, with an enlargement of the field joint with the O-ring. (Feynman, 1988)

said that the O-rings in the field joints were "a potentially major problem affecting flight safety and program costs." Cook acted as liaison between the different specialists, reporting back on everything that might, at some time in the future, represent a threat to the finances of the Space Transportation System.

The engineers had told him that the joints were unsatisfactory and that there was every chance that one day they might fail to function — with disastrous consequences. And no matter what happened it looked as though they would soon have to change the booster design — with all the extra cost that this would entail.

The second memo, dated February 3, had been written immediately after the *Challenger* explosion. In this, Richard Cook wrote that the general feeling among NASA experts was that the accident had been caused by a field joint burning through. Cook continued:

> It is also the consensus of engineers in the propulsion division, Office of Space Flight, that if such a burn-through occurred, it was probably preventable and that for well over a year, the solid rocket boosters have been flying in an unsafe condition. Even if it cannot be ascertained with absolute certainty that a burn-through precipitated the explosion, it is clear that the O-ring problem must be repaired before the shuttle can fly again.[2]

Cook also said that, the way he saw it, not enough had been done at the Marshall Space Flight Center to resolve the problem and that senior personnel at NASA headquarters would not even give a hearing to worried engineers from the booster division. Even after the disaster, they had not been brought in to help with the investigation.[3]

The New York Times ran the story on the front page. This was the first public intimation that the disaster had not been the tragic but necessary price of progress referred to by the President in his words to the bereaved just two weeks earlier. That it was not an unforeseeable malfunction. Instead, by all accounts, this disaster could have been, and should have been, prevented.

The newspaper described its source as a solid rocket booster expert. The journalist denied that Cook had leaked the two memos. Nevertheless, William Rogers was infuriated by the story. For one thing, the article contained information which the commission should have been the first to hear of — instead of having to read about it in the newspaper. Second, it was unbelievable that a resource analyst in the comptroller's office — a bookkeeper — could give the impression that he knew something about space shuttle technology and be taken seriously.

But, infuriated or not, Rogers could not ignore this piece of information. He would have to bother the commissioners after all, by calling them to Washington on Monday to examine Richard C. Cook.

The meeting on Monday was held behind closed doors. Rogers, clearly riled, subjected Cook to a veritable cross-examination. Cook willingly described how, at the Marshall Center, it had been a standing topic of discussion: How to get the joints to function properly without throwing half a million dollars' worth of solid rocket booster segments away. New segments had to be ordered at least thirteen months in advance, so they would be faced with a serious dilemma — in terms of the schedules as well — if existing material had to be scrapped.[4]

Rogers delved into Cook's motives, especially in writing the second memo:

ROGERS: *You were doing it in terms of the budget? I mean, was*
             that the purpose of writing the memorandum after the
             accident?
COOK: *Yes.*
ROGERS: *Did you have reason to think that efforts would not be*
             made after the accident to investigate it thoroughly?

COOK:	*No, in fact, I knew that across the street [at NASA headquarters] they were doing the same analyses we were doing.*
ROGERS:	*Then will you explain again why you wrote the memorandum?*
COOK:	*To document what I felt were all the budgetary implications of the situation.*
ROGERS:	*Only budgetary? Was there any other purpose? Well, if you'd rather not answer that's all right. I'm just curious about why you wrote the memorandum. I mean, it doesn't sound as if you had budgetary considerations in mind. It sounds differently. But I wanted you to have an opportunity to tell the commission why you wrote it.*

Cook, clearly ill at ease, had to go through it all one more time:

COOK:	*I wrote it because I felt it was a serious enough situation that I didn't think that until these various issues were resolved with the SRBs that I had been involved with — that they had to be taken care of before the shuttle could safely continue.*
ROGERS:	*Did you think your engineering experience based upon the short time you had been with NASA improved your ability to pass judgment on what others had decided? Well, here again, I don't really want to press you. Do you have anything else to tell the commission?[5]*

Cook was allowed to leave the witness stand. He had been shocked by the tone of the hearing. *He* had not leaked his own documents, and yet he was being treated like the defendant in a criminal case — or, at the very least, like a public servant who has displayed a serious lack of loyalty.

Lunch was past before Rogers cooled down and announced that he had spoken with Cook again and now understood that, in fact, it had been Cook's boss who had asked him to write about the disaster. That being so, he had absolutely no objections to make and could only, on behalf of the commission, thank him for his testimony and his responsible behavior. Both of his memoranda would be published in their entirety. In addition, a number of documents dealing with the history of the solid booster rockets would be made available to the press. Finally,

Richard Cook was asked to repeat his testimony the following day at a public sitting of the commission.

The fuss had barely died down before the commission received a fresh shock. Allan J. McDonald, a forty-eight-year-old senior engineer from Morton Thiokol, who had represented the company at the Kennedy Space Center for the *Challenger* launch, presented himself to them, saying that he had come forward of his own free will. He believed the commission ought to know that on January 27, the eve of the *Challenger* launch, his company had put a veto on igniting their solid rocket boosters in temperatures lower than 53° F. Managers at NASA had said they were "appalled" by Thiokol's veto. For several hours the matter had been debated in a teleconference between Thiokol in Utah, the Marshall Space Flight Center in Alabama, and the Kennedy Space Center in Florida. Eventually, Thiokol had given in and accepted the idea of a launch, even though the temperature would be at least 20° F lower than that which, just a few hours previously, they had claimed to be the lowest warrantable temperature.

But McDonald had refused to accept his own company's change of heart. In the heat of the moment he had said, "I sure wouldn't want to be the person that had to stand in front of a board of inquiry to explain why we launched this outside the qualification of the solid rocket motor."[6]

Now it was William Rogers' turn to be shocked. If what McDonald said was true, then the commission was being led up the garden path by people at Thiokol and, not least, by principal figures in the launch decision chain at NASA. In both places, any number of people must have known about the lengthy teleconference that was going on as the temperature dropped on the night before the launch. And yet, until now, it had been thought necessary to inform the commission only that a brief discussion had been held regarding "a concern by Thiokol over low temperatures"; which, however, was soon concluded when "Thiokol recommended that the launch proceed." And that definitely did not tie in with McDonald's version of the proceedings.[7]

Rogers had been counting on NASA's own willingness to get to the bottom of things. He had expected to work on close and straightforward terms with NASA and had never dreamt that anything might have been swept under the rug. This was one reason why he had reacted so strongly to Cook's revelations — blackening, as they did, the name of the civilian space agency. Now his trust had received a serious blow, and he was angry. From now on, the inquiry into the *Challenger* disaster would have to take account of much more than just the possible technical malfunctions. It would also have to delve into the workings of the launch decision chain at all levels in NASA. If there was any suggestion of patent dishonesty or attempts to cover up the true course of events, then they would have to consider instituting criminal proceedings against those chiefly responsible. And there were increasing indications that Richard Cook, whom William Rogers had attacked that morning for disloyalty to NASA, had had more than enough reason to raise the alarm.

Remedial Action —
None Required

William Rogers had opposed Feynman's idea of dividing up the work of the commission because he regarded NASA's own internal investigating teams as an extension of the commission. The commissioners would receive technical material from these teams and augment this later by holding hearings as they saw fit. Now, however, Rogers could no longer trust NASA implicitly. The commission itself would have to clamp down and insist upon overseeing every stage of the investigation. On February 15, 1986, after two days at the Kennedy Space Center, followed by preliminary closed-door hearings, the chairman issued a sensational press release:

> In recent days, the Commission has been investigating all aspects of the decision making process leading up to the launch of the *Challenger* and has found that the process may have been flawed. The President has been so informed.
>
> Dr. William Graham, Acting Administrator of NASA, has been asked not to include on the internal investigating teams at NASA persons involved in that process.
>
> The Commission will, of course, continue its investigation and will make a full report to the President within 120 days.[1]

William Graham then had to withdraw from the investigating teams a number of directors and managers from the various NASA field centers. In principle, they all now stood accused.

Three days after issuing the press release, Rogers made it known that the commission would be split up into subcommittees, which would visit the various NASA field centers and the premises of the major contractors to direct on-the-spot investigations of all relevant data. Exactly the solution which Richard Feynman had wanted at the outset.

Not surprisingly, the inquiry ended up revolving, to a very great extent, around the solid rocket boosters. An understanding of their construction was essential, including the question of how one could assemble a 149-foot rocket comprised of several different segments and still be sure that the joints between the individual segments would be absolutely airtight — thus leaving only one possible exit open to that tremendous pressure: the nozzle at its foot. The history of the booster design would also have to be taken into account; and all of the documents and certificates completed before the initial approval of the design, along with the subsequent documentation of the individual flights, would have to be gone through with a fine-tooth comb.

The commission would have to examine the interaction between three parties: the major contractor (Morton Thiokol in Utah), the NASA center with special responsibility for the solid rocket boosters (the Marshall Space Flight Center in Huntsville), and, lastly, NASA's senior administration in Washington, which had overall responsibility for the Space Transportation program. It very soon became clear that the interests of these three parties did not necessarily coincide.

As we have seen, Morton Thiokol had landed the lucrative booster contract in November 1973, in the face of protests from Congress's General Accounting Office — which was not at all convinced that the handling of bids for the contract had been entirely fair and impartial.[2] Aside from the Utah connections of the NASA Administrator at the time, the argument that Thiokol's proposal seemed to be the cheapest in terms of both construction and operating costs had played a not insignificant part in NASA's deliberations. Particular mention had been made of Thiokol's proposal for two O-ring seals situated immediately above one another in the joint. A small leak-check port allowed compressed nitrogen to be blown between the O-rings, thereby pressure-testing the seal without having to pressurize the entire booster casing.

It might be the case, as their competitor Aerojet maintained, that a booster casing with no joints would be the best solution from a safety point of view. But this argument had been evaluated back in 1973 when the space shuttle contracts were up for bids. Morton Thiokol had won the booster contract over Aerojet's design — which was certainly safer but also decidedly more expensive. So NASA had considered Thiokol's solution to the question of pressure-testing a segmented booster both neat and innovative. It would save time and money before each launch.[3]

Morton Thiokol had given the impression that their solid rocket

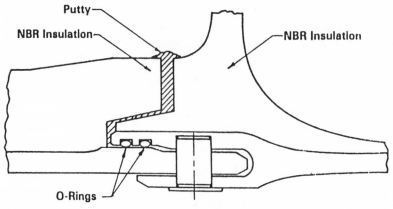

Putty

NBR Insulation

NBR Insulation

O-Rings

FIELD JOINT WITH PUTTY AND O-RINGS. (LEWIS, 1988)

booster would be a relatively straightforward, improved version of the highly dependable Titan III — albeit on a somewhat larger scale. The new booster would be the biggest American solid rocket motor to date. With the Titan rocket too, booster segments had been mounted one on top of the other and at no point had any problems with leaking seals between the individual segments arisen. And the Titan rocket had only one O-ring in each joint to ensure that it was leakproof; Thiokol would use two, thereby building in an extra safeguard — just in case anything should go wrong. Marshall's technicians approved the construction plans and informed the NASA administration that it all seemed sensible and sound.

Thiokol was not quite so forthcoming about the one crucial difference between the joints in these two rockets. Granted, the Titan segments were fitted with one rubber seal, but only as backup. Normally, one particular design element ensured a very tight interlock between the insulation of one rocket segment and the next, thus allowing it alone to bear the brunt of the fierce combustion, without the O-ring ever being affected by it. The O-ring was capable of taking the strain if anything went wrong, but under normal circumstances it was not meant to be subjected to any strain whatsoever. Thiokol, on the other hand, had constructed their model in such a way that the O-rings alone had to contain the pressure, with just a small amount of asbestos-filled insulating putty between them and combustion gases burning at thousands of degrees.

The idea was for the 30 atmospheres of the combustion gases to squeeze against the putty, which would then compress the air in the gap between it and the primary O-ring. In this way, the compressed air would pressure-activate the rubber seal. In other words, it would be forced into the groove in which the rubber seal lay and distribute the pressure around it, thereby pushing it hard up against the facing cylinder wall. In the highly unlikely event that the primary O-ring should fail, the secondary O-ring would be there as backup.

The reasoning behind this sounded convincing. Nevertheless, there was that one vital difference between the space shuttle booster and the Titan: Here, the pressure was contained and the joints held airtight solely by the O-rings and the putty. They did not constitute a reserve system, backing up the first line of defense — the insulation. They *were* the front line.[4]

Furthermore, for practical reasons, the Thiokol plant had to assemble the booster segments horizontally when putting them to the test in Utah. In contrast, the Titan III was always assembled vertically, allowing each segment to be slipped neatly into the one below. This method would be employed later at Cape Kennedy, but at the plant Thiokol had to put up with conditions that were less than perfect. And so the design was altered, making the gap between the two segments somewhat larger. This would facilitate the assembly of components weighing several tons. But so that the gap would not be too big, it was then reduced by the addition of thin metal shims. Thereafter, the O-rings should be able to handle the remaining opening.[5]

One last, crucial difference between the two rockets was, of course, so obvious that no one ever thought to mention it. The shuttle boosters would have to be reused up to twenty times. Not only would the booster segments be subjected to terrific expansion under pressure. They also had to be transported thousands of miles, and during these journeys the pressure of their own weight caused them to become ever so slightly oval in shape. And after every trip they had to be reassembled quite precisely, to close tolerances.

In September 1977, a "hydroburst test" was carried out. A rocket segment was filled with oil under great pressure to simulate the pressure of the liftoff. It was during this test that engineers first discovered, to their surprise, that instead of narrowing under pressure, the gap in the joint widened. The tang of the joint and the inner clevis bent away from each other instead of toward each other and, in so doing, reduced, rather than increased, the pressure on the O-rings. The extra-long tang on the upper

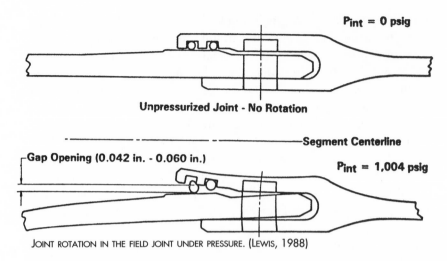

JOINT ROTATION IN THE FIELD JOINT UNDER PRESSURE. (LEWIS, 1988)

of the two cylinders—which had been necessary to allow room for two O-ring grooves—only added to the problem. The tang from the upper segment shifted around inside the clevis of the lower segment, because of the way the relatively thin casing swelled like a balloon between the joints. The joint "rotated" and the O-rings drew away from the sealing surface. The system was reacting quite contrary to expectations.

Thiokol's engineers were taken aback by this, but they were also working on a very tight schedule. Everything now hinged on whether these new findings would render the joint design impracticable, in which case they had a serious problem. In the estimate of experts at Thiokol this unforeseen joint rotation seemed hardly likely to present any great problem, practically speaking; so they forwarded the documentation on the hydroburst findings to their opposite numbers at the Marshall Space Flight Center.

At Marshall, however, a much more serious view was taken of this phenomenon. Glenn Eudy, chief engineer at Marshall for the solid-fuel rocket motors, described it as a "very critical SRM issue." He rejected various proposals from Thiokol for temporary solutions involving shimming to reduce the rotation. In a memo written on September 2, 1977, he said:

> I personally believe that our first choice should be to correct the design in a way that eliminates the possibility of O-ring clearance.[6]

In a report dated October 21, another engineer at Marshall, Leon Ray, described it as "unacceptable" for the tang not to be redesigned. Otherwise, in a worst-case scenario, the gap in the joint could become so wide that—either due to the joint rotation or because of two segments not being perfectly cylindrical—the joint would spring a leak, just a few feet away from a tank containing vast amounts of liquid explosive. In a memo written in January 1978, John Q. Miller of the Marshall Space Flight Center described such a leak as potentially disastrous and added:

> We see no valid reason for not redesigning the joint to accepted standards.[7]

A year later, on January 19, 1979, John Q. Miller wrote another memo to Glenn Eudy, complaining about the line taken by Thiokol on the problem.

> We find the Thiokol position regarding design adequacy of the clevis joint to be completely unacceptable.

He supported this statement by pointing out that the primary O-ring would now extrude into a gap which would be widened when it ought to be tightening. According to Miller, this method was "forcing the seal to function in a way which violates industry and government O-ring application practices"—whereby the O-ring was expected to adhere to the tang and clevis surface right from the start. O-rings were normally designed to seal by compression, not extrusion. In addition to this, joint rotation allowed the secondary O-ring to become "completely disengaged from its sealing surface on the tang." The clevis joint secondary O-ring seal had been verified by tests to be "unsatisfactory."[8]

In February 1979, Glenn Eudy and one of his engineers from the Marshall Space Flight Center visited several suppliers of the Viton rubber seals and asked for their views on this problem. Although guarded in their replies, these suppliers did say that the solid rocket motor O-ring extrusion gap was greater than anything they had experienced in other designs. The official report of the visit to Precision Rubber Products Corporation in Lebanon, Tennessee, stated that the company believed "that the O-ring was being asked to perform beyond its intended design and that a different type of seal should be considered."[9]

The Rogers commission could find no evidence of these internal NASA documents ever being forwarded to Thiokol. At any rate, nothing came of these serious objections. The design of the joints remained unchanged and hence—in the opinion of certain specialists at the Marshall center—quite unacceptable. And yet—amazing as it may seem—on

September 15, 1980, the solid rocket boosters were officially approved. The only explanation must be that by this time senior management at NASA were fast losing patience. One limit after another had been exceeded in every branch of the Space Transportation System and now the absolute limit had obviously been reached. NASA's Administrator was acutely aware of the fact that both the politicians on Capitol Hill and the American people would very much like to see *Columbia* launched in the very near future.

As we know, the first *Columbia* flight was a great success. But on the space shuttle's second flight (STS-2) on November 12, 1981, the first serious problems with those vital field joints were encountered. Later analysis of the used boosters showed that the primary O-ring on one of the boosters had been subjected not only to pressure but also to burning gases. The insulating putty had not provided enough protection and the hot gases had melted and vaporized the primary O-ring on the right booster's aft field joint to a depth of 0.053 inch — about one-fifth of the O-ring's diameter. But — this erosion notwithstanding — the rubber seal had served its purpose. Now NASA had proof not only of the joint rotation but of the fact that, contrary to all expectation, the rubber seals had to withstand both the pressure and the fierce heat of the combustion. And yet Marshall technicians said not one word about this problem at official meetings with personnel from headquarters and other NASA field centers held prior to the next launch. They preferred to mull over the problem behind closed doors at Huntsville.[10]

Soon it became impossible to go on perpetuating officially the idea that the secondary O-ring was actually there as backup. As John Q. Miller had already pointed out, it was true that this ring protected the primary O-ring, but only for a fraction of a second, before the joint rotation disengaged it from the sealing surface. From then on, for the rest of its more than two minutes of flying time, it was of no help. So, in fact, you were left with just the one O-ring in the booster joints. And hence, in all fairness, the design certificates would have to be altered accordingly.

This meant that the joint could no longer be classified as Criticality 1R (R for redundant) but had to be moved into the Criticality 1 category. Everything depended on one rubber ring, 0.280 inch thick.

The new certificates contained a section entitled "Failure Effect Summary." Under this heading details were given of what would happen if the component in question—in this case the primary O-ring—were to fail. In this section NASA had to write that, as far as could be gathered, the result would be:

> Loss of mission, vehicle and crew due to metal erosion, burn-through and probable case burst resulting in fire and deflagration.[11]

The certificate was verified and certified on December 17, 1982, more than three years before *Challenger* would, quite graphically, demonstrate the practical application of this phenomenon.

Despite this reclassification as Criticality 1, it was a long while before experts at both NASA and Thiokol abandoned the notion of the secondary ring still effectively providing a backup. As far as they were concerned, the reclassification was nothing more than an overcautious formality. Others believed that the secondary ring had never been envisaged as a reserve component. It was only there to facilitate pressure testing of the primary O-ring before launching. Hence there was nothing disturbing about the fact that it disengaged after liftoff. Still others—the majority—never heard a word about the new classification.[12]

Around the same time, NASA was making plans for a whole new generation of lightweight boosters for the space shuttle system, to be constructed from a new and much lighter material—carbon filament. During the developmental work on these boosters an engineer from a company named Hercules Inc. stumbled across an idea which looked as though it might well solve part of the joint rotation problem. For the new boosters, Hercules Inc. had come up with a new joint design involving a special capture feature that locked the joint components tightly into place, thus obstructing rotation. The new joint would also be fitted with an extra O-ring.

Both Hercules Inc. and Morton Thiokol submitted bids for production of the new, lighter boosters. Hercules had built the capture feature into their field joints, and in so doing had most likely solved the problem; Thiokol stuck to the old, unsatisfactory design. There were those who believed that this obstinacy could be ascribed solely to the company's reluctance to make any changes whatsoever that might be interpreted as an admission that the company's existing hardware was not up to scratch.

In May 1982, NASA awarded the contract to Hercules Inc., and the first models of the new, safer design were scheduled to be ready for testing in 1986, six months after the *Challenger* explosion.

The new joint design could also have been applied to the existing steel segments but, for one thing, this would have entailed a "halt in production" and, for another, it would have meant a weight increase of 600 pounds on each booster—neither of which NASA needed.[13]

NASA was eager to save weight wherever possible, in order to increase the cargo-carrying capacity of the space shuttle and, hence, make it more cost-effective. In 1983 NASA switched to using booster cases that were between 0.02 and 0.04 inch thinner. At the same time a more powerful fuel mixture was introduced along with an improved nozzle design, intended to increase the power of the booster. These improvements had, however, the side effect that the joints now rotated even more than before.[14]

A year later, in February 1984, scorched O-rings were discovered again, after mission 41-B. On this occasion the rings were eroded to a depth of approximately 0.050 inch. Not even then did anyone take steps to issue a directive prohibiting the space shuttle from flying until the problem had been solved. Instead, in March 1984, the Marshall Space Flight Center produced an official system report that rendered it possible and defensible to continue with the launches:

> Possibility exists for some O-ring erosion on future flights. Analysis indicates max. erosion possible is .090 inch according to Flight Readiness Review findings for STS-13. Laboratory tests show sealing integrity at 3,000 psi using an O-ring with a simulated erosion depth of .095 inch. Therefore, this is not a constraint to further launches.[15]

The language is dry and technical but the message is clear enough. Scorching of the O-rings was now accepted as being part of the design. They even went so far as to set a margin of safety which they felt sure could not be exceeded. Instead of insisting that the possibility for combustion gases to reach the thin rubber seals at all was totally unacceptable, the official line was taken that a certain amount of erosion could be accepted. The Marshall Problem Assessment System concluded: "Remedial action—none required."

It was at this time that the pressure testing of the joints became a moot point among technicians at both Thiokol and Marshall. It so happened that they now had even more proof that the hot gases were penetrating to the O-rings through blowholes in the asbestos-filled insulating putty. These leak paths might have been caused by incorrect distribution of the packing prior to the assembly of the booster segments, but there was no denying that they could also have occurred during the actual prelaunch pressure testing. If the pressure from this broke through the primary ring it would blow tiny pinholes in the putty, which could then be used by flames moving in the opposite direction. At a temperature of 6,000° F, jets of hot gas would tear through these channels to home in on the O-rings like a welding torch.

It was, however, impossible to forgo this pressure test. The joints had to be checked for leaks before liftoff, since the integrity of the joints was a crucial design factor. Nor was it altogether irrelevant that the injection of pressure between the O-rings blew these outward—in which position they would supposedly find it easier to seal the gap later on during the launch. And so, as problems with the joints began to crop up, the pressure was increased from 50 psi to 100 psi and, finally, to 200 psi.

The findings of fresh studies made in April 1984 suggested, however, that the increased pressure had only exacerbated the problems with the O-rings. At the time, the technicians involved could not quite see the connection, but now, looking back on it, there can be no doubt. On the first eight flights, with the pressure at 50 psi, the incidence of observed O-ring anomalies was 12 percent. For the next seven flights, the pressure during testing was increased to 100 psi, and the incidence of anomalies rose to 56 percent. On the last ten flights a pressure of 200 psi was employed, and the result was not long in emerging: an incidence of 88 percent O-ring anomalies—blowbys, scorching, or both.[16]

The increase in pressure had been introduced to prevent leaks. NASA wanted to be sure that the packing was not masking a leak in an O-ring. Now it looked as though this precautionary measure had actually helped aggravate the problem dramatically. The greater pressure blew holes in the packing from the outside, leaving the O-rings unprotected later when the hot gases hit the packing from the inside. The technicians involved in this process must have noted this connection and yet, strangely enough, they still regarded the blowholes only as a sign that the pressure testing had been effective.

The commissioners found it hard to see the logic in this. One of them, Dr. Walker, examined Project Manager Lawrence Mulloy from the Marshall Space Flight Center:

WALKER: *Do you agree that the primary cause of the erosion is the blowholes in the putty?*

MULLOY: *I believe it is. Yes.*

WALKER: *And so your leak check procedure created blowholes in the putty?*

MULLOY: *That is one cause of blowholes in the putty.*

WALKER: *But in other words, your leak check procedure could indeed cause what was your primary problem. Didn't that concern you?*

MULLOY: *Yes, sir.*[17]

Almost a year to the day before the *Challenger* disaster, mission 51-C had been launched. On that occasion, the temperature had been 51° F, the coldest to date for a launch of the space shuttle. When the solid rocket boosters were fished up and taken apart, erosion was found in the O-rings of both boosters. Still worse, for the first time it was observed that gases had escaped past the primary ring and covered a 100 degree arc to reach the secondary ring. Technicians described this phenomenon as a race between the flame and the O-ring. All of the rubber on its surface had been eroded while the joint was opening up and the O-ring was trying to expand and contain the combustion. In this case the O-ring had won the race, but the surface behind the primary O-ring was jet black and covered with soot and grease.[18]

Technicians at the Marshall Space Flight Center and Morton Thiokol did not like what they saw at all. On January 31, 1985, Marshall asked Thiokol for a complete review of analysis findings from previous flights. For the first time, in the reply submitted eight days later, Thiokol mentioned the temperature as being a contributory factor.

> . . . low temperature enhanced probability of blowby — [flight] 51-C experienced worst case temperature change in Florida history.

The contractor predicted that the next flight could be expected to experience the same phenomenon, but still this was not believed to present any impediment to the launch.

> The condition is not desirable but is acceptable.[19]

Fortunately, when flight 51-B was launched in April 1985, the weather at Cape Kennedy was no longer cold. Yet even on that occasion, an erosion was observed in a nozzle joint, this time to a depth of 0.171 inch — two-thirds of the diameter and almost twice as deep as what had previously been described as the greatest possible erosion. In this instance

the primary O-ring lost the race. The secondary ring had saved the day
— by a hair — and had managed to get a grip, but even this ring was
eroded by 0.032 inch.[20]

Lawrence Mulloy now had to take some action. So he placed a
"launch constraint" on the space shuttle. This would prohibit Marshall
from approving a launch until the problem had been solved or, at the
very least, thoroughly appraised before each launch. This prohibition
took effect six months before the *Challenger* disaster, but Mulloy rou-
tinely wrote and signed waivers for every launch after July 10, 1985 — to
keep the schedules from being compromised. In other words, not only
was he the one who could place this constraint on the system; he also
had the power to issue dispensations, as he saw fit. Lawrence Mulloy in
no way believed that he was doing anything wrong in this. For one thing,
the problems with the O-rings could no longer be characterized as unforeseen; and for another, until now every flight had been executed without any serious technical breakdown occurring. Phenomena such as
scorching and blowbys had imperceptibly become part of the standard
setup, giving no cause for concern. Later, Mulloy put it like this:

> Since the risk of the O-ring erosion was accepted and indeed ex-
> pected, it was no longer considered an anomaly to be resolved be-
> fore the next flight.[21]

At NASA, no one in the management levels above the Marshall Space
Flight Center knew anything about any launch constraint. Again, at von
Braun's old workplace, they stuck to their habit of keeping their problems
to themselves.[22]

That a more covert awareness of the extent of the problem did exist
was demonstrated in July 1985, when Marshall ordered seventy-two new
booster segments from Thiokol. The contract clearly stressed that all of
the new segments were to be supplied with the Hercules capture feature,
as a first step toward a solution, whether Thiokol liked it or not. The
new segments would be ready for delivery in August 1988. In the meantime, NASA proposed to go on flying with the old ones.[23]

Thiokol's annual report for 1985 sported a splendid painting of the space
shuttle soaring into the sky on its front cover. Under the picture it said:

> Twin Morton Thiokol solid propellant rocket propulsion motors
> have performed flawlessly for nineteen space shuttle flights.[24]

On July 31 that same year, having addressed the problems that had arisen on flight 51-B, a worried senior scientist and acknowledged rocket seal expert at Morton Thiokol, Roger Boisjoly, wrote a letter to his boss, Robert K. Lund, the vice president of engineering:

> This letter is written to insure that management is fully aware of the seriousness of the current O-ring erosion problem in the SRM joints from an engineering standpoint.
> The mistakenly accepted position on the joint problem was to fly without fear of failure and to run a series of design evaluations which would ultimately lead to a solution or at least a significant reduction of the erosion problem.

Boisjoly pointed out that the latest development had highlighted the inadequacy of such an approach. This time, fortunately, the serious erosion had occurred in a joint close to the nozzle, where it was not so dangerous, but . . .

> If the same scenario should occur in a field joint (and it could), then it is a jump ball as to the success or failure of the joint because the secondary O-ring cannot respond to the clevis opening rate and may not be capable of pressurization. The result would be a catastrophe of the highest order — loss of human life.

He closed his letter by saying:

> It is my honest and very real fear that if we do not take immediate action to dedicate a team to solving the problem, with the field joint having the number one priority, then we stand in jeopardy of losing a flight along with all the launch pad facilities.[25]

To the Rogers commission, Boisjoly described the consternation at Thiokol when they saw the erosion on 51-B:

> . . . the erosion was considerably more than we had ever seen before. There was probably eighty percent of the [primary] O-ring gone, the entire section was actually vaporized. So at that point we said, "My God, if that ever happens in a field joint, we've bought the farm because the secondary seal might not be capable of holding it and sealing."[26]

No matter how much the Marshall center might have wanted it, there was no longer any way that headquarters could be kept out of this. A manager from headquarters had been present at the examination of the charred boosters in Utah. He had seen how disturbed the engineers were by what they found. And he recommended that Marshall and Thiokol be asked to provide an evaluation at a meeting of all three parties in August 1985.

The meeting took place on August 19. NASA's senior management had to look at a great many slides marked with a great many bullets, while being told, for the first time, about the problems. The presentation adopted the tone that, while the situation was of course serious, both the contractor and the Marshall center were aware of the problems and had the situation under control. If they hadn't, then one of the points made on the slides might have sounded pretty serious:

- The lack of a good secondary seal in the field joint is most critical and ways to reduce joint rotation should be incorporated as soon as possible to reduce criticality.

Four bullets further down, however, the managers could relax again:

- Analysis of existing data indicates that it is safe to continue flying existing design as long as all joints are checked with a 200 psig stabilization pressure, are free of contamination in the seal areas and meet O-ring requirements.

Although they did add:

- Efforts need to continue at an accelerated pace to eliminate SRM seal erosion.[27]

So in fact they were still abiding by the increased pressure which had exacerbated the problem, and still being reluctant to issue a flight veto that would have given them the time to get to the root of the design problems. Flight schedules had to be met, and the turnaround time reduced. Work on the field joints would have to be carried on simultaneously and "at an accelerated pace"; in any case, they would soon have seventy-two new segments at their disposal. In other words, they proceeded with the very approach which Roger Boisjoly from Thiokol had criticized so strongly.

Jesse Moore, the Associate Administrator for Space Flight, was to have attended this meeting but he never turned up. Moore would later be the man who "pressed the button" to send *Challenger* on its way, a man who would afterward claim that he had known nothing about any problems with the field joints.

Thiokol now appointed a special "seal team" to concentrate on the field joints. The company was acutely aware that NASA's patience was fast running out. Management did not sleep any better for knowing that NASA — in compliance with the U.S. Competition in Contracting Act — proposed to put the contract out for bids. For the first time ever, Thiokol would soon be facing tough competition for its order to supply boosters to NASA.

On the one hand, therefore, the company had to defer to NASA and show decisiveness and drive in facing up to these problems. On the other hand, they had — above all else — to stand by their assertion that the design was still basically sound and, hence, that no risk was involved in using the Thiokol solid rocket boosters.

It may be that the Thiokol management was capable of making this dual strategy work. But the engineers and technicians who, like Roger Boisjoly, were working at a hectic pace to produce forty-three new designs for the field joints without any great support from the people at the top, did not find it so easy.[28]

Even though their work was supposed to have top priority, time and again members of the seal team were called away to deal with other tasks. They had to wait for equipment, and getting their tests run entailed fighting a constant battle against the demands of the day-to-day routine. In October 1985, Roger Boisjoly wrote again to his bosses and tried to get them to listen:

> I might add that even NASA perceives that the team is being blocked in its engineering efforts to accomplish its tasks. NASA is sending an engineering representative to stay with us starting Oct. 14th. We feel that this is the direct result of their feeling that we (MTI) are not responding quickly enough on the seal problem. . . .
>
> Finally, the basic problem boils down to the fact that ALL MTI problems have #1 priority and that upper management apparently feels that the SRM program is ours for sure and the customer be damned.[29]

When questioned by the Rogers commission interviewer, Roger Boisjoly was even more direct in his explanation of why he had to write this memorandum:

> This so-called seal team was, for all intents and purposes, nonexistent. So I said, "Oh no, they ain't going to do this to me. I ain't going to take this load on my shoulder. There ain't no way." So I wrote my memo, my company private memo, and I says, "Folks aren't going to say that they're not aware of this because I'm going to get action." And that's why I wrote it, because I was scared to death that we were just going to take an extended period of time to fix this problem, and in the meantime this same thing could happen on a field joint that happened on this nozzle joint and we'd buy the whole thing.[30]

Another senior technical employee at Thiokol, Robert V. Ebeling, wrote to his boss, Allan McDonald, around the same time:

> HELP! the seal task force is constantly being delayed by every possible means . . .
> The allegiance to the O-ring investigation task force is very limited to a group of engineers numbering 8–10. Our assigned people in manufacturing and quality have the desire, but are encumbered with other significant work. Others in manufacturing, quality, procurement who are not involved directly, but whose help we need, are generating plenty of resistance. We are getting more instructional paper than engineering data. We wish we could get action by verbal request but such is not the case. This is a red flag.[31]

Naturally the Rogers commission also summoned Ebeling, to hear how he had viewed the situation back then:

EBELING:
> *My experience on O-rings was and is to this date that the O-ring is not a mechanism and never should be a mechanism that sees the heat of the magnitude of our motors . . . and in all cases, keep the heat of our rocket motors away from those seals. Whatever it is, you do not need chamber pressure to energize a seal.*

COVERT: *In this regard, then, did you have an increasing concern as you saw the tendency first to accept thermal distress and then to say, well, we can model this reasonably and we can accept a little bit of erosion, and then etc. etc.? Did this cause you a feeling of, if not distress, then betrayal in terms of your feeling about O-rings?*

EBELING: *I'm sure sorry you asked that question.*

COVERT: *I'm sorry I had to.*

EBELING: *To answer your question, yes. In fact, I have been an advocate—I used to sit in on the O-ring task force and was involved in the seals since Brian Russell worked directly for me, and I had a certain allegiance to this type of thing anyway, that I felt that we shouldn't ship any more rocket motors until we got it fixed.*

COVERT: *Did you voice this concern?*

EBELING: *Unfortunately, not to the right people.*[32]

In December 1985, Morton Thiokol made a formal request to NASA to have the problem with the O-rings closed. Later, the company maintained that a director at Marshall had in fact asked that such a request be made. Thiokol admitted that the problems had not been resolved but did say that every effort was being made to get them sorted out. They saw no reason, therefore, for repeatedly having to account for these problems, as they would have to do as long as Marshall's computers were registering the problem as "open"—that is, unsolved.

Again, the commission found it hard to understand how a problem that was not solved could be "closed." Astonishment and disbelief were evinced in the examination of Brian Russell of Thiokol:

ROGERS: *Can I interrupt? So you're trying to figure out how to fix it, right? And you're doing some things to try to help you figure out how to fix it.*

Now, why at that point would you close it out?

RUSSELL: *Because I was asked to do it.*

ROGERS: *I see. Well, that explains it.*

ROBERT W. RUMMEL: *It explains it, but really doesn't make any sense. On one hand you close out items that you've been reviewing flight by flight, that have obviously critical implications, on the basis that after you close it out, you're going to try to fix it.*

> *So I think what you're really saying is, you're closing it out because you don't want to be bothered. Somebody doesn't want to be bothered with flight-by-flight reviews, but you're going to continue to work on it after it's closed out.*[33]

On January 23, 1986, five days before the *Challenger* flight, the problem was closed at the Marshall Space Flight Center. The computers no longer recognized any problem with the field joints and O-rings, even though the problem was, in effect, wide open.

The Rogers commission had no difficulty in clearly expressing its dissatisfaction. Phenomena such as joint rotation, blowbys, and erosion were quite inconceivable when the field joints were designed. When they manifested themselves anyway, these phenomena were accepted to the point where, at the routine review held after every flight, they were deemed to present no risk. Arbitrary safety margins were set, only to be hastily extended when overtaken by developments.

In some respects, the interests of the Marshall Space Flight Center and Morton Thiokol conflicted. It was Marshall's job to keep Thiokol on their toes, while Thiokol endeavored to tell their watchdog no more than was absolutely necessary. They became masters in the art of procrastination—and the flights proceeded. But the interests of both parties were undoubtedly served by keeping NASA headquarters at arm's length. In Thiokol's case, to preserve the possibility of new and lucrative contracts; in Marshall's, to preserve a reputation won way back in von Braun's great days at Huntsville—for quality workmanship "better than the finest watch."

The harshest criticism was voiced by Richard Feynman:

> [In the flight reviews they] agonize whether they can go even though they had some blowby in the seal or they have a cracked blade in the pump of one of the engines, and then it flies and nothing happens.
>
> Then it is suggested, therefore, that the risk is no longer so high for the next flight. We can lower our standards a little bit because we got away with it last time . . . an argument is always given that the last time it worked.
>
> It is a kind of Russian roulette. You got away with it and it was a risk. You got away with it. But it shouldn't have been done over

and over again like that. When I look at the reviews, I find the perpetual movement heading for trouble.[34]

On twenty-four launches a risk had been taken, to keep the "production line" moving, and each time the astronauts had escaped with their lives. On mission number 25, *Challenger* 51-L, their luck ran out.

36. Feynman's Experiment

The sheaves of documents testifying to the fact that engineers had been working on the field joint problem for some time were not just lying around waiting to be picked up. None of these documents were marked confidential, but they were well hidden among mountains of paperwork at the Thiokol plant, the Marshall center, and NASA headquarters — immersed in the "sea of information" that seems to be an inevitable facet of life in a big organization. Only after months of work was it possible to piece together the picture as it has been presented in the previous chapter.

Richard Feynman did not have the temperament for sifting through piles of paper. He took an active part in the hearings, but left it to others to track down the documentation.

On the other hand, he could not resist moving in on another front. At that very first private sitting on Monday, February 10, 1986, when Cook and McDonald dropped their bombshell, Feynman conceived an idea for a striking and instructive demonstration of the technical problem which, in his opinion, had been the space shuttle's downfall. As yet, he knew nothing about the documents but he had grasped the concept of the joint design from his chat with the NASA experts on Saturday, and he had received a strange phone call the previous day, Sunday, from the Pentagon representative on the commission, Major General Kutyna, who told him:

> "I was working on my carburetor this morning, and I was thinking: the shuttle took off when the temperature was 28 or 29 degrees. The coldest temperature previous to that was 53 degrees. You're a professor; what, sir, is the effect of cold on the O-rings?"
>
> "Oh!" I said. "It makes them stiff. Yes, of course!"
>
> That's all he had to tell me. It was a clue for which I got a lot

of credit later, but it was his observation. A professor of theoretical physics always has to be told what to look for. He just uses his knowledge to explain the observations of the experimenters![1]

The next morning, Monday, Feynman asked his high-ranking former student Bill Graham to furnish him with any available material on the elasticity of the O-rings at different temperatures. While Feynman was attending the session at which Cook and McDonald made their astonishing disclosures, Graham's people were working fast. No sooner had the session come to a close than Feynman was handed a note from July 1985 detailing the results of a test at the Thiokol plant. Experiments had been made with compressing a piece of O-ring as it would be compressed in the joint before liftoff. After that, the gap had been opened up at the same speed and to the same extent as it would when the rocket was ignited. And finally the rubber's ability to regain its original shape after compression at different temperatures had been measured. The most telling passage in the note stated:

> At 100° F the O-ring maintained contact with both tang and clevis surfaces. At 75° F the O-ring lost contact for 2.4 seconds. At 50° F the O-ring did not reestablish contact in ten minutes, at which time the test was terminated.[2]

Feynman was arrogant enough to feel less than satisfied with this note. He could not have cared less about timespans of 2.4 seconds and 10 minutes. He had been hoping for data involving reaction times measured in milliseconds. The rocket reached its full thrust in just 0.6 second. As far as he was concerned, any timespan greater than this was of little interest.

Sitting there in his hotel, feeling as though he was wasting his time, he conceived the idea of experimenting with the O-rings and cold himself. All he lacked was an O-ring. Graham could not help him. Yes, they did have O-rings, but they were at the Kennedy Space Center in Florida. He could get hold of a test sample but that would take some time. While they were talking, however, it suddenly occurred to Graham that he had a sectioned model of a field joint which was to be passed around the commissioners the following day—while Lawrence Mulloy from the Marshall Space Flight Center gave them all the facts about field joints that Feynman already knew. This section also contained two small pieces of O-ring that he might be able to extract.

Feynman's spirits immediately lifted. Now here was an amusing exercise to get his teeth into. He arranged to meet Graham the next morning at the NASA chief's office before the commission hearing, which this time would be held in public. A number of television networks would also be present.

Early in the morning of Tuesday, February 11, Feynman was standing in the snow outside his hotel trying to hail a taxi. Having managed to grab one, he asked the driver to find him a hardware store, fast. At first, the driver did not catch what he had said. This was a fancy hotel. Usually he drove fares from here to Capitol Hill, the airport, or — once in a while — the White House. But a hardware store?

Feynman explained that he needed some tools. After driving around they found a hardware store. It was now 8:15 — the hardware store opened at 8:30. Being more used to California weather, Feynman had no overcoat and had to hang around stamping his feet in the snow until then. When he finally got in, he asked for some tapered pliers and the smallest C-clamp they had. It looked like it was going to be a good day after all.

Back to NASA headquarters. On the way to Graham's office Feynman began to worry that his C-clamp might not be small enough to dip into the glass of ice water he had in mind. So he went to see the NASA doctors to ask them for a surgical clamp like the ones they used on their rubber tubes. They had no such clamps, but they helped him check that his C-clamp could be lowered into the glass.

At Graham's office the pliers and C-clamp were brought out and a sample of rubber seal was easily extracted from the model. Feynman felt a little ashamed of himself, but he could not resist cheating a bit and trying out the experiment before performing it before his audience. Then he slipped the piece of rubber back into the model and left the office, whistling to himself and reminding Graham to bring the model to the hearing.

Feynman had the pliers in one pocket and the C-clamp in another. At the previous sitting a glass of ice water had been set out at each place. This time there was none. Feynman had to ask one of the attendants for a glass of ice water. "Certainly, sir." The hearing got underway. Larry Mulloy began by describing the booster joints. Feynman could not concentrate. He was still missing one vital prop for his act. Impatiently he signaled again, only to receive a reassuring nod. His water was on its way.

Feynman used the interim to question Mulloy, thus setting the scene perfectly for the subsequent experiment. While all this was going on, the model of the joint was working its way around the table.

FEYNMAN:	*During a launch, there are vibrations which cause the rocket joints to move a little bit — is that correct?*
MULLOY:	*That is correct, sir.*
FEYNMAN:	*And inside the joints, these so-called O-rings are supposed to expand to make a seal — is that right?*
MULLOY:	*Yes, sir. In static conditions they should be in direct contact with the tang and clevis and squeezed twenty-thousandths of an inch.*
FEYNMAN:	*Why don't we take the O-rings out?*
MULLOY:	*Because then you would have hot gas expanding through the joint . . .*
FEYNMAN:	*Now, in order for the seal to work correctly, the O-rings must be made of rubber — not something like lead, which, when you squash it, it stays.*
MULLOY:	*Yes, sir.*
FEYNMAN:	*Now, if the O-rings weren't resilient for a second or two, would that be enough to be a very serious situation?*
MULLOY:	*Yes, sir.*[3]

Feynman had no further questions. The course was now marked out. If Mulloy had already been familiar with Feynman's way of working, he would have noticed that the professor was now playing his favorite role — that of the naive individual who wants to be quite sure that he has understood these very complex facts correctly. At the same time, Feynman had shown Rogers that he could also play the part of the astute lawyer imperceptibly tightening the net around his unsuspecting victim.

Now the model had reached Feynman. The pliers were whipped out of the jacket pocket, and he extracted the O-ring section. But he still did not have the ice water.

At long last the door opened and a young woman sidled quietly in carrying a tray with ice water for everyone. Naturally, she started with the chairman and the vice-chairman and then proceeded around the table. By the time Feynman was handed a glass he already had the rubber sample wedged in the C-clamp. Quickly, he lowered the whole thing into the ice water and left it there for a few moments.

Major General Kutyna, who was seated next to Feynman, had followed all these preparations with interest and immediately grasped what Feynman was doing. When Feynman could control himself no longer and was about to press the button on his microphone, Kutyna grabbed his arm and held him back. He explained to him in a whisper that he would gain maximum effect if he waited until Mulloy's lecture brought him to one particular slide, a copy of which the commissioners had in front of them.

Feynman waited, although it was not easy. When this slide came up, he was ready. He pressed his button and switched on his microphone:

> We spoke this morning about the resiliency of the seal and if the material weren't resilient, it wouldn't work in the appropriate mode but would be less satisfactory, in fact it might not work well. I took this stuff that I got out of your seals and I put it in ice water and I discovered that when you put pressure on it for a while and then undo it, it doesn't stretch back. It stays the same dimension. In other words, for a few seconds at least—and more seconds than that—there is no resilience in this particular material when it is at a temperature of 32° F. I believe that has some significance for our problem.[4]

While he was speaking, Feynman unscrewed the C-clamp, so that the press photographers and television cameras could get their pictures. Feynman was well aware that he had just generated a new anecdote to add to the string that his students never grew tired of relating. Single-handed, he had got the better of this high-tech space agency and produced an accurate diagnosis just fourteen days after the disaster, with nothing other than common sense, pliers, a C-clamp and a glass of ice water.

Rogers hemmed and hawed and said that, on the face of it, this seemed like a most interesting experiment. But they had to move on. The commission would return to this question at a later date, when they came to the hearing on weather conditions. At that time Mr. Mulloy would no doubt wish to comment more fully on this matter.

Larry Mulloy was not exactly thrilled by the turn his lecture had taken, but Feynman was not letting him off the hook:

> I just wanted to know, before the event, from information that was available . . . was it fully appreciated everywhere that this seal would become unsatisfactory at some temperature and was there some sort of suggestion of a temperature at which the SRB shouldn't be run?

The question was clear-cut, scientific, and could hardly be misunderstood. The same cannot be said of Larry Mulloy's reply:

Yes, sir. There was a suggestion of that, to answer the first question. First, the data that was presented, it was the judgment that under the conditions that we would see on launch day, given the configuration that we were in, that seal would function at that temperature. That was the final judgment.[5]

37. *Teleconferences*

Next, the Rogers commission wanted to take a closer look at the decision-making process as it had operated on January 27, the day before the *Challenger* launch. According to Allan J. McDonald's testimony, NASA had forced through a launch despite a clear veto from the solid rocket booster contractors, Morton Thiokol. The commission took such a serious view of these allegations that three days were set aside at the end of February for public hearings on this matter.

First of all, however, the commissioners had to acquaint themselves with all the ramifications of NASA's launch decision chain, which worked on a basically hierarchical principle. The process began at the bottom, at what was known as Level IV. Here, all of the space shuttle prime contractors had to attest that components supplied by them and their subcontractors were in good working order and that they saw no impediment to the launch. Obviously, before these testimonials were certified and verified, a series of meetings were held with each contractor — until NASA was certain that all outstanding problems had been solved.

Testimonials and certificates from this nethermost level were then passed on to the respective project managers at the Marshall, Johnson, and Kennedy centers, who constituted Level III. Every single contractor had already been assigned an opposite number at a NASA center to whom he reported regularly and with whom he discussed any problems. Now the contractor had to confirm with his opposite number that he was ready. The Marshall Space Flight Center had responsibility for the main engines and fuel tank as well as the solid rocket boosters. Project managers for each of these departments (among them Lawrence Mulloy) reported back to the manager of the Marshall Shuttle Projects office, Stanley R. Reinartz, that they were ready to launch. Thereafter, a brief formal meeting was held with the Marshall Center director, William

STS Organization

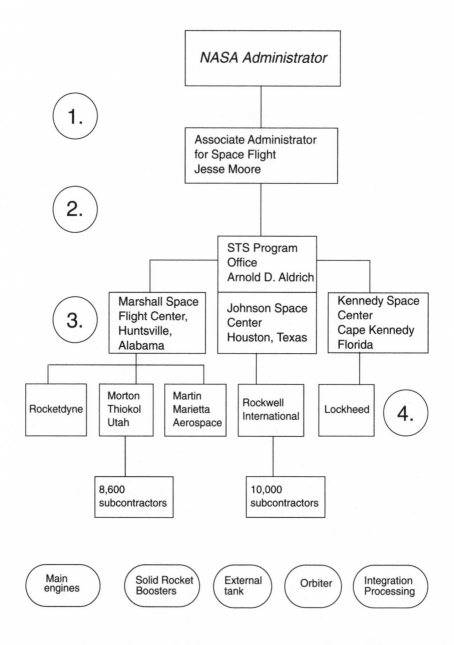

Lucas, at which it would be established, once and for all, that—in the opinion of both the contractors and NASA—all components were now operational. The same process was, of course, also taking place simultaneously at the Johnson and Kennedy Space Centers regarding those parts of the space shuttle system for which these centers acted as watchdogs.

The process then moved up to Level II—assuming that all of the NASA Level III centers had reported A-OK to the manager of NASA's National Space Transportation Systems Program Office, Arnold D. Aldrich. This coordinating office had its physical location at Johnson Space Center in Houston, but had no special link with that center. This was where all the threads were drawn together, and only now was it possible to gain an overall view of any remaining problems that needed to be sorted out before the launch.

Eventually, Aldrich was able to advise the Associate Administrator for Space Flight, Jesse Moore, at NASA headquarters—Level I—that everything was set for the Flight Readiness Review on January 15, 1986.[1]

This system might seem rather regimented. The idea behind the hierarchical structure was that it should prevent the decision makers at the top from being swamped by minor details that ought to have been dealt with long before further down the chain. Only major problems with some bearing on the launch found their way to the upper levels, and it became a point of honor with the individual NASA field centers to sort out any problems themselves, so that the two topmost levels would not be inconvenienced by failures and delays at their center.

The Flight Readiness Review held for *Challenger* on January 15 was an undramatic affair. Jesse Moore presided over a teleconference at which managers and middle managers from the relevant NASA centers gave the all clear for mission 51-L. Project Manager Larry Mulloy from Marshall—responsible for the solid rocket boosters—stated: "no major problems or issues,"[2] and Moore was free to move on to the next question.

Neither Moore (Level I) nor Aldrich (Level II) received any information whatsoever about the problems with the O-rings leading to a "launch constraint" which Mulloy (Level III) had now personally waived for six consecutive flights before 51-L. Nor had these problems left any trace of themselves in certificates or testimonials bearing the necessary verification from Morton Thiokol (Level IV).[3] Both Level III and Level IV believed they had the problems under control.

Minor hitches were resolved then and there. With the TDRS-B satellite on board, *Challenger* would exceed the permitted weight for any potential forced landing on one of two possible emergency runways. Agreement to issue a waiver was reached without any trouble. A few other problems would have to be resolved before the next meeting, less than fourteen days away.[4]

On January 25, 1986, the L-1 review was held. Once again, at this meeting the Marshall people from Huntsville sat across the table from representatives of the STS Program Office in Houston, with Jesse Moore from headquarters in Washington in the chair. The pecking order was quite clear. Marshall's delegation, led by Stanley Reinartz, reported to Arnold Aldrich, who in turn reported to Jesse Moore. Also present was the director of the Marshall center, William Lucas — although officially with no part to play in the decision-making process.

All outstanding problems had now been resolved. The last "open" matters could be pronounced "closed." The only major talking point was provided by the weather conditions. There was some concern that it might rain. *Challenger* had been standing out on its launch pad for thirty-five days, and during that time it had been exposed to seven inches of rain — almost three times the normal rainfall for such a period.[5] The ten-man delegation from Marshall, led by William Lucas himself, had almost been late for the meeting because bad weather had made it impossible for their plane to land at Kennedy.[6] Because of its tiles, the Orbiter could not, under any circumstances, be launched in the rain.

Other than the rain, the meeting touched briefly on the cold front which, according to the meteorologists, was moving up behind it. Not even at this juncture did anyone mention the possibility of any problems with the solid rocket boosters.

As we saw earlier, on Monday, January 27, an attempt was made to launch *Challenger*. This attempt had, however, to be abandoned just after midday because of strong crosswinds over the space center runway. Directly after the scrub, at 1:00 p.m., the Mission Management Team, led by Arnie Aldrich, met to hear whether any of the experts at other NASA centers had any objection to a launch twenty-four hours later than first planned. At this point, the meteorologists reported that the temperature would drop to the low twenties on Monday night. A decision was made to switch on the heating elements situated at various points on the Orbiter and to leave the faucets running on the launch pad; but nothing was said about any problem with the boosters — much less the O-rings. The experts from the Marshall Space Flight Center were still saying "Go" to a launch the following morning.

It was at this point that one of the Morton Thiokol representatives at the Kennedy Space Center rang home to the plant in Utah — just to be on the safe side. He wanted to ask experts there for their feelings about the low temperature. At Utah, senior engineers, among them Robert Ebeling and Roger Boisjoly, were immediately called in for a hastily improvised conference. The engineers were alarmed to hear about the cold conditions. These temperatures were far below the existing "data base" from previous flights.

Later that afternoon, Bob Ebeling made a call to Thiokol's Allan J. McDonald, stationed at Kennedy. Ebeling wanted to tell him that the engineers in Utah were taking this situation very seriously, and to ask McDonald to obtain more information on anticipated temperatures in the hours up to launch time.

McDonald now learned that the temperature would drop below freezing just before midnight and that it would fall to around 22° F in the early hours. At the time scheduled for liftoff (9:38) it would be around 26° F.[7]

He rang these figures through to Ebeling in Utah. McDonald told Ebeling that he had better round up his own people and notify the vice president of engineering, Robert Lund. It was vital, McDonald stressed, that any eventual decisions be taken by management on the *technical* side, rather than by the project management people from Thiokol's senior administration.

Allan McDonald also tried to get hold of Lawrence Mulloy, to notify the people at Marshall of Thiokol's concern. Mulloy, however, could not be reached. Instead McDonald got through to Cecil Houston, another Marshall man at Kennedy. Cecil Houston listened to what he had to say and promised to arrange a teleconference between Levels III and IV at which the people from Kennedy could discuss the problems with Thiokol and the experts at the Marshall Space Flight Center. Cecil Houston would inform his boss, Stanley Reinartz at Marshall, to make sure that he too would be listening in on the net.

This first teleconference commenced at 5:45 p.m., while McAuliffe and the other six astronauts, all-unwitting, were resting their stiff limbs after all the wasted hours spent that day on their backs in the Orbiter.

At the teleconference, Thiokol presented their arguments for postponing the launch until the temperature had risen. Mention was made of the cold 51-C launch and the widespread damage to the O-rings on that occasion. Stan Reinartz told the Rogers commission that his feeling had been not that the people at Thiokol were advocating a postponement but rather that they wanted to have a full and frank discussion about these problems before proceeding with the countdown.[8] Reinartz's second-in-command, Judson Lovingood, however, was in no doubt: Thiokol recommended a postponement. Lovingood had therefore suggested contacting the launch decision chain's Level II and Level I — Arnold Aldrich and Jesse Moore — to put them in the picture and confront them with Thiokol's concern.

Reinartz wanted to notify his ultimate boss at the Marshall Space Flight Center, William Lucas, but he saw no reason whatsoever to inform higher levels within NASA of the situation.[9]

The members of the Rogers commission found it incomprehensible that a problem of this magnitude had not been referred to a higher authority. Instead, considerable effort had been made to keep it a Level III matter. Here was a textbook example of inexplicable behavior where it would undeniably be of help to know the underlying rules or laws. The commission's understanding was helped a little, though, by the arrival — some way into the investigation — of an anonymous character sketch of William R. Lucas, who had inherited not only the Marshall Space Flight Center but also his management style from Wernher von Braun. This anonymous communication said:

> The Marshall Space Flight Center is run by one man, William R. Lucas. His style of management can best be described as feudalistic. In his ten year tenure as center director, he has established a personal empire built on the "good old boy" principle. The only criteria for advancement is total loyalty to this man. Loyalty to country, NASA, the space program mean nothing. Many a highly skilled manager, scientist or engineer has been "buried" in the organization because they underestimated the man's psychopathic reaction to dissent.[10]

Reinartz had to put his loyalty to William Lucas before any thought for Aldrich, a man working at the rival Johnson Space Center. It had long been a source of irritation and jealousy — and not only at Marshall — that this should have been the center chosen to house the coordinating STS Program Office. Many people at Huntsville felt that Houston had already received more than its fair share of the credit for NASA's Manned Space Flight. All things considered, Lucas would not be happy about Marshall having to take its problems up to Level II for airing. Lucas's style — like von Braun's—involved playing his cards very close to his chest when dealing with other NASA field centers or with headquarters in Washington. He had come to Cape Kennedy to see his men do what he expected of them. The anonymous source told the commission:

> It has been apparent for some time that the Flight Readiness Review process developed by Lucas and other senior NASA managers simply was not doing the job. It was not determining flight readiness. Rather, it established a political situation within NASA in which no center could come to a review and say that it was "not ready." To do so would invite the question "If you're not ready, then why are you not doing your job? It is your job to be ready." . . . As a consequence, each center gets up and basically "snows" Headquarters with highly technical rationales that no one but the immediate experts involved can completely judge. Lucas made it known that under no circumstances is the Marshall Center to be the cause for delaying a launch.[11]

William Lucas prided himself on the fact that at no point during the twenty-five Flight Readiness Reviews up to and including *Challenger* had the Marshall Space Flight Center been forced to raise any objections to a launch.[12] Immediately after the first teleconference, he heard Reinartz talking about Thiokol's "concern" regarding the cold and the O-rings. He also heard that another "conference" was to be held at 8:45 p.m. He was feeling quite confident, convinced that his people would get this problem sorted out without there having to be any thought of scrubbing the launch. And so it was obviously quite unnecessary to advise Arnold Aldrich and Jesse Moore.

Between the two teleconferences, Thiokol engineers had time to prepare various charts and graphs to be faxed to the other participants at Kennedy and Marshall before the meeting resumed. These documents showed the previous instances of eroded O-rings and the blowby, whereby hot gases had penetrated past the rings.

When the second teleconference commenced at 8:45, Roger Boisjoly presented those involved with a chart detailing his main concerns in technical language and employing those inevitable oversized bullets. In fact, this chart had already been used at the meeting at headquarters in August 1985.

Launch Decision Chain Regarding the Solid Rocket Boosters

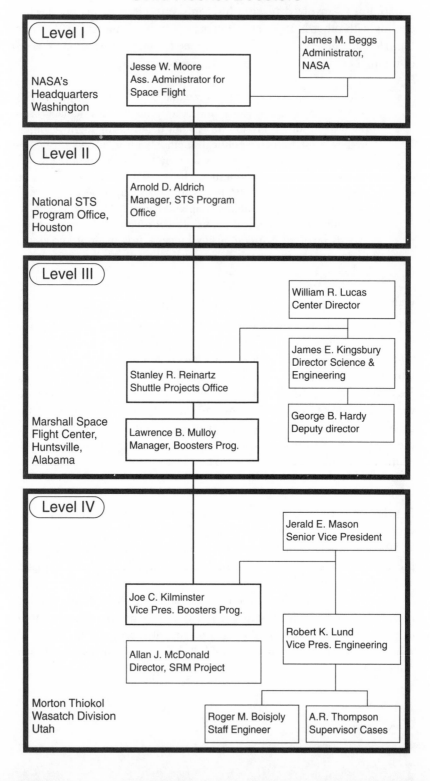

PRIMARY CONCERNS

- FIELD JOINTS — HIGHEST CONCERN

- EROSION PENETRATION OF PRIMARY SEAL REQUIRES RELIABLE SECONDARY SEAL FOR PRESSURE INTEGRITY

  - IGNITION TRANSIENT (0-600 MS)
    - (0-170 MS) HIGH PROBABILITY OF RELIABLE SECONDARY SEAL
    - (170-330 MS) REDUCED PROBABILITY OF RELIABLE SECONDARY SEAL
    - (330-600 MS) HIGH PROBABILITY OF NO SECONDARY SEAL CAPABILITY

  - STEADY STATE (600 MS — 2 MINUTES)
    - IF EROSION PENETRATES PRIMARY O-RING SEAL — HIGH PROBABILITY OF NO SECONDARY SEAL CAPABILITY
      - BENCH TESTING SHOWED O-RING NOT CAPABLE OF MAINTAINING CONTACT WITH METAL PARTS GAP OPENING RATE TO MEOP (Maximum Engine Operating Pressure)
      - BENCH TESTING SHOWED CAPABILITY TO MAINTAIN O-RING CONTACT DURING INITIAL PHASE (0-170 MS) OF TRANSIENT[13]

Here, at long last, were the data in milliseconds that Feynman had been looking for. During his presentation Boisjoly translated his chart into everyday language, so that even the nontechnicians on the line could follow it. His main point was that if a primary O-ring broke down while flying under full pressure, the secondary O-ring would not be able to come to its rescue. Boisjoly's chart stressed the fact that the O-rings were not protected by any secondary system. They were, in fact, Criticality 1. Under normal conditions they could, however, get away with running such a risk, because the ring sealed the gap within the first 170 milliseconds, while there was still every chance of it being safeguarded by the secondary O-ring — before that was disengaged by the joint rotation.

But in this case, because of the low temperatures, conditions were far from normal. Boisjoly's biggest worry was that the O-ring would react more slowly this time around, and thus be pushed down into that more problematic gray area after the 170 milliseconds — when it was highly unlikely that the secondary O-ring would still be functioning, and when hot gases would burn off the surface of the O-ring just as fast as it was trying to get into position.

To underline the connection between the low temperatures and a slower O-ring, Boisjoly had also brought along another chart, rather more didactic in its wording.

JOINT PRIMARY CONCERNS

- A Temperature Lower Than Current Data Base Results in Changing Primary O-Ring Sealing Timing Function
- SRM (Solid Rocket Motor) 15A
  (used on the 51-C flight in January 1985)
  80° ARC Black Grease Between O-Rings
- Lower O-Ring Squeeze due to lower temp.
- Higher O-Ring shore hardness
- Thicker grease viscosity
- Higher O-Ring pressure actuation time
- If actuation time increases, threshold of secondary seal pressurization capability is approached
- If threshold is reached then secondary seal may not be capable of being pressurized[14]

Roger Boisjoly adopted an even more didactic tone with the Rogers commission when asked to elaborate upon his chart. He simply explained that as anyone could see, it was easier to squeeze a sponge into a crack than it was to force in a brick. In the cold, as Feynman had also demonstrated in his experiment, the rubber rings were turning into bricks. The grease, being colder, would also grow rigid and viscous and help keep the O-rings stuck in their grooves longer.

At the teleconference, Boisjoly was asked to quantify his concern. Could he produce any figures showing how great the probability was of things going as badly wrong as he was claiming? He could not. And there was one good reason why not. No Orbiter had ever been launched in such cold conditions. Boisjoly could only say that he was convinced "that it was away from goodness in the current data base."

The validity of the findings from the coldest launch so far (51-C) were also contested. Had soot not also been observed in the grease on 61-A, which had been launched in October 1985 at a temperature of 75° F? Quite obviously, the temperature could not be the only factor involved. Boisjoly insisted that the damage had been far more drastic on 51-C.

The grease had been "jet black," he said. Mulloy asked wearily when they had started using criteria based on the color of the grease.

Then it was Thiokol's vice president of engineering, Robert K. Lund's turn to speak. The heads of Thiokol's engineering division would now have to come to a decision on the basis of their own engineers' calculations. Having spent the entire afternoon discussing the problems with his staff, Lund was in no doubt. *Challenger* would have to stay on the launch pad until the temperature rose again. As a basis for his recommendation he cited the temperature of just over 53° when 51-C had been launched a year earlier—another textbook example, in fact, of the Russian roulette principle. On that occasion, the blackened grease between the O-rings notwithstanding, everything had gone well. So there was no reason not to risk another shot at it. Nevertheless, he went against his principles enough to advise, in the strongest terms possible, against running below the 51-C temperature. Conditions had to be the same as last time (or better). Otherwise it would be better to wait. They ought not to tempt fate any more than they were already doing. Bob Lund handwrote Thiokol's recommendation personally, in block capitals:

RECOMMENDATIONS:

• O-RING TEMP MUST BE ≥ 53° F AT LAUNCH

DEVELOPMENT MOTORS AT 47° F to 52° F WITH PUTTY PACKING HAD NO BLOW-BY

SRM 15 (51-C) (THE BEST SIMULATION) WORKED AT 53° F

• PROJECT AMBIENT CONDITIONS (TEMP & WIND) TO DETERMINE LAUNCH TIME[15]

Marshall's Deputy Director for Science and Engineering, George B. Hardy, found it almost impossible to hear Bob Lund out. This was not merely a cautious suggestion that a postponement might be most advisable. This was a veto. A damn contractor was presuming to decide that *Challenger* should be grounded. Hardy said he was "appalled" by Thiokol's recommendation. Lawrence Mulloy, also from Marshall, interrupted to ask indignantly whether Thiokol really wanted him to wait until "next April" before launching *Challenger*. He for one could find nothing whatsoever in Thiokol's material to shake his belief that the primary

O-ring would do its job, just as it had done twenty-four times previously.[16]

Hardy changed tack. Then again, he said, the way he saw it, if the contractor had issued a veto, then of course there could be no question of launching the space shuttle. Mulloy then asked Thiokol vice president Joseph C. Kilminster, one of the links in the official launch decision chain, whether he agreed with his colleague Robert Lund. Kilminster replied that he would have to abide by the expertise of his engineers and, hence, their recommendation. Present in the room while all this was going on was Thiokol's senior vice president, Jerald E. Mason—both Lund's and Kilminster's ultimate boss. Until now he had kept silent. Now he stirred, prompting Kilminster to quickly ask for a five-minute caucus, to allow the Thiokol people to consult among themselves. Evidently, a pause for thought was called for. It was 10:30, the debate had been running back and forth for almost two hours, and the *Challenger* astronauts were now bedding down for the night, to be ready to cope with the demands that would be made on them early the next morning.

Instead of taking five minutes, the caucus lasted almost half an hour. At Thiokol, Jerald Mason began by making it clear that from now on whatever line the company took would be "a management decision." Neither staff engineer Roger M. Boisjoly nor his supervisor Arnold R. Thompson had anything to do with management. Nevertheless, they tried to get through to their superiors, in the hope of preventing a launch which, from an engineering point of view, would have to be regarded as indefensible. In Boisjoly's words:

> Arnie actually got up from his position, which was down the table, and walked up the table and put a quad pad down in front of the table, in front of the management folks, and tried to sketch out once again what his concern was with the joint, and when he realized he wasn't getting through, he just stopped . . .
>
> I went up and discussed the photos once again and tried to make the point that it was my opinion from actual observations that temperature was indeed a discriminator and we should not ignore the physical evidence that we observed.[17]

Not a single engineer in the room supported a launch. Opposition was unanimous among the technical personnel involved in the conference. Jerry Mason repeated that it was now up to the management of the company to reach a decision. He ran his eyes over his staff and asked rhetorically, "Am I the only one who wants to fly?"[18] He turned to Robert Lund, who had written out Thiokol's veto, and asked him to "take off his engineering hat and put on his management hat." He ought to be assessing the situation as a responsible company manager. By now, the engineers had realized that management was going to make the decision to launch *Challenger*, come hell or high water. They respected their management's right to make such a decision. For their part, there was nothing more they could do. But they could not help seeing this decision as a searing defeat.

Boisjoly told the Rogers commission how the situation at this meeting had imperceptibly altered — in a complete "turnaround from other pre-flight meetings." Usually, at such meetings, it was the job of the engineers and technicians to prove, beyond a shadow of a doubt, that this or that technical solution was warrantable. This time, the boot was on the other foot; the "burden of proof" had shifted. The engineers had to prove, with one hundred percent certainty, that something would go wrong if the flight was not stopped. And, again, there was a good reason why they could not. After all, there were still all those instances of flights on which the O-rings had not been damaged. They could not deny that *Challenger might* also make it. And since they could not be certain that "mission, vehicle and crew" would be lost, Thiokol's management agreed to reverse Lund's decision. Naturally, the commission examined Robert Lund on this question.[19]

ROGERS: *How do you explain the fact that you changed your mind when you changed your hat?*

LUND: *. . . We had to prove to them that we weren't ready. And so we got ourselves in the thought process that we were trying to find some way to prove to them it wouldn't work, and we were unable to do that. We couldn't prove absolutely that the motor wouldn't work.*

ROGERS: *In other words, you honestly believed that you had a duty to prove that it would not work?*

LUND: *Well, that is kind of the mode we got ourselves into that evening. It seems like we have always been in the opposite mode. I should have detected that, but I did not, but the roles kind of switched.*[20]

I guess one of the big things that we really didn't know [was] whether the temperature was the driver or not. We couldn't tell. We had hot motors that blew by and cold motors that blew by and some that did not. The data was inconclusive. And so I had trouble justifying it in my own mind and saying, by golly, temperature is the factor.[21]

Roger Boisjoly told the commission that he had been "crucified" more than once at meetings with people from the Marshall Space Flight Center for using phrases such as "I think," "I feel," or "I expect." He had learned the hard way with William Lucas — never open your mouth unless your technical documentation is 100 percent in order. Anyone could be "judgmental" but it had taken more than hot air to put Neil Armstrong on the moon. The only thing that counted at Marshall was "engineering-supported statements."

> When people go in front of Dr. Lucas, they know full well that if they use words like that or if they use engineering judgment to try to explain a position, they will be shot down in flames. And for that reason, nobody goes to him without a complete, fully documented, verifiable set of data.[22]

Robert Lund's problem was that he and his engineers had no such data because they were looking only at those flights on which problems had arisen with the O-rings. And these had happened at temperatures ranging all the way from 53° F to 75° F. So in that sense the findings were not conclusive. If, however, anyone had thought to cast an eye over all of the flights — something which, in the heat of the moment, never occurred to anyone — then the picture became much clearer. Only three instances of O-ring thermal distress had been observed on the twenty flights that had taken place at 66° F or above. All four flights staged at a temperature of 63° F or below had shown heat damage. So in fact there were some strong indications that damage to the O-rings was not only likely but almost a dead certainty at anything below 65° F or thereabouts.[23]

Obviously the people at the Kennedy Space Center also spent the break debating this point. Allan McDonald from Thiokol told Larry Mulloy from Marshall that, according to the documentation, the solid rocket boosters were only qualified for use between 40° F and 90° F. So he could understand that Mulloy found it hard to accept a new benchmark of 53° F; on the other hand, he just did not know how he could allow himself to go below 40° F. Mulloy said he would never dream of doing so but the 40° F figure was actually a set limit for the mean bulk temperature of the 500 tons of solid fuel inside the boosters. While they were talking, this was still registering a temperature of 55° F. So there was no problem.

This was the first time McDonald had come across this argument about the mean bulk temperature. The way he saw it, the temperature limit had to be based on "ambient temperature." Nor could he make sense of the fact that, according to Mulloy, the space shuttle system as a whole was certified down to 31° F. By his way of thinking, the lowest limit would have to be 40° F — on the principle that a chain is only as strong as its weakest link. Besides, even hours and hours of incredibly severe frost would never bring the mean bulk temperature of all those tons of fuel down that far. Exactly, said Mulloy, which is why it was also safe to fly. It was at this point that McDonald gave up and said that he was not prepared to be party to a decision to fly at temperatures for which the boosters were not qualified.[24]

Boisjoly described his impression of the mood at the meeting to the Rogers commission:

BOISJOLY: *I felt personally that management was under a lot of pressure to launch and that they made a very tough decision, but I didn't agree with it.*

WALKER: *Do you know the source of the pressure on management that you alluded to?*

BOISJOLY: *Well, the comments made over the [net] is what I felt, I can't speak for them, but I felt it — I felt the tone of the meeting . . .*[25]

Larry Mulloy disagreed totally. He did not believe that Thiokol had been pressured in any way whatsoever. The mood had been matter-of-fact and businesslike:

MULLOY: *It has been suggested, implied or stated that we directed Thiokol to go reconsider these data. That is not true. Thiokol asked for a caucus so that they could consider the discussions that had ensued and the comments that Mr. Hardy and I and the others had made.*[26]

The crux of his own counterstatement was that, at the eleventh hour, Thiokol had suddenly come up with some completely new "launch commit criteria" for the solid rocket boosters. This was unacceptable and unnecessary. The existing benchmarks had worked perfectly well so far and there was no need to change them now.

Rogers could not suppress a touch of irony:

ROGERS: *Do you remember any other occasion when the contractor recommended against launch and you persuaded them they were wrong and had them change their mind?*

MULLOY: *No, sir.*[27]

At 11:00 the meeting was resumed. Over the net, Thiokol vice president Kilminster said that they had taken another look at all the available data and discussed the situation and they had to admit that the data was inconclusive. Thiokol would, therefore, no longer oppose a launch. On the contrary, they now recommended that *Challenger* be launched as planned. Furthermore, the Thiokol management had managed to draw up a new rationale, which Kilminster now read out:

MTI ASSESSMENT OF TEMPERATURE CONCERN ON SRM-25 (51-L) LAUNCH

- CALCULATIONS SHOW THAT SRM-25 O-RINGS WILL BE 20° F COLDER THAN SRM-15 O-RINGS
- TEMPERATURE DATA NOT CONCLUSIVE ON PREDICTING PRIMARY O-RING BLOW-BY
- ENGINEERING ASSESSMENT IS THAT:
  - COLDER O-RINGS WILL HAVE INCREASED EFFECTIVE DUROMETER ("HARDER")
  - "HARDER" O-RINGS WILL TAKE LONGER TO "SEAT"
    - MORE GAS MAY PASS PRIMARY O-RING BEFORE THE PRIMARY SEAL SEATS (RELATIVE TO SRM-15)
      - DEMONSTRATED SEALING THRESHOLD IS 3 TIMES GREATER THAN 0.038" EROSION EXPERIENCED ON SRM-15
  - IF THE PRIMARY SEAL DOES NOT SEAT, THE SECONDARY SEAL WILL SEAT
    - PRESSURE WILL GET TO THE SECONDARY SEAL BEFORE THE METAL PARTS ROTATE
      - O-RING PRESSURE LEAK CHECK PLACES SECONDARY SEAL IN OUTBOARD POSITION WHICH MINIMIZES SEALING TIME

- MTI RECOMMENDS STS-51L LAUNCH PROCEED ON 28 JANUARY 1986
  - SRM-25 WILL NOT BE SIGNIFICANTLY DIFFERENT FROM SRM-15

[Signed:]
JOE C. KILMINSTER, VICE PRESIDENT
SPACE BOOSTER PROGRAMS MORTON THIOKOL INC. Wasatch Division[28]

Here, while faithfully reiterating all of his engineers' arguments, Kilminster managed to reach the very opposite conclusion. He justified this by asserting that the O-rings had scope for three times the erosion observed on 51-C. Erosion three times as great had, in fact, been encountered before with no disastrous outcome. Consequently there ought to be some latitude for *Challenger* — even if the cold did have the effect that the engineers claimed. Furthermore, the pressure testing between the O-rings in each joint pressed the O-rings outward in their grooves. When the joint rotated, the secondary O-ring would therefore be in the position where the gap between it and the opposite wall was narrowest. Kilminster said nothing about the fact that this same pressure testing pushed the primary O-ring up to the upper rim of its groove, into a position where the gap would be at its widest. Contrary to the general feeling among the company's engineers, he was convinced that the secondary O-ring would provide a reliable safety net if the primary O-ring proved to be as slow as they claimed.

With Thiokol's senior vice president, Jerald Mason, Richard Feynman took up the question of whether the O-rings really had a safety factor of three. How could they be sure, he asked, that they couldn't get more erosion than the amount they got on 51-C? Mason could not say for sure.

FEYNMAN: *What made you think that 51-C was the maximum erosion that you could possibly expect?*

MASON: *We felt that the factor of three times was not likely to be exceeded.*[29]

Larry Mulloy and George Hardy listened with indulgent interest as Kilminster recited his arguments. Now they were seeing more eye to eye with him and the contractor whom he represented. As a matter of form, George Hardy did, however, ask him to sign his rationale and fax it to Kennedy and to Marshall—so that it would be quite clear that the contractor with sole responsibility for the *Challenger* solid rocket boosters wholeheartedly recommended a launch the next morning.

The meeting was at an end. At Cape Kennedy, where *Challenger* hung, upended and gleaming snow white—a national icon bathed in the light of scores of spotlights—it was 11:15. A little later, all over the base, klaxons sounded the warning for everyone to clear the launch pad, so that the job of pumping in the 800 tons of liquid explosive could begin all over again.

In Utah, Roger Boisjoly returned to his office. Feeling empty inside, he sat down at his desk, switched on his light, and wrote the following diary entry, for no one's eyes but his own:

> I sincerely hope this launch does not result in a catastrophe. I personally do not agree with some of the statements made in Joe Kilminster's written summary stating that SRM-25 is okay to fly.[30]

After the teleconference, Marshall's Stanley Reinartz asked those present whether they now saw any other outstanding problems. Brief mention was made of the ice on the launch pad, but the ice team would attend to that. This left just the weather conditions in the Atlantic. The two special ships cruising offshore were reporting very high seas, which might make it difficult to retrieve the boosters after use. At any rate, they had better allow for the possibility of losing the parachutes and booster nose cones, which disengaged on splashdown.

Larry Mulloy rang Arnold Aldrich (Level II) and told him about the stormy conditions in the Atlantic. Aldrich asked him what exactly they might stand to lose—in financial terms. When advised by Mulloy that it would be somewhere in the region of "only" a million dollars,[31] Aldrich felt that they would just have to accept the loss of the parachutes and nose cones, if they wanted to get *Challenger* off the ground. Mulloy saw no reason to mention one word about the hours they had just spent in a teleconference with Thiokol. Rogers asked him why not.

MULLOY: *At that time, and I still consider today, that was a Level III issue, Level III being an SRB element or an external tank element or an Orbiter. There was no violation of the Launch Commit Criteria. There was no waiver required in my judgment at that time and still today.[32]*

Major General Kutyna was still not quite clear on this point:

KUTYNA: *Larry, let me follow through on that, and I'm kind of aware of the launch decision process, and you said you made the decision at your level on this thing. If this were an airplane, an airline, and I had just had a two-hour argument with Boeing on whether the wing was going to fall off or not, I think I would tell the pilot, at least mention it.*

MULLOY: *I did, sir.*

KUTYNA: *You did?*

MULLOY: *Yes, sir.*

KUTYNA: *Tell me what levels above you.*

MULLOY: *As I stated earlier, Mr. Reinartz, who is my manager, was at the meeting, and in the morning, about five o'clock in the operations support room, where we all were, I informed Dr. Lucas of the content of the discussion.*

KUTYNA: *But he is not in the launch decision chain.*

MULLOY: *No, sir. Mr. Reinartz is in the launch decision chain, though.*

KUTYNA: *And he is the highest level in that chain?*

MULLOY: *No. Normally it would go from me to Mr. Reinartz to Mr. Aldrich to Mr. Moore.*[33]

But, according to their testimonies, neither Aldrich (Level II) nor Moore (Level I), who was to "push the button," had heard a word about this lengthy discussion. Once again, the Marshall Space Flight Center had sorted things out itself with Morton Thiokol. How they did this was no business of the upper levels. The main thing was that, once Thiokol had recommended "STS-51L launch proceed," the problem no longer existed. The commission found it hard to believe that—even though no official notification had been made—word had not even reached senior management on the grapevine that thrives in every large organization. But both Moore and Aldrich denied, throughout the hearings, having heard so much as a whisper.

The director of the Marshall Space Flight Center, William R. Lucas, was, however, told—although not until early the next morning. He felt satisfied that his people had solved the problem. So the center at Huntsville could not be blamed for any delay this time either.

Lucas assured the commission that he had heard nothing between around eight in the evening and five o'clock the next morning. About the conversation early that morning, he said:

LUCAS: *[Reinartz] told me, as I testified, when I went into the control room, that an issue had been resolved, that there were some people at Thiokol who had a concern about the weather, that that had been discussed very thoroughly by the Thiokol people and by the Marshall Space Flight Center people, that it had been concluded agreeably that there was no problem, that he had a recommendation by Thiokol to launch and our most knowledgeable people and engineering talent agreed with that. So from my perspective, I didn't have — I didn't see that as an issue.*

ROGERS: *And if you had known that Thiokol engineers almost to a man opposed the flight, would that have changed your view?*

LUCAS: *I'm certain that it would.*[34]

William Lucas took his seat in the control room alongside Arnold Aldrich and Jesse Moore, but it did not occur to him to tell them what he had just heard. Like von Braun, Lucas took a military view of chains of command and he — the director-in-chief of the Marshall Space Flight Center — did not need to say anything to Arnold Aldrich, who had such close ties with the Johnson Space Center in Houston. If anything had to be passed on to Aldrich, he had people to see to that.

ROGERS: *I gather you didn't tell Mr. Aldrich or Mr. Moore what Mr. Reinartz had told you?*

LUCAS: *No, sir. That is not the reporting channel. Mr. Reinartz reports directly to Mr. Aldrich. In a sense, Mr. Reinartz informs me as the institutional manager of the progress he is making in implementing his program, but I have never on any occasion reported to Mr. Aldrich.*[35]

And Reinartz had not reported anything to Level II, no more than Mulloy had done. The senior management could make their decision regarding the launch without having to be bothered by details about the cold, the rubber, or the lack of elasticity in some lowly booster joints.

Rogers found it hard to conceal his annoyance, even with Reinartz:

ROGERS: *In the Navy we used to have an expression about going by the book and I gather you were going by the book. But doesn't that process require some judgment? Don't you have to use common sense? Wouldn't common sense require that you tell the decision makers about this serious problem that was different from anything in the past?*

Reinartz replied that he believed that "the Thiokol and Marshall people had fully examined that concern and that it had been satisfactorily dispositioned."[36]

At seven o'clock that same morning, B. K. Davis from Stevenson's ice team worked his way around the launch pad, under the Orbiter, measuring the temperature with a pyrometer — an instrument capable, by means of infrared beams, of measuring surface temperatures at a distance simply by being pointed at the desired area. When he pointed it at the left booster, the surface temperature read 33° F. Astonishingly, the right booster was considerably colder. Davis reckoned this disparity had to be due to the fact that the right booster had been exposed to an icy wind chilled even further by a supercooled fuel tank containing subzero liquid hydrogen. The instrument showed a surface temperature, on the nethermost section of the right booster, of 19° F.[37] This was 34° F below the limit stated in Thiokol's veto. But Davis knew nothing about this veto and no one had asked him to report the surface temperatures of the boosters, so he carried on with his work while writing down the temperatures for subsequent analysis.

38. Rockwell's Warning

When the commission moved on to the question of ice on the launch pad, they were to hear the same theme repeated—only on different instruments.

This time, Rockwell International played the Thiokol part opposite NASA. Rockwell was responsible for the actual Orbiter and had to approve the conditions under which the space shuttle system was launched. This company was based in California, so, with a time difference of three hours between themselves and Florida, their managers and technicians had to get up very early to be ready, with time to spare, before launch time. The first that Rockwell's director of the space shuttle project, Dr. Rocco Petrone, heard about the icicles on the tower was when he arrived at his office at 4:30 a.m. local time. His staff told him that the Kennedy Space Center had requested an "input" from Rockwell by 6:00 (9:00 in Florida), the time of the last meeting before the launch. Experts at Rockwell could keep abreast of the ice situation on television screens showing film of conditions on the launch pad in Florida.

The experts were partly concerned that large chunks of ice would be shaken loose when the motors started to reverberate—and might strike the Orbiter's fragile shell. Mention was also made of a phenomenon known as aspiration. Previous liftoffs had shown that the huge solid rocket boosters took a deep breath just as the fierce combustion inside them was triggered off. For this reason, the ice team on the pad were instructed to removed every pebble or screw that they might come across. Any stray odds and ends could be sucked up into the thrusters and do damage before being spat out again by the terrific force of the combustion. And in this case, chunks of ice constituted massive odds and ends, far larger than any tiny screw. Petrone told the commission:

The prime thing we were concerned about was the unknown baseline. We had not launched in conditions of that nature, and we just felt we had an unknown.

I then called my program managers over in Florida at 5:45 [PST] and said we could not recommend launching from here, from what we see. We think the tiles could be endangered, and we had a very short conversation. We had a meeting to go through, and I said, let's make sure that NASA understands that Rockwell feels it is not safe to launch, and that was the end of my conversation.[1]

Earlier in the night, the ice team had radioed in a report from the launch pad on conditions there. The transcripts of these conversations were not made public until much later. Here, the leader of the ice team, Charles Stevenson, is talking to his immediate superior, the Director of Shuttle Engineering, Horace Lamberth, and the Launch Director, Gene Thomas:

THOMAS: *Hey, we gotta come out of there [the countdown hold] when you guys [are] telling us you're pretty sure that water system's gonna work.*

LAMBERTH: *Yeah, you feel comfortable with what you see out there, Charlie, now?*

STEVENSON: *We have a lot of ice, if that's what you mean. I don't feel comfortable with that on the FSS [Flight Service Structure].*

LAMBERTH: *Then what choices we got?*

STEVENSON: *Well, I'd say the only choice you got today is not to go. We're just taking a chance of hitting the vehicle.*[2]

At the nine o'clock meeting, Rockwell's representative in Florida, Robert Glaysher, explained that the ice constituted an unknown factor and that Rockwell could not predict where any shards might fall or how the Orbiter would cope, in the event that it was struck. Glaysher told the commission about his recommendation:

My exact quote — and it comes in two parts. The first one was, Rockwell could not 100 percent assure that it is safe to fly which I quickly changed to Rockwell cannot assure that it is safe to fly.[3]

Another Rockwell man, Martin Cioffoletti, told Aldrich that they had no data base from which to make a safe assumption that the ice would *not* hit the spacecraft. They could not calculate the trajectory of ice falling from the tower as the Orbiter passed it on its way up. At this point the commission chairman himself intervened in the examination:

ROGERS: *But I think NASA's position probably would be that they thought that you were satisfied with the launch. Did you convey to them in a way that they were able to understand that you were not approving the launch from your standpoint?*

CIOFFOLETTI: *I felt that by telling them that we did not have a sufficient data base and could not analyze the trajectory of the ice, I felt he understood that Rockwell was not giving a positive indication that we were for a launch.*[4]

Not surprisingly, this was not the way NASA had seen it. Several NASA centers had also put forward their own arguments for believing that the ice would not, in fact, hit the Orbiter. Arnold Aldrich described how he had interpreted Rockwell's warning:

> [Glaysher's] statement to me, as best I can reconstruct it to report to you at this time, was that, while he did not disagree with the analysis that JSC [the Johnson Space Center] and KSC [the Kennedy Space Center] had reported, that they would not give an unqualified go for launch as ice on the launch complex was a condition which had not previously been experienced, and thus this posed a small additional, but unquantifiable, risk. Mr. Glaysher did not ask or insist that we do not launch, however.
>
> At the conclusion of the above review, I felt reasonably confident that the launch should proceed.[5]

On this occasion, Level II (Aldrich) could not plead ignorance of serious objections raised further down the chain. Rogers therefore asked Aldrich whether he could recall a single instance when a main contractor had opposed a launch but the countdown had continued all the same. Aldrich replied by saying that he had viewed Rockwell's recommendation as an intimation of concern and not a veto against the launch. Rogers was growing exasperated:

> If the decision-making process is such that the prime contractor thinks he objected and testified under oath that they took a position it was unsafe to launch, and you say that it was not your understanding, that shows us serious deficiencies in the process.[6]

At an executive session, held before the public hearing, Sally Ride had remarked that everything must hinge on determining whether, at the close of the meeting, Rockwell had said, "We don't want to launch," and Rogers had added:

> That's exactly it. If Rockwell comes up in a public session and says we advised NASA not to launch and they went ahead anyway, then we have got a hell of a problem.[7]

Regardless of whether Rockwell had made its concern clear to Aldrich, by the time the nine o'clock meeting was over, Petrone knew that his company's point of view did not carry enough clout to stop the buildup to the launch. Neil Armstrong cross-examined Rocco Petrone on this point:

ARMSTRONG: *Now clearly when they resumed the count, you knew that your recommendation essentially had been either considered and overruled or dispositioned in some other way.*

PETRONE: *That's right, sir.*

Armstrong asked if Petrone had expressed an opinion about that:

PETRONE: *Mr. Armstrong, I felt that we had expressed our opinion to the proper level on the proper occasion of the meeting that had been set up for it. I felt that I had done all I could.*

RIDE: *Did it surprise you that NASA picked up on the count?*

PETRONE: *I was disappointed that they did, yes.*[8]

39. The Reaction

John Fabian had himself been into space along with Sally Ride, in June 1983, on the mission code-named 31-C. That was when she had become the first American woman in space. They had actually flown with *Challenger* on that occasion. Later, he had taken part in mission 51-G with *Discovery*, six months before *Challenger* exploded. He had since left NASA to work under Major General Donald J. Kutyna on certain top secret military satellite programs, and had been assigned to the Rogers commission as aide to Sally Ride.

Fabian saw how difficult it was for her to maintain her businesslike demeanor and keep a check on her emotions as the hearings revealed what had actually taken place and all of the bureaucratic brinkmanship involved. She had felt particularly close to her female successors in the astronaut corps, and the *Challenger* disaster had cost two of them their lives. All the signs were that much more than a malfunction had occurred; and that the accident could have been avoided.

John Fabian later described his own feelings:

> It just unraveled like Watergate. We felt betrayed. It was one thing to understand the technical reasons for the solid rocket explosion. But NASA had always put its people above everything. To hear how they put off and covered up the needed repairs and to know that it killed your friends is a little hard to take.[1]

In the middle of March 1986, Sally Ride told *Time* magazine:

> I'm not ready to fly again. I think there are very few astronauts who are ready.[2]

Among the public at large, the reaction had begun to set in. The most important hearings were televised. Until these hearings people had almost felt sorry for NASA, that enthusiastic, dynamic agency — so unfairly struck by such a tragedy, even when everything humanly and technologically possible had been done to prevent it. Now the American people too were feeling betrayed. Out of the blue, the spotless, eternally youthful organization that John F. Kennedy had set in motion — and which, many felt, perpetuated his dream — had been exposed as what could only be described as a disgusting con game. A number of those who had testified received anonymous threatening phone calls; some chose to go underground for a spell, taking their families with them.[3] One reader wrote to *Time* at the beginning of March to vent his wrath:

> Why should the mother of two young children be sent hurtling into outer space aboard a rocket ship filled with explosive fuel? Certainly there are more suitable subjects with which to experiment.[4]

In the same edition, the magazine reported that Rogers, the commission chairman, had informed the President that he was appalled by the picture that was emerging of the decision-making process carried out prior to the launch. From Congress, Senator Ernest Hollings said that now, unfortunately, the disaster seemed "like an avoidable accident rather than an unavoidable one." So it was no great help for NASA's Jesse Moore to tell the newsmagazine: "We grieve at NASA. We grieve every day."

Prior to this, *Aviation Week and Space Technology* had always stood by NASA through thick and thin. This was the journal which had written, just a few days after the accident, that no matter what a commission investigation might unearth, one thing had to be established right at the start: NASA's planners, engineers, and technicians had done everything in their power to make the space shuttle as safe as possible.

As the investigation progressed, the journal could no longer be so certain of this. It admitted that what had been discovered, beneath NASA's professional veneer, was "a hidebound space agency fraught with lax management oversight, intramural turf battles between headquarters and key field centers and a tendency towards compartmentalized bureaucratic thinking that in the aftermath of the accident has generated self-serving responses."[5]

The most unexpected criticism, however, came from another quarter. On March 4, 1986, John Young — NASA's chief astronaut — put his name to a memorandum which he then forwarded to his boss at headquarters, with copies to all of his astronaut colleagues at NASA. Young was fully aware that this was an unprecedented step in the relationship between the astronauts and NASA's senior management. He also realized that, having been the commander on *Columbia*'s first flight five years earlier, he of all people was under a particular obligation to play the role of ambassador for the space shuttle system. Nevertheless, he could not remain silent any longer. In his memorandum he wrote:

> From watching the Presidential Commission open session interviews on television, it is clear that none of the direct participants have the faintest doubt that they did anything but absolutely the correct thing in launching 51-L at every step of the way. While it is difficult to believe that any humans can have such complete and total confidence, it is even more difficult to understand a management system that allows us to fly a solid rocket booster single seal design that explosively, dynamically verifies its Criticality 1 performance in its application. . . . There is only one driving reason that a potentially dangerous system would be allowed to fly: launch schedule pressure.[6]

Young had enclosed a long list of other critical systems and conditions pertinent to the space shuttle. He mentioned the wheels and the brakes, which had come very close to failing catastrophically on several occasions, and cited risky landings at Florida, where the weather could change more quickly and more unpredictably than in California. He also drew attention to thirty-four other critical points in the space shuttle system — the landing systems, main engines, reaction thrusters, fuel cells, etc. Although the astronauts were not aware of the problems with the O-rings, they were familiar with most of the points mentioned by Young. But it had taken Young to see them all in context and conclude:

> On an individual basis, they were not big enough to slow or stop the launch rates. But totally, this list is awesome. The list proves to me that there are some very lucky people around here.[7]

At the beginning of April 1986, the Rogers commission summoned John Young and a group of other space shuttle astronauts. For the first time in the history of American space exploration, the astronauts were given the opportunity to voice their own concerns and wishes regarding the space program in public to an independent body.

During this session, the astronauts reiterated that they had never been told about the problems with the solid rocket boosters.

This hearing, however, ended up focusing on the absence of abort systems in the Orbiter. If the space shuttle came in at the wrong angle for landing, and even if they knew as much as an hour beforehand that the computers would bring them down somewhere without a landing strip, then the astronauts had no chance of getting out of the cabin before "landing." And if two main engines failed early on in the proceedings, the space shuttle would not have enough inertia to make it back to Cape Kennedy. In which case, the astronauts would have to make an emergency landing in the Atlantic. They carried life jackets on board and they had trained in a floating model of a cabin. The newspapers had printed pictures of a smiling Judith Resnik extricating herself from the model and climbing down into the NASA dinghy. But the astronauts no longer believed they would survive an emergency landing in the ocean. One of them, Henry Hartsfield, put it like this:

> When you talk about smacking the water at 200 knots with an airplane that is basically an airliner-type design, I'm convinced it's going to break up. If you've got a 60,000 pound payload behind you, it's probably going to come into the cockpit with you.[8]

John Young brought up the subject of the explosive charges which the space shuttle had to carry so that it could be blown up from earth if it went out of control. He described how uneasy the astronauts felt—understandably enough—knowing that during every launch there was a man at Cape Kennedy with his finger on a button that could kill them simply by transmitting a radio signal. Young had taken it as a tacit and unwritten part of the "deal" that these charges would be removed after the first four test flights, when the crew were left without the ejection seat option. But this had not happened and the armed forces' safety officers who took responsibility for this at the Kennedy Space Center still refused to remove them.

The astronauts told the commission about the hazardous landing procedure which necessitated a lengthy training both in the simulator and in a special aircraft built for the purpose. NASA had wanted to cut back on the training time to save money and push more crews through the bottleneck—in order to chalk up more flights per year. The simulators were out of date, using technology that had been state-of-the-art fifteen years before. The contractor supplying computers for the Orbiter and the simulator had recently informed NASA that they could count on spare parts only for another three years for computers that had long since gone out of production. Feynman, the computer expert, found it amusing to see how the NASA flagship, instead of keeping pace with contemporary chip technology, was still stuck in the Stone Age with ferrite core memories.

In another move to save time and money, attempts had been made on several occasions to force a landing at Kennedy in unwarrantable weather conditions — just so that the space shuttle could "fly" itself home, instead of a week having to be taken up with transporting it from California. The problem here was that you had to know what the weather was like on the runway one and a half hours before landing, and very often it was impossible to predict weather conditions in Florida that far in advance. Young explained about the risk involved if contact was not maintained with the shuttle until the last minute:

> Once you've been given the go for de-orbit, if you lose communication or if you don't have communications right up to the time you de-orbit — we waved off Crippen one time three minutes prior to de-orbit on 41-C. They reported to us that the weather was going to be clear at the time of landing. At the time of landing, there was 11,500-foot rain showers over the end of the runway. . . . I think we were about three minutes away from having Crip land in some pretty interesting rain showers.[9]

Feynman was not very popular with the astronauts when he asked whether it would not be best to automate the entire landing procedure, to ensure that an injured or unconscious crew could be brought down safely. But here he was running up against their holy of holies: the right stuff. If the pilots did not have control over the space shuttle on the approach to the runway, then they would feel that they no longer had control of the situation, that they could *do* something. Instead they would be forced back into the passive role of the early Mercury astronauts in their little tin can. They never said it outright but it must have crossed their minds that, in that case, even Ham the chimpanzee could steer the space shuttle.

Nor could Feynman understand why the astronauts should be the ones to decide when the landing gear was lowered. Surely it would be safer to leave that too to a computer? On this point also the astronauts begged to differ, although they were unable to offer any good argument against it. They were probably afraid of losing the last vestige of control over an almost totally automated system. At any rate, Feynman was not convinced by their arguments and stuck to his somewhat simplistic picture of the astronauts' role during the flight:

> I found that very odd — a kind of silliness having to do with the psychology of the pilots: they're heroes in the eyes of the public; everybody has the idea that they're steering the shuttle around, whereas the truth is they don't have to do anything until they push that button to lower the landing gear. They can't stand the idea that they really have nothing to do.[10]

There was no consensus either when Feynman asserted that the space shuttle would never be capable of flying on any kind of healthy commercial basis — a role that had been a fallacy from the start. Now, with just three space shuttles left, some people were already saying that the shuttles were too important for more of them to be risked on launching the odd commercial satellite. Feynman put the case quite clearly:

> The question is whether we should be sending up things on a manned shuttle when we could have done it without it.[11]

Again he was encroaching on the astronauts' self-esteem. They had been training for years and could not help but take the view that they carried out vital assignments in outer space and did a far better job than any unmanned system could. Feynman did not disagree with them on this; he was simply not sure whether such a risky and extravagant solution was necessary every time a communications satellite had to be put into orbit.

NASA's original advertising campaign for the space shuttle system had promised "safe, cost-effective and routine access to space." By now, everyone agreed that unrealistic financial targets had been set for the system. As far as the other factors were concerned, everyone — astronauts and commissioners alike — agreed with astronaut Hartsfield's conclusion:

> The space shuttle transportation system is certainly not an operational system in the traditional sense, and it can't be. It will never be routine, and there will always be risks associated with it, flying in space. We will fix all these problems and go fly again, but it is still going to be a risky business, and I think everyone should remember that.[12]

40. Slow Motion

Alongside the examination of decision makers and astronauts, parallel with the scrutiny of files at NASA and the Thiokol plant, ran another, technical investigation — a second-by-second breakdown of how the disaster unfolded. Some commissioners may well have found it a relief to leave behind the murky world of contradictory testimonies and immerse themselves instead in ice-cold data. There had been no disagreement among the computers. They had faithfully sent back scores of reports per second from all of their various sensors, just as the numerous cameras had captured what they could see from their vantage points. Now the technological jigsaw puzzle could be assembled, piece by piece.

Challenger's flight could be followed right up until the last moments, with the latest computer enhancement techniques being used to bring out details that could never have been caught by the naked eye. The pictures and the computer data could then be compared in minutest detail and, taken together, they made it possible to rerun the entire sequence of events, only this time with the speed taken down to the silent world of milliseconds — even more fragmented than television's slow-motion replays of moves in a fast-paced football game.

As we saw earlier, *Challenger's* solid rocket boosters were ignited at 11:38:00:010 on January 28, 1986. So the investigators did have a fixed point from which to measure everything else that occurred. The ambient temperature was 36° F — 15° F colder than it had been for any previous mission. On liftoff, the temperature in the joint on the right booster was estimated at 28° F ± 5° F.

At 6.6 seconds before this very precise liftoff time, all three main engines had been ignited, tilting the whole stack over to one side, with the boosters still bolted to the launch pad. So far, their combined thrust had

0/360°
045°
315°
090°——————270°
135°   225°
180°

180°
225°
135°
270°——————090°
315°
045°
360/0°

LEFT   EXTERNAL   RIGHT
SRB   TANK   SRB

POSITIONING OF SOLID ROCKET BOOSTERS MEASURED IN DEGREES. (*COMMISSION REPORT*)

merely been absorbed, as a "twang" motion, within the construction. The boosters arched; just for a moment the circular field joints became slightly oval in shape. As the boosters sprang back into their upright position, they were ignited, the detonator bolts were discharged, and T = 0 — that magical moment that everyone had been working toward for such a long time — had arrived.

At T plus 0.25 second the cameras registered the first upward movement. Newton's third law in practice: reaction following upon action. *Challenger's* flight had begun. A microscopic instant later, at T plus 0.678 second, the photographic evidence showed, first, off-white smoke, then a puff of gray smoke spurting from the vicinity of the aft field joint on the right SRB. The two cameras on the launch pad, which had the best view of the smoke, both failed to function, but others registered it from a slightly more oblique angle. It was determined that the smoke was issuing from the sector between the 270 and 310 degree points on the booster's circumference, directly opposite the fuel tank. This puff of smoke seemed to indicate that the hot gases (temperature: 5,600° F) had first encountered ice in the joint (the white smoke) and that the O-rings had not slipped into place and sealed the surface but had instead been burned and eroded away along with the putty and grease in the joint.

Puffs of smoke continued to issue from the joint at a rate of three or four per second — at approximately the same rate as the "twang" motion stowed away inside the structure. The last puff of black smoke was observed at T plus 2.733 seconds. It was not possible to establish whether the smoke had stopped pouring out or whether it had simply been dispersed by the accelerated speed and become indistinguishable from the exhaust from the SRBs when viewed from below at an angle. The smoke could last be picked out at T plus 3.375 seconds.

The commission examined photographs and film from earlier launches but could find no evidence of similar phenomena.

At T plus 4.339 seconds, as scheduled, the main engines received the signal to throttle up to 104 percent of normal power. All systems were functioning normally, and they reacted as they were supposed to. At T plus 7.724 seconds, *Challenger* slowly went into its programmed roll maneuver. As planned, the engines dropped down to 94 percent of normal power. The roll was completed by T plus 21.124 seconds.

At this point in the launch, *Challenger* would very soon reach that level in the atmosphere where the craft would be subjected to the greatest aerodynamic stress. So at T plus 35.379 seconds the engines were given the order to reduce power to 65 percent. Although the force from the SRBs, which provided three-quarters of the combined thrust, could not be influenced, a solution had been arrived at. The hollow core inside the boosters was molded in such a way that the thrust would be greatest at the very start, when it was needed most; then it fell, to synchronize with the reduced power from the main engines during that same interval; and then, after a while, it rose again, in line with the engine power.

From T plus 36.990 seconds *Challenger* was passing through a level at which it encountered wind shear — extremely turbulent atmospheric conditions. This turbulence was greater than anything experienced on previous flights, but the computers and the navigational equipment on board were reacting as intended, passing the appropriate course adjustments to the five big bell-shaped nozzles. Although more radical than on any previous flight, these adjustments still fell within all design limits and did their job of keeping the Orbiter on course.

Both the main engines and the boosters now increased power once more. At this point, at T plus 58.788 seconds, a computer enhancement of a single frame of film revealed the first tiny flicker of flame on the right booster. It was pinned down to a spot estimated as lying close to the 305 degree point on the circumference, next to the aft field joint. Just one frame further on in the roll of film, computer enhancement was no longer necessary. At T plus 59.262 seconds, it was no longer a case of a flicker of flame, but of a constant, clearly defined plume issuing from the back of the solid rocket booster and running along its side.

At that same moment, the Orbiter computers registered a minor differential between the chamber pressures of the two boosters. The right-hand booster had lost a little of its pressure due to the fact that now a little hole had appeared in its side.

At T plus 60.248 seconds, the plume of flame had grown and spread toward the lower of the two rings bolting the booster to the fuel tank. The flames altered shape, curving around until they were impinging on the strut linking the external tank and the booster. They were also starting to lick at the tank. The Orbiter was now pitching about, having hit the roughest patch of air turbulence in the launch. At T plus 62.484 seconds the unequal compressive stress caused by the flame was so great that the Orbiter's guidance systems started to compensate for it. The big nozzle on the left booster gimballed sideways to counter the imbalance brought about by the drop in pressure on the right side.

At T plus 64.660 seconds, the first signs were observed of the flames having breached the external tank and penetrated to that mass of liquid hydrogen. Both the shape and the color of the flames altered as they mixed with the stream of hydrogen pouring from the tank. A flickering reflection could now be discerned on the dark underside of the Orbiter. At T plus 66.764 seconds, this breach was confirmed by a pressure drop in the hydrogen system.

After this, things happened so fast that it would suffice to say that the Orbiter exploded after 73 seconds of flight. But one could also stay with this extremely slow-motion replay and focus on those last, fatal seconds. At T plus 72.204 seconds the computers registered the fact that the two boosters were no longer flying parallel. The flames had now severed the lower strut and the right-hand booster was rotating around the upper attachment strut.

At T plus 73.124 seconds a halo of steam was observed around the bottom dome of the external tank. Shortly thereafter, the whole of this dome caved in and fell off. All at once a fresh surge of power was imparted to the craft from the vast amounts of hydrogen which had suddenly been provided with an escape route. This surge — which was equal to the force of a extra solid rocket booster — pushed the hydrogen tank upward into the intertank structure. Around the same time, a sweeping blow from the upper strut rammed the nose of the right-hand booster sideways, into the lower segment of the oxygen tank. At T plus 73.137 seconds the photographs revealed a cloud of vapor between the two fuel tanks. Oxygen was also starting to pour out. All the conditions for a huge explosion were now present.

At T plus 73.191 seconds a searing flash of light was visible on the film between the Orbiter and the bottom section of the external tank. And again at 73.213 seconds in the forward section. The explosions merged at T plus 73.327 seconds. The engine sensors registered a drop in pressure in the fuel hoses and tried to increase the pressure by opening the valves still further. The combustion was growing weaker and weaker as its fuel, the hydrogen, gave out and the oxygen was left to burn more or less on its own. This caused the temperature in the turbopumps to increase. As the temperature climbed beyond the redline limit, the con- scientious computers shut down the engines — one at a time as the tem- perature was exceeded.

The last radio signal from *Challenger* was picked up at T plus 74.130 seconds. At T plus 74.587 seconds came a final blinding flash in the vicinity of the Orbiter's forward section. The time was now 11:39:14:597.

Challenger was enveloped in a cloud of flame, its color changing to ocher as the fuel in the orbital maneuvering engines also ignited. The Orbiter somersaulted and was blown apart: the tail with its three main engines still firing away; the crew module with a tail of wires and hoses trailing behind it; a wing.

It was still traveling at almost twice the speed of sound, at an altitude of 46,000 feet and eighteen miles off the Florida coast. Thanks to their own inertia and Newton's first law, pieces of wreckage reached an alti- tude of 122,000 feet before plummeting downward into the Atlantic.[1]

No part of this sequence could have been altered by anyone in the control center or by Commander Scobee — even if anyone had been aware that something was wrong. At no point did any of the technicians on the ground notice anything amiss. No one saw the smoke around the launch pad until afterward. All communication between the spacecraft and Houston had been perfectly normal, professional, and composed. Smith's "Uh-oh!" from the Orbiter at T plus 73 seconds had been the first indication that he realized something was terribly wrong.

The timetable detailed above was pieced together from data from the Orbiter, together with photographic evidence. As each piece of wreckage was fished up from the ocean floor, one fact after another was substan- tiated. And on April 13, at long last, the crucial missing section of the right booster was located and brought to the surface.

This fragment was scarred, battered, and rusty. It consisted of the aft section of the booster cylinder surrounding the presumed breach in the field joint. Between the 294 and 316 degree points on the circumference investigators discovered an oblong-shaped area, fifteen inches high by twenty-eight inches long, in the metal covering the joint. The center of the hole lay somewhere around the 307 degree position. There was no doubt about it, they were looking at *Challenger*'s cause of death. At the worst possible spot, two feet from the external tank, the solid rocket booster's casing had burned through — thus sparking off the disaster. Later, the fragment of booster which had been lying in the same position, immediately below the first, was found. Here, too, burn-through was observed at the same spot.

All the pieces of the jigsaw puzzle had now fallen into place, and the commission was able to establish, beyond a shadow of a doubt, the technical cause behind the loss of *Challenger*:

> A combustion gas leak through the right Solid Rocket Motor aft field joint initiated at or shortly after ignition eventually weakened and/or penetrated the External Tank initiating vehicle structural breakup and loss of the Space Shuttle *Challenger* during STS Mission 51-L.[2]

41. Evaluating the Risk

Richard Feynman found it annoying that the different working parties within the commission never had to meet to discuss each other's findings. On more than one occasion, Feynman himself had sent synopses of various test results to the commission secretariat for circulation to the other commissioners. On almost every occasion, he discovered that the secretariat had not passed on his reports. Rogers wanted no further debate at this stage. It was time to write the report, and each working party would have to concentrate on its own section. Feynman had better confine his activities to Kutyna's working party and keep his mind on the technical causes of the disaster.

But Feynman, the Nobel Prize-winning commissioner, would not accept such limitations. He wanted to look at the interplay of all the factors involved and attempted to write a report of his own, in which he brought together everything he had learned from his own and the commission's investigations. When the other members of the commission did eventually get to see the draft of Feynman's report, they considered it significant enough to warrant incorporation into the final commission report. But first Feynman wanted to discuss it with the commission as a whole. Rogers did not feel they had time for this, and suggested that Feynman's report be included as a supplement or an appendix to the main report. Several of the other commissioners advised Feynman against accepting this solution, since it would mean that his report would be detached from the main report and would not appear until several months later — along with all the other supplementary material. Feynman did not have the strength to go on arguing indefinitely, so he agreed to the idea of an appendix — not least because such a form would allow him a greater say in the actual content.

Feynman entitled his report "Personal Observations on the Reliability of the Shuttle." Very early on, he had been struck by what seemed to be a recurrent theme in the commission's investigations: the disparity between the engineer's evaluation of the risks involved in different systems and that of management.

On one particular occasion, during the course of his work, Feynman arranged a meeting between three engineers from NASA's shuttle main engine department and their boss, Judson Lovingood. During this meeting, Lovingood became infuriated by Feynman's habit of going into everything in minutest detail. Lovingood did not feel that this was leading anywhere. So Feynman came straight to the point:

> "In order to speed things up, I'll tell you what I'm doing, so you'll know where I'm aiming. I want to know whether there's the same lack of communication between the engineers and the management who are working on the engine as we found in the case of the booster rockets."
>
> Mr. Lovingood says, "I don't think so. As a matter of fact, although I'm now a manager, I was trained as an engineer."
>
> "All right," I said, "here's a piece of paper each. Please write on your paper the answer to this question: what do you think is the probability that a flight would be uncompleted due to a failure in this engine?"

A secret ballot was then held. When the slips of paper were unfolded, two had written 1:200, one had written 1:300, and there was one odd note:

> Cannot quantify. Reliability is judged from:
> • past experience
> • quality control in manufacturing
> • engineering judgment[1]

Feynman announced that he had been handed four slips of paper but that one of them had not answered the question. The bullets alone were enough to convince him that this must have been Lovingood. The latter did not, however, believe that he had not answered the question. Feynman confronted him head-on:

> "You didn't tell me *what* your confidence was, sir; you told me *how* you determined it. What I want to know is: after you determined it, what *was* it?"

He says, "100 percent" — the engineers' jaws drop, my jaw drops;
I look at him, everybody looks at him — "uh, uh, minus epsilon*!"[2]

When the silence became too embarrassing for Lovingood, he
changed his reply, saying instead that, in his opinion, the risk was around
1:100,000. Feynman, who had the engineers' slips in front to him, was
then able to inform him that the difference between the engineers' es-
timates and that of management amounted to a factor of more than 300.[3]

We saw earlier how George A. Keyworth, former presidential adviser —
having recognized the possibility of a space shuttle being lost — worked
on alternative launch systems geared especially toward meeting the Pentagon's needs, even while the whole country would be reeling from the
shock of a disaster in which astronauts had perished. Evidently Keyworth
had not estimated the risk at 1:100,000.

After the *Challenger* calamity, yet another report from the Rand think
tank came to light. In this report, prepared long before the accident,
experts at Rand had predicted that allowance would have to be made for
the probability of losing one out of the four Orbiters in an accident and
that such an accident would ground the remaining Orbiters for a prolonged length of time. Rand noted that fifty-two anomalies had occurred
on *Columbia*'s first flight. The think tank calculated that the space shuttle
system would have to fly eighty faultless missions before achieving the
reliability of the expendable rockets.[4]

Feynman was amazed that anyone in a position of responsibility at
NASA could believe the risk to be as low as 1 in 100,000 launches.

> Since 1 part in 100,000 would imply that one could launch a shut-
> tle each day for 300 years expecting to lose only one, we could
> properly ask, "What is the cause for management's fantastic faith in
> the machinery?"[5]

He pointed out that, if management was right, and the probability of
failure was as low as 1 in 100,000, it would take an inordinate number
of tests to determine it: you would get nothing but a string of perfect
flights, without a single hitch.

If, however, management was wrong, and the actual risk was greater
than they professed to believe, one would, instead, be faced with a sce-

* In science, epsilon denotes a minuscule quantity.

nario in which the majority of flights displayed various faults, major and minor, and abnormal occurrences, all of which had simply not conspired — so far — to bring about an utter catastrophe. And this was precisely the scenario with which they had been presented. Orbiters had been just a hairbreadth away from disaster on several occasions, and the design had proved to be anything but flawless — even if everything had run smoothly until the twenty-fifth flight.

> It would appear that, for whatever purpose — be it for internal or external consumption — the management of NASA exaggerates the reliability of its product to the point of fantasy.[6]

Feynman spoke to the range safety officer, who had persistently refused to remove the destruct charges from the space shuttle, on the grounds that it still entailed a greater risk than NASA claimed. In order to determine the risk involved in the space shuttle system's solid rocket boosters, this officer had studied all the information gleaned from working with American solid-fuel rockets. He discovered that, out of 2,900 launches, 121 had failed — a risk factor of approximately 1:25. But this figure also included the very first flights, on which one would have to expect a greater number of failures. He therefore reckoned that the figure could reasonably be put at 1:50. If, in addition, the most meticulous inspections and quality control checks were carried out — bearing in mind that the system did have to carry human beings — it might be possible to attain a factor of just over 1:100. He did not believe, however, that technology had yet reached the stage where it could produce a solid-fuel system with a risk factor of 1:1,000. And since the space shuttle system employed two such rockets, the risk, obviously, had to be doubled.[7]

NASA had assured him, right to the end, that the risk was 1:100,000 and hence, to all intents and purposes, nonexistent. This was how NASA justified its nonchalant plans to fly with plutonium and liquid fuel in the cargo hold on the two missions scheduled for later in 1986. The safety officer stuck by his figures and refused to remove the destruct charges, which might save residential areas from being hit by a stray space shuttle. Feynman agreed with him.

In his appendix, Feynman also criticized the sliding scale used to define safety limits — a method he had previously referred to as Russian roulette. Again and again, NASA accepted a risk simply because they had flown with it before and nothing had gone wrong. Quite unexpected and unacceptable phenomena such as erosion and blowby had gradually been accepted as a natural part of the design. Then they had started inventing design limits for these unacceptable phenomena and put forward apparently logical arguments to justify continuing with the flights.

Feynman took a particularly dim view of the argument presented at the conferences preceding the *Challenger* launch: that the O-rings embodied a safety factor of three. A certain amount of erosion had occurred on 51-C, but there was "scope for" three times as much erosion — so said the people from both NASA and Thiokol. This gave a safety factor of three; and that made the 51-L mission warrantable.

> This is a strange use of the engineer's term "safety factor." If a bridge is built to withstand a certain load without the beams permanently deforming, cracking or breaking, it may be designed for the materials used to actually stand up to under three times the load. This "safety factor" is to allow for uncertain excesses of load, or unknown extra loads, or weaknesses in the material that might have unexpected flaws, et cetera. But if the expected load comes onto the new bridge and a crack appears in a beam, this is a failure of design. There was no safety factor at all, even though the bridge did not actually collapse because the crack only went one-third of the way through the beam. The O-rings of the solid rocket boosters were not designed to erode. Erosion was a clue that something was wrong. Erosion was not something from which safety could be inferred.[8]

Feynman regarded the problems surrounding the main engines as a sort of control experiment. He wanted to check whether the same lax attitude toward safety measures had been adopted with them as the program progressed.

He was impressed by the advances the engineers in this branch of the space shuttle project had made in getting the better of the main engines — which were so much more complex than the SRBs had been. Many a prototype engine had blown up in the process but — one step at a time — most of the problems had been eliminated. Although often without the nature of the problem having been thoroughly understood.

When, for instance, hairline cracks were discovered in the turbine blades of the high-pressure oxygen turbopump, no one knew whether these were due to a flaw in the metal, oxygen saturation in the atmosphere around the turbopump, heat stress on starting and stopping, vibration or resonance. Nor was it known how long it would take such a crack to develop into a proper fracture; nor which loads would put the greatest strain on the engine. Instead, the turbine blade was replaced and the engine sent on its way once more. This failure to fully understand individual components was an inevitable consequence of the fact that NASA had built the engine from the top down, instead of building incrementally from the bottom upward, using tried and tested components.

Originally, what they had been looking for was an engine design that could withstand fifty-five missions — a projected life span of 27,000 seconds in all. Instead they had found themselves having to replace oxygen turbopumps after every fifth or sixth launch, and hydrogen pumps after every third or fourth. In other words, at the very most, only 10 percent of the original design specifications had been fulfilled; and yet they went on flying.[9] It was the same story with the engine bearings and valves.

When Feynman put his question to Rocketdyne, the contractors for the main engines, they judged the risk to be 1:10,000; as we have seen, the NASA engineers said 1:300 or 1:200; the management said 1:100,000. An independent consultancy firm quoted Feynman a risk factor of 1 or 2:100.[10]

Feynman also looked at the Orbiter's computer systems, on which he was particularly well qualified to pass judgment. As previously mentioned, he was astonished to find that NASA was still flying with fifteen-year-old ferrite core technology in computers which had long since gone out of production. Since the memories of these computers were very limited, the astronauts had to load the programs for ascent and descent and the payload programs something like four times during flight. Feynman learned that it would have cost far too much in terms of both money and time to convert these thousands upon thousands of program lines for modern computer systems. So they had had to do without faster and much more dependable modern hardware.

Nevertheless, Feynman cited the computer programmers' approach to "bugs" in their line of work as an example to staff in other branches of the space shuttle project. Because the programmers started at the bottom. Only when a program module had been thoroughly tested and debugged could it take its place alongside the other modules and a fresh test be embarked upon, to reveal new bugs. Finally, a rival computer team were asked to "play customer" and subject the system to every diabolical challenge they could think of, before "buying it." Then it was subjected to further tests involving exhaustive run-throughs in the simulator, where every fault was considered to be serious as it would have been if it had occurred during a real flight. Only after all that was the system released for use.[11]

> To summarize, then, the computer software checking system is of highest quality. There appears to be no process of gradually fooling oneself while degrading standards, the process so characteristic of the solid rocket booster and space shuttle main engine safety systems.[12]

The programmers could, however, reveal how they had recently been told by their bosses that those higher up in the chain considered these extremely time-consuming tests to be too expensive and quite unnecessary. They would have to find a simpler way of doing things, the programmers were told. After all, the space shuttle system was no longer under development.

On the basis of his investigations, Feynman was able to conclude that the space shuttle system had most probably been flying with a risk factor of approximately 1:100.

> If a reasonable launch schedule is to be maintained, engineering often cannot be done fast enough to keep up with the expectations of the originally conservative certification criteria designed to guarantee a very safe vehicle. In such situations, safety criteria are altered subtly—and with often apparently logical arguments—so that flights can still be certified in time. The shuttle therefore flies in a relatively unsafe condition with a chance of failure on the order of 1 percent. (It is difficult to be more accurate.)
>
> Official management, on the other hand, claims to believe the probability of failure is a thousand times less. One reason for this may be an attempt to assure the government of NASA's perfection and success in order to ensure the supply of funds. The other may be that they sincerely believe it to be true, demonstrating an almost incredible lack of communication between the managers and their working engineers.[13]

One very serious consequence of this strategy, or lack of understanding on the part of NASA's management, had been that ordinary citizens, like Christa McAuliffe, had been encouraged to join flights into space, as though it was just a matter of an innocent little jaunt in a space shuttle that was every bit as safe as a standard passenger plane. Feynman believed that the test pilots, with their military training, knew the risk involved, and he admired them for the courage they showed as astronauts. But enticing ordinary American citizens on board was practically criminal.

Feynman knew that the Rogers commission was in the process of preparing a list of recommendations for the President and for NASA. He could not, however, close his own appendix without presenting his personal conclusions regarding the necessary changes.

> Let us make recommendations to ensure that NASA officials deal in a world of reality, understanding technological weaknesses and imperfections well enough to be actively trying to eliminate them. They must live in a world of reality in comparing the costs and utility of the shuttle to other methods of entering space. And they must be realistic in making contracts and in estimating the costs and difficulties of each project. Only realistic flight schedules should be proposed — schedules that have a reasonable chance of being met. If in this way the government would not support NASA, then so be it. NASA owes it to the citizens from whom it asks support to be frank, honest and informative, so that these citizens can make the wisest decisions for the use of their limited resources.
>
> For successful technology, reality must take precedence over public relations, for Nature cannot be fooled.[14]

42. Crisis in Conscience

For the first time in years, while sitting in enforced retirement just before the lunar landing, NASA's old boss James Webb found time to catch up on his reading. And, as we have seen, he decided to write about his experience with big organizations and large-scale endeavors. While working on his book, in the early spring of 1968, he read an article in the *Harvard Business Review* which troubled him deeply. This article, "Crisis in Conscience at Quasar," outlined a situation that had arisen in a large company, a state of affairs which Webb described as every company president's nightmare — namely, one where he cannot trust what he is being told by his departmental managers; where subsidiary companies start, systematically, to pull the wool over the eyes of their parent company — with disastrous consequences.

Although Webb did not go into this article in depth in his book, he did make particular reference to it and summarized the essence of it:

> A situation where trusted associates are so deeply committed to the line of action they have decided upon that they begin to cover up and buy time for a hopeful, but increasingly remote, solution in order to avoid loss of face in the organization.[1]

The author of this article, John J. Fendrock, was the head of a large electronics company which had done work for NASA. He outlined an imaginary situation — a case — based on a true story from the world of business. The gist of this case was that middle managers often found themselves in a situation where they had to do things which were morally and ethically cloudy, questionable from a business point of view or possibly even of doubtful legality. Theoretically they could of course refuse to do these things, but usually they did not. They obeyed orders and said nothing about what they knew—either to the authorities or to the parent company.

Webb found this disturbing because, as we have seen, the key factor in his organizational charts was the constant flow of feedback to the head office which enabled senior management to step in like a shot and straighten out any tricky situations further down in the organization. Just as the huge Saturn booster would be helpless without the instantaneous feedback from its thousands of sensor points.

But if this feedback could not be trusted, or if management was simply not informed of critical situations, then the company could not possibly be managed from the top. In that case, its course would be determined, not by management, but by a long succession of unrelated, randomly interacting incidents, all far beyond management's control.

This was not how Webb had viewed NASA during the development of the Apollo project. On the contrary, he had had a sense of thousands of employees working, loyally and enthusiastically, toward the same goal. Scientists and engineers had worked closely, at all times, with flight controllers and astronauts; the universities had shared their knowledge with American industry. At its best, the public sector had been indistinguishable from private enterprise. Everything had run like a dream.

But maybe the dream was coming to an end. Maybe even the NASA system would not be immune to the problems that were now beginning to crop up, more and more often, in other large-scale corporations and governmental bodies—where senior management felt increasingly isolated from the people below.[2]

In Fendrock's case, a large company with a great many subsidiaries suddenly discovers that the optimistic financial statement which it has just presented in good faith to its shareholders is totally false. One of their subsidiaries became involved in some very risky ventures, which completely miscarried. This leaves the parent company in the unexpected situation of having to suffer a very substantial loss, one which it had had no chance to prevent by means of management intervention. For some time, the subsidiary company has been submitting misleading and increasingly spurious reports to the upper echelons, in the ever more desperate hope that some miraculous turn for the better will make the figures tally again.

The parent company is swift to react and wastes no time in firing the managers responsible in the subsidiary company. And yet, a niggling doubt remains. Maybe the problem is not that easily resolved. It is therefore decided that a series of hearings should be held, to interview members of the subsidiary's middle management, who must have known what was going on and yet failed to blow the whistle. The aim of these interviews was, of course, to gain some insight which might prevent a similar situation from arising in the future.

Again and again, the fictional hearings described in Fendrock's case unearth the same line of reasoning. The middle managers were well aware of what was going on but they would never have dreamed of "tattling" to the men at the top. As far as they were concerned, their first loyalty was to their immediate bosses within the subsidiary company. Anything beyond that seemed hazy and remote. Their knowledge of the parent company and the head office was drawn mainly from the glossy annual report. These were unreal, abstract entities, obviously deserving of respect and correct behavior, but in their everyday lives they worked for their own managers. And it would be tantamount to mutiny and suicide to turn your back on them when the chips were down. In a typical quote from these interviews, one middle manager says:

> Frankly, at no time did it occur to me that I had a greater responsibility than the one I had to John [his supervisor]. Perhaps this is wrong, but I have always felt that I owe more loyalty to my supervisor than to the company. And besides, I'm not certain to what degree personal morality should enter into business decisions.[3]

Another was asked why he could not tell the head office that for months on end he had been asked to falsify accounts.

> Your question implies that I was *unable* to do this. Actually, it was always possible, but I was not *required* to do it. However, I was expected to give my observations to John [his supervisor] and to support him in any decision he made as to how the information was to be handled.[4]

In every interview, employees seemed appalled by the idea of going behind the backs of their immediate superiors. Such methods could only have an undermining effect. One employee put it like this:

> I honestly feel that what gets reported back to headquarters can only reflect what the president [of the subsidiary] sees fit. I would hit the ceiling if I found out one of my project managers was reporting directly or indirectly to the president. By the same token, the president shouldn't have to guard against insurgency in his ranks.[5]

Clearly, the employees of the subsidiary see themselves as a self-contained system within the greater system. It seems only natural for them to give their principal allegiance to their own local place of work and their immediate superiors. Basic common sense dictates that the maintenance of a good working environment, close to home, comes before everything else. And they have the distinct impression that it would do them no good to undermine the authority of their local boss.

As children they were taught to despise anyone who ran telling tales to the grown-ups. And even now, in their adult lives, they would never dream of going over someone's head. Obviously, then, they find themselves on the horns of a dilemma if they are aware of deliberate dishonesty in their workplace. In every case, however, they seem to have distinguished quite sharply between their own personal morality, as they would apply it to situations in their private lives, and the rules of business, which often can, and have to be, rather more shadowy. It would take almost unprecedented courage on the part of an employee to take on the company for whom he works, armed with his personal, down-to-earth sense of morality and decency. And there can be little doubt that the person concerned would then have both management and his colleagues at his throat.

The last man from the subsidiary company to be examined is an engineer by profession, a man who, although not particularly interested in the administrative side of things, has worked his way up to a post as vice president of engineering — and is proud of it.

> I don't pretend to be a business manager; rather, I am an engineering manager. The tangibles of engineering are something I grasp and manipulate readily, but the intangibles of business are quite another thing.[6]

In Fendrock's case he has been equipped with a pipe, which he draws on thoughtfully between questions.

> Q. You were aware, were you not, that the reports sent to headquarters distorted conditions at Quasar to such an extent that the status of projects was inaccurately reported, actual and projected earnings were blatantly inflated, and the entire status of the operation was totally misrepresented. How could you have accepted such a situation?

A. If only I could answer you in a manner that might express my feelings at the time. Was I aware of what was going on? Yes, of course I was. But I didn't *want* to know about it. I will go so far now as to say that I tried *not to know* what was being done. Realistically, once I accepted the basic decision to ride the thing out, I felt stuck with the consequences. There was nothing, as I saw it, that I could do to alter the course taken.[7]

Only one employee in Fendrock's case will not accept the assumption, implicit in the head office's question, that the rot is confined to the subsidiary, which has grossly misled the parent company in its efforts not to lose face. He believes, rather, that the subsidiary would never have got itself into such a situation if the head office had not continually been putting pressure on its subsidiaries, prompting them to display aggressiveness and optimism and to get results—fast. In so doing, the head office has indirectly encouraged its subsidiary's gradual involvement in these risky ventures as it struggled to maintain its standing within the organization. And it was solely because of this constant pressure that they had taken such big commercial risks.

Like the gambler at the roulette wheel, we plunged deeper—with about the same odds—and lost.[8]

When the losses began to show up, they had no choice but to run fresh risks, in the hope of covering up the initial loss—all with more and more catastrophic consequences. Finally, the inevitable happened, the whole affair came to light and the head office set about finding a scapegoat for a state of affairs which they themselves had initiated—though no blame could ever be pinned on them.

To me, what happened was that headquarters decided on a set course of action, passed the word down, and then—when it became impossible for us to follow through—they looked for scapegoats.[9]

There was something about this account—fictitious though it was—that Webb could not get out of his head. Basically, it tallied too well with human nature. So well that there was no way of telling how large organizations could ever guarantee that a management's picture of its organization was—to all intents and purposes—accurate. Or be certain that every branch of the organization was working toward the same goal. It might be that every managing body set its own limits, consciously or unconsciously, for the information it wished to receive. And company directors might, therefore, also be confronted, in the stream of reports from the lower echelons, with their own impression of how things stood and, hence, find themselves locked, so to speak, inside a hall of mirrors, surrounded by their own reflections.

If, instead of being fictitious, Fendrock's case represented the actual state of affairs inside all large organizations, one might begin to doubt whether such organizations can be directed and supervised with any degree of certainty. It seemed that—beneath the well-organized boxes and arrows of the organizational charts—anarchy, jealousy, and infighting prevailed at all levels within the system. On paper, the organization's goals might have been defined by the senior administration, but the actual course was determined by the quite unpredictable sum of all the psychological and commercial conflicts and alliances further down in the system.

Eighteen years later, James Webb read the report on NASA and *Challenger* with disbelief. By then he was an old man, suffering from Parkinson's disease and able to get around only with difficulty. He may well have spared a thought for Fendrock's article when he came to the commission's piece on the Marshall Space Flight Center:

> The Commission is troubled by what appears to be a propensity of management at Marshall to contain potentially serious problems and to attempt to resolve them internally rather than communicate them forwards. This tendency is altogether at odds with the need for Marshall to function as part of a system working towards successful flight missions, interfacing and communicating with the other parts of the system that work to the same end.[10]

43. Recommendations

The time had come for the Rogers commission to assemble the findings of their investigations for the final report on the *Challenger* disaster. The stipulated 120 days were just about up. The commission subcommittees had examined 160 individuals, and these interviews alone had generated 12,000 pages of copy. The actual hearings, both public and private, had added another 2,800 pages. In addition, the investigators had examined 6,300 documents, amounting to 122,000 pages in all, and keyed them into the commission data bases.[1] This vast store of material would henceforth be held for the nation in the National Archives in Washington.

The report shed light on events leading up to the disaster and unraveled the tangled web of underlying causes. It concluded by making a number of recommendations to the President — recommendations which would have to be implemented by NASA if the space shuttle was ever to fly again.

The most urgent recommendation dealt, not surprisingly, with the SRB field joints, which would have to be redesigned to function according to the original design specifications. The joint rotation had to be eliminated and the joints fabricated in such a way that they would no longer be temperature-sensitive. Nor ought they to be affected by transportation and handling or assembly procedures, much less by being reused over and over again.

The commission did not give any indication of how NASA was supposed to accomplish this objective, but they did express their lack of confidence in NASA's ability to fulfill it single-handed, by appointing an Independent Solid Rocket Motor Design oversight committee — just to keep an eye on things.

The commission did not confine itself to technical problems. In a number of key recommendations, they dealt with the changes which would have to be made in the management of the space shuttle program. The commission wished to make sure that control over all the ramifications of the space shuttle program would be centralized, to prevent any recurrence of those obscure conflicts of loyalty between an employee's responsibility to the managers of his own NASA center and his obligations to the big bosses in Washington.

The commission also wanted to see astronauts being promoted to top jobs within NASA. Who better than the "aeromedical test subjects" to guard against anyone ever turning a blind eye to safety procedures again?

In another move to reinforce the safety aspect, the commission recommended the setting up of an Office of Safety, Reliability, and Quality Assurance, authorized to oversee every link in the NASA chain and report directly to NASA headquarters. It would be impossible to eliminate completely the problems pinpointed by the Fendrock case, but it was felt that such a safety office could clarify matters to the point where headquarters would be faced with genuine problems rather than cover-ups and wishful thinking.

The commission recommended specific changes in the space shuttle program routines. NASA would have to develop a system that would enable astronauts to escape from the space shuttle before touchdown in the event of a forced landing. The commission, however, did not believe it would have been possible to develop a system that might have saved the lives of the *Challenger* crew. But the astronauts ought to have the possibility of bailing out during the final approach.

As far as approach and landing were concerned, the commission would prefer that, from now on, this be carried out in California every time. NASA was urged to abandon the idea of scheduled landings in Florida, because of the unstable weather conditions. The problems with the Orbiter landing gear, undercarriage, and brakes referred to by the astronauts would also have to be resolved before the space shuttle system's launch permit would be renewed.

The Marshall Space Flight Center did not get off scot-free. The commission recommended changes in the personnel, the organizational structure, and the basic attitude at Marshall. From now on, all important prelaunch reviews were to be tape-recorded, and the flight crew commander for that particular mission would take part in these meetings, to assure himself that everything was as safe as it ought to be.

The American taxpayers had not been amused to read in *The New York Times* that in 1976 government auditors had discovered fourteen warehouses at Marshall packed to the gunwales with brand-new equipment and appliances, ordered by employees who had either left the space center long before or died. Since that time, all this expensive equipment had been lying there gathering dust.[2]

Both the Marshall Space Flight Center and Morton Thiokol were strongly criticized in the report for being uncooperative and too prone to cover up serious problems. Shortly before the publication of the report, *Aviation Week and Space Technology* had revealed that two Thiokol employees, Allan J. McDonald and Roger Boisjoly (exactly as set out in Fendrock's case), had been subjected to harassment by the company and transferred to inferior posts — purely on account of their testimonies.

These disclosures evoked a fierce response from U.S. congressmen. Representative Edward J. Markey told the magazine:

> I have never seen such blatant, unethical and potentially dangerous corporate behavior as is exhibited here by MTI. The supplier of the suspect rocket booster is stripping its most conscientious engineers of their staff and responsibilities. This deserves the strongest condemnation and calls for immediate action.[3]

Following a bit of congressional string pulling, Allan J. McDonald was assigned to a senior post with the division detailed to sort out the field joint problems. Only when he was satisfied would Thiokol be given the green light for their new design. Roger Boisjoly was on sick leave, unfit to work. Unable to rid himself of the feeling that he could have done more to prevent the disaster, he had sought psychiatric help.[4]

Richard P. Feynman was not happy with all of the commission's recommendations, his main concern being that an Office of Safety would simply add another layer of bureaucracy to a setup that was already confusing enough. Nor did he approve of the haste with which the report had been written. Feynman had been hoping for some prolonged and frank discussions with the other commissioners before the final section was written up. He had imagined, perhaps a little naively, that they had put the job sharing behind them and now they would have time to discuss the overall picture — something which none of them were capable of discerning alone. But time was short. What mattered now were the commission's recommendations. Rogers turned up for meetings armed with ready-made suggestions and demanding concrete amendment proposals on the table then and there if any changes were to be made.

But this last-minute rush did not stop the chairman from taking time for an elaborate debate on the color of the report jacket. As far as Feynman could recall, most of the commissioners preferred the red. But blue it was.

A certain armed neutrality prevailed between Feynman and Rogers. The chairman was not happy about Feynman's appendix, even though it had been relegated to the supplementary section of the report. On several occasions he tried to have particular pieces removed from it, saying that, basically, these had already been covered in the main body of the report. Each time, Feynman reluctantly complied, albeit with mounting frustration. New, revised versions of his appendix kept appearing, each one shorter than the one before. On a couple of occasions Feynman's text was totally wiped off the commission data base and had to be painstakingly reconstructed.

Feynman irritated the chairman with his habit of harking back to the fact that senior managers at NASA did not seem to have been telling the truth in pleading ignorance of the critical proceedings on the eve of the launch. Feynman found it difficult to understand how the commission could just accept their protestations of innocence — simply because it could not be proved that they *had* known anything.

Their mutual animosity came to a head one day in the chairman's office. Feynman was told by Rogers that he believed the report might be viewed as very negative. So he wanted to suggest a tenth and more positive recommendation, to offset the nine critical points. He handed Feynman a slip of paper containing his proposal:

> The Commission strongly recommends that NASA continue to receive the support of the Administration and the nation. The agency constitutes a national resource and plays a critical role in space exploration and development. It also provides a symbol of national pride and technological leadership. The Commission applauds NASA's spectacular achievements of the past and anticipates impressive achievements to come. The findings and recommendations presented in this report are intended to contribute to the future NASA successes that the nation both expects and requires as the 21st century approaches.[5]

Feynman was furious. The commission had never discussed questions of such a political nature. The other nine recommendations were the conclusions of technical and organizational investigations. The impartial findings of impartial investigations. This, on the other hand, was a political recommendation. Rogers tried to counter by arguing that this was, after all, a presidential commission, so it would be only fitting to close with such a patriotically worded statement. Feynman was still against it.

A few days later, Rogers rang to assure Feynman that he understood his misgivings, but that all the other commissioners had agreed to this statement. Feynman was surprised that he had not been given the chance to put his point of view. But, things being as they were, he saw no reason to oppose it any longer.

Later he discovered, purely by chance, that only a handful of the other commissioners had heard about this new recommendation. Rogers had tricked him. Taking the bull by the horns, Feynman now threatened to resign from the commission. He sent the chairman a telegram:

> PLEASE TAKE MY SIGNATURE OFF THE REPORT UNLESS TWO THINGS OCCUR:
> 1) THERE IS NO TENTH RECOMMENDATION, AND
> 2) MY REPORT APPEARS WITHOUT MODIFICATION FROM VERSION #23.[6]

This move prompted Rogers to accept Feynman's appendix in its unaltered form. In return, Feynman offered to go along with the disputed statement, as long as it was not presented as a recommendation on a par with the other nine. Rogers was welcome to call it a "concluding thought," in which case it would no longer be possible to use the phrase "The Commission strongly recommends . . ." Instead, Feynman suggested "The Commission urges . . ." Rogers gave it one last try. Could they write "strongly urges . . ."? Feynman was adamant — "*urges*". And "urges" it was.

President Reagan forwarded the report to James Fletcher whom he had recalled to run NASA — the same Fletcher who had seen to it that his home state of Utah won the vital contract for the solid rocket boosters. Fletcher was to submit a plan to the President of how NASA proposed to implement the recommendations made in the report.

James Fletcher made his reply in a press release — most diplomatically and very circumspectly. It was quite obvious that he was trying to underplay the report's significance for NASA:

> We have been pressing on, despite the pain, seeking answers to difficult questions; beginning carefully to make changes where they are needed. We have been at work. Yet, like all Americans, we have awaited the Rogers Commission report, hoping to learn from it as well.[7]

Fletcher promised to study the report's conclusions "with great care, an open mind and without reservations."

Reagan's political instincts told him they were going to need more than "an open mind." Four days after NASA issued its press release, he made it clear to the space agency that the whole idea of these recommendations was for them to be followed to the letter. Fletcher was given a month in which to send in his plan and was told to report on his progress after one year. At long last the message had got through to NASA. Now Fletcher was saying:

> Where management is weak, we will strengthen it. Where engineering or design or process need improving, we will improve them. Where our internal communications are poor, we will see that they get better.[8]

To the press, William Rogers emphasized that he did not wish criminal charges to be brought against any of those responsible. It ought to be enough that the whole course of the affair had been brought into the open and that particular individuals had been forced to appear in public to give an account of themselves. The report's prime aim must be to prevent this from happening again. Besides, the responsibility had to be placed not only on individuals but also on organizational patterns, processes, and structures within NASA and its subcontractors.

But several congressmen were not prepared to let those responsible off that easily. Senator Hollings, in particular, pulled no punches. Referring to Lawrence Mulloy, he told journalists:

> I find that gross negligence. I don't think he was trying to kill astronauts, but let's be blunt about it. I think that conduct was . . . in the nature of willful, gross misconduct.[9]

William Lucas too was given something to think about:

> He showed no remorse [during the hearings], no misgivings, no understanding of individual responsibility.[10]

Richard P. Feynman's involvement with the Rogers commission would prove to be his final public victory. He only just managed to tape-record his experiences and his work on the commission before losing his fight against cancer in February 1988.

Two Minutes and
Forty-five Seconds

Both in the main body of its report and in its synopses, the Rogers commission touched only very lightly on the fate of the *Challenger* crew. But of course the same scientific investigative methods had been employed when it came to reconstructing the last seconds of the crew's lives. Some of the findings of this investigation were classified, some were tucked away in the supplements to the report, which were not published until some months later.

The crew module had been located in the first weeks of March, lying at a depth of ninety feet twenty miles off the Florida coast. The entire flight deck section of the Orbiter appeared to have come down in one piece. At the beginning of April, under the tightest security, the astronauts' bodies were retrieved. All superfluous personnel and all members of the press had been cleared out of the area, and the salvage ships did not dock until darkness had fallen.

Specialists then set about the required postmortem examinations, and finally, at the end of April, the seven coffins were handed over to the astronauts' families at a ceremony held on the Kennedy Space Center runway. The managers of every NASA center were present, along with astronaut colleagues of the deceased—all there to pay their last respects to the *Challenger* Seven.

Time magazine was able to announce that Christa McAuliffe would be buried on a hilltop close to her high school in New Hampshire. The two pilots, Francis Scobee and Mike Smith, would be given heroes' burials at Arlington National Cemetery in Washington. Onizuka was to be buried in the military cemetery on Hawaii; McNair in his hometown; and Jarvis's ashes would be scattered over the Pacific. Judith Resnik's parents told *Time* that they wanted her ashes to be scattered over *Challenger*'s splashdown point in the Atlantic.[1]

These funerals cast a pall over the celebrations arranged by NASA for the twenty-fifth anniversary of the flight into space, in 1961, of America's first astronaut, Alan Shepard.

On August 4, 1986, two months after the publication of the Rogers report, *Aviation Week* carried a shocking headline that was to reverberate throughout the American press:

SHUTTLE CREW SURVIVED BREAKUP
Began Emergency Procedures

The magazine told its readers how the Rogers commission had established that the crew cabin had apparently survived the explosion more or less intact. This part of the Orbiter system had been designed as an independent unit, pressure-sealed to preserve the crew's oxygen supply. It was reckoned that the force of the actual explosion would not have been strong enough to contribute directly to the death or serious injury of the crew.

Of course, the explosion might have cut off the oxygen supply to the cabin, but the crew's emergency personal egress air pack cylinders had been found in the wreck. Of the four cylinders retrieved, three had been used—among them the one belonging to Mike Smith. The pilot's oxygen supply could only be connected up with the help of the astronaut immediately behind him. In this case presumably Judith Resnik.

At any rate, in the seconds after the explosion some of the crew had been active and conscious. It was possible to gauge the amount consumed from the individual oxygen cylinders at between three-quarters and seven-eighths of their contents. In other words, a number of the crew had been breathing for the two minutes and forty-five seconds it took for *Challenger* to impact with the sea at 204 mph, or a force of 200 g—at which death would, of course, be instantaneous.

According to NASA's doctors, the oxygen consumption could not be taken as proof that the crew had been conscious. If the module had been damaged in such a way that it was no longer airtight, the crew would have been subjected to decompression, which in turn would have led, after between fifteen and eighteen seconds—to dizziness and subsequent loss of consciousness. But the shattered wreck made it impossible to ascertain whether this had, in fact, been the case. So one could not discount the possibility that, in the worst case, the crew had been fully conscious right up until they hit the sea.

Dr. Joseph P. Kerwin of NASA's Johnson Space Center did, however, cite three factors which suggested that the crew had slipped rapidly into merciful oblivion:

> 1. The accident happened at 48,000 feet. The crew cabin was at that altitude or higher for almost a minute. Without an oxygen supply, loss of cabin pressure would have caused rapid loss of consciousness, and it would not have been regained before water impact.
> 2. Personal air pack activation could have been an instinctive response to unexpected loss of cabin pressure.
> 3. If a leak developed in the crew compartment as a result of structural damage during or after breakup, breathing air would not have prevented rapid loss of consciousness.

Dr. Kerwin also had a fourth factor, which he presented in the same matter-of-fact tone, but which touched more closely on how the crew would have been likely to react had they been aware — during that almost three-minute descent to the ocean waves — of the fate that awaited them. Dr. Kerwin clung to the reassuring fact that when the crew were brought to the surface, they were still lying in position in their seats.

> 4. Crew seats and restraint harnesses showed patterns of failure that demonstrate that all seats were in place and occupied at water impact, with all harnesses locked. This would likely be the case had rapid loss of consciousness occurred, but it does not constitute proof.[2]

Although the Rogers commission did not wish to hold any of the key people involved in the *Challenger* disaster legally liable, there was a tacit assumption that radical changes would be made at both senior and middle management levels.

Often, these changes did not seem like firings as such, but more of a discreet tap on the shoulder or a considerate suggestion that, after all the strain that the *Challenger* affair must have put on them, a transfer to a less demanding job might be in order. On occasion such moves might even masquerade as promotions.

Veteran NASA man James Fletcher moved in to take over the supreme post at NASA, as Administrator. In the intervening years Fletcher had been working as a consultant for the Pentagon — a fact that led several observers to interpret this appointment as tipping the scales even further in favor of those military interests now waiting in the wings of the American space program. Senator Al Gore tried to block Fletcher's appointment. He pointed out that Fletcher had deceived the American people and Congress in the early 1970s when the cost of the space shuttle system was being debated. And it had been during Fletcher's time at NASA that cutbacks had been made in quality control and safety trials. Gore maintained that NASA had saved all of a half million dollars on such vital aspects of the space shuttle system, and that safety checks had been reduced by 70 percent. In 1970, NASA's quality assurance department had employed a staff of 1,689; in 1986, only 505 were left. The most drastic cuts had been made at the Marshall Space Flight Center, from 615 to just 88 quality control staff.[1] But none of this hindered Fletcher's appointment.

As his second-in-command, Fletcher had Dale D. Myers, the man who, around 1970, had been so exceptionally adroit at procuring lucrative space shuttle contracts for North American Rockwell.[2]

The Associate Administrator for Space Flight, Jesse Moore, from NASA headquarters, was the man who had, officially, been ultimately responsible for the launch. While still maintaining that he had never been informed about the two teleconferences, he nonetheless handed in his resignation and took a job in the aerospace industry.[3] In accordance with the Rogers commission's recommendations, his position at Level I of the decision chain was taken over by an astronaut, Richard H. Truly. Thus, an astronaut made it right to the top of the NASA organization for the first time.

Arnold D. Aldrich from Level II in the decision chain was promoted to senior manager of the entire space shuttle system and transferred to headquarters in Washington. He would not, however, have anything to do with critical decisions pertaining to specific launches or landings. Such decisions would be made by his newly appointed second-in-command, veteran space shuttle astronaut Robert Crippen.[4]

The fatal decisions concerning *Challenger*'s last flight were, as we have seen, taken at Level III. And here the ax fell with a vengeance. At the Marshall Space Flight Center, Lawrence Mulloy, who had mocked the worried Thiokol engineers, was transferred to other work. He got the message and handed in his resignation soon afterward. A few months earlier, George Hardy, who had been directly responsible, on behalf of the Marshall center, for the solid rocket boosters, also left of his own free will. Mulloy and Hardy's boss, Stanley Reinartz, requested a transfer on health grounds. And the all-powerful director of the Marshall center, William Lucas, had his resignation approved just a few weeks after publication of the Rogers commission report, which, with its extensive criticism of his management style and the atmosphere at Marshall, left him no alternative.[5] It seemed a very long time since the newspapers had been hailing the Heroes of Huntsville.

Inevitably, the repercussions of the report also hit Morton Thiokol. The company's chief executive, senior vice president Jerald Mason, who had maintained to the bitter end that the O-rings had a safety factor of three, was granted early retirement. Vice president J. C. Kilminster, who had signed the telefax authorizing the *Challenger* flight at Mason's behest, was transferred to another division within Thiokol. The vice president of engineering, Robert K. Lund, who had started by backing up his engineers and then later changed his mind when asked to put on another hat, was also given a job elsewhere.[6]

Thiokol's Allan J. McDonald, the only manager to openly protest against the launch, was promoted to manager of Thiokol's SRB design division.[7] At long last, the company had realized that McDonald might well provide the sole key to any future dealings with the U.S. Congress.

Thiokol's contract with NASA stated that a reduction of $10 million in incentive fee earned would ensue if a fault in an SRB led to "loss of life and mission." It also stated that, in such a case, the contractor would have to sign a document admitting responsibility for the resulting accident, adding social stigma and legal liability to financial loss. The signing of such a document could affect the company's chances of obtaining new government contracts and would also expose it to the risk of suits by private parties.

Thiokol promptly acquiesced to the loss of the $10 million but was very reluctant to sign any document that hinted at "legal liability." For its part, NASA had no wish to terminate its working arrangement with Thiokol, only to have to start all over again with a new contractor—a process that would undoubtedly lead to more delays. They therefore came to an arrangement whereby Thiokol would "voluntarily accept" the $10 million reduction in the incentive fee.

In return, Thiokol would not have to sign the document concerning legal liability. This agreement aimed to avoid "litigation and keep priority on returning the shuttle to flight, and bypass the question of the company's liability for the accident." Over and above this, Thiokol undertook to carry out work to the tune of $505 million—at no profit to themselves—on redesigning the field joint and replacing the reusable hardware lost in the *Challenger* accident.[8]

> In 1988, Morton Thiokol's CEO, Charles S. Locke, revealed to *Business Week* that he had been given support and good advice throughout the whole affair by Warren E. Anderson, who had been CEO of the Union Carbide Corporation when a cloud of toxic gas spewed out of its plant in Bhopal, India, killing 2,000 people.[9]

That is how the big boys tackle the job of rapping knuckles and apportioning blame—in proper, pinstriped fashion. The relatives of the *Challenger* crew found it hard to understand how the protagonists in this drama could get off so lightly. A month or so after the report had been published, *Aviation Week* was able to report that Michael Smith's family had filed a suit against NASA personnel Mulloy, Hardy, and Moore, among others, demanding damages of $15 million. Smith's widow, June, told the press that she had purposely made no comment until the Rogers report appeared. She had now read it and discovered:

> Incredibly terrible judgment, shockingly sparse concerns for human life, instances of officials lacking the courage to exercise the re-

sponsibilities of their high office and some very bewildering thought processes.[10]

The family's lawyers cited exacerbating circumstances, alleging that Smith had been tossed around the cabin in the seconds before his death and that he had known that he and the other crew members were doomed.[11]

The families of the other astronauts soon followed suit. Betty Grisson, the widow of Gus Grisson, who had been killed in the fire in the Apollo capsule, strongly urged the *Challenger* wives to sue. "They don't care anything about you," she told them — "they" being NASA and the aerospace companies.[12]

This could have developed into an extremely unpleasant situation and both the government and NASA dreaded the thought of the publicity and the bitterness that could be stirred up by long-drawn-out court cases. The government therefore hurriedly entered into an out-of-court settlement with the bereaved families, whereby the latter promised not to disclose details of the sums awarded. According to *Aviation Week*, Thiokol had been forced into making a substantial contribution to these damages.[13] In 1988, *The New York Times* reported that the average compensation package had amounted to between $2 and $3.5 million per family.[14]

Roger Boisjoly, the Thiokol engineer who had fought harder than anyone to make his own and the Marshall center managers realize why the cold would be fatal for *Challenger*, was haunted by a feeling of powerlessness. He sued his employers for $2 million in damages, claiming that both his career and his health had suffered.[15] While this case was dragging on, he filed an additional suit against Thiokol, for attempting to mislead the American government. Boisjoly claimed that as far back as 1985 Thiokol had been keeping NASA in the dark about test results that were less than positive, for fear of losing orders for more SRBs. This time, Boisjoly's lawyers demanded that Thiokol pay $2 billion in damages to the U.S. Treasury. According to a new statute, if the company was found guilty, Boisjoly would be awarded 25 percent of this amount.[16]

Boisjoly's suit was rejected in September 1988,[17] by which time he had been undergoing therapy for "Post Traumatic Stress Disorder" for two years. He was only just getting back on his feet. The *Challenger* disaster had shaken him to the very core of his being. Long after the disaster, he still could not comprehend how an engineer's arguments, charts, and graphs could just be swept aside with a wave of the hand, as though they were of no relevance whatsoever. Psychologically, it had done him good to lecture on his experiences to engineering students. In all of his lectures, he asked these engineers of the future to be prepared for the day when their professional evaluations would clash head-on with their employers' interests. He had no doubts that they would all, at some time, find themselves in that situation.

Boisjoly was now fifty years old and unemployed. There was only one thing he was sure of: he would never again apply for a job with an aerospace firm.[18]

When, in the late autumn of 1986, Henry S. F. Cooper visited the Johnson Space Center, he received a clear impression of the impact the *Challenger* explosion had made within NASA. The people at Houston had lived and worked with the *Challenger* Seven. Story Musgrave, an astronaut who had himself made two flights with *Challenger*, described his feelings thus:

> For the first week or so, there was a sort of paralysis. We were like zombies — no one saying much to anyone. You were there, in the office, but you were just sitting in a chair looking out of the window.[19]

Apart from the loss of close friends, what really hurt was that the fault had been so appallingly basic. This was low-tech stuff, the sort of fault that could just as easily have occurred in a factory making washing machines. But this was not a washing machine factory; this was NASA, with all that that entailed. Seen in that light, the accident was almost unbearable.

This was probably how an aging James Webb was feeling when he quietly told Joseph J. Trento:

> There was an organization that was regarded as being perfect, that suddenly doesn't do the simplest thing. My point of view was that you had to do all the simplest things as well as the complicated things if you are going to succeed. That's the only way you can have an agency that can year after year deliver the goods.[20]

At the time when Apollo 13 was almost lost in space, it had been incredibly difficult to predict such an accident, and so one could live with it. As Story Musgrave told Cooper:

> Before the launch of Apollo 13, there was a series of errors. But nothing and nobody is perfect, and we all had the feeling that the fuel-tank rupture was an unfortunate incident in which good people had given it their best shot but missed something. With the *Challenger*, though, the trail goes on and on and on, and it turns out that the trouble is endemic to a major part of the organization.[21]

Cooper asked several employees whether they felt betrayed by their colleagues in the organization. The majority would not say so in so many words. What they would say, however, was that they unquestionably felt a sort of controlled anger. Musgrave:

> Normally, there is an intellectual honesty around here. I will say that part of NASA let the rest down. Yes, I'm disappointed that part of NASA did not do its job.[22]

Cooper also spoke to Gene Krantz, the legendary flight controller who guided the *Eagle* safely down onto the lunar surface and who had brought Apollo 13 back to earth in one piece. Krantz would not blame Jesse Moore for "pressing the button":

> Consider their position. There is enormous pressure to launch. You can see how they might say, "We have set up elaborate mission rules, which we try to satisfy. We try to satisfy weather conditions here, and at Edwards Air Force Base, and at various abort sites. There is in fact no mission rule preventing a launch at low temperatures. We have the entire system ready to go. Also, we have slipped this mission several times already, and if we slip it again all the crews—the launch crew, the flight controllers, the astronauts—will be more tired. Every time we off-load and refuel, we run risks. And if we delay again we will be passing scheduling and cargo problems on to future missions. So I can see an argument for launching. A launch decision is one of the most complex management decisions in the whole world.

Here he took a deep breath, before going on to say:

> It's harder for me to put myself in the booster guys' shoes. To this day, I cannot figure out what went wrong with their reasoning.[23]

On the first anniversary of the *Challenger* disaster, seventy-three seconds silence was observed at all NASA centers throughout the country.[24] A few days later 120 tons of salvaged and carefully numbered pieces of wreckage from *Challenger* were lowered into a couple of empty Minuteman rocket silos at Cape Kennedy.[25] The silos were then sealed. Thus, the technological remains of *Challenger* also found a final resting place.

Sally K. Ride was appointed as Special Assistant to the Administrator and asked to conduct an analysis of the future for the American space program. Eleven months later she published her report, in which she recommended that NASA abandon all thought of grandiose projects along the lines of Kennedy's lunar program and the projected Mars program. Instead, the organization ought to pursue a more moderate course, take things a bit easier, consolidate, and develop individual aspects a little at a time.

Ride stressed that the space shuttle system should never become a goal in itself. The sole purpose of the space shuttle system was to provide a means of attaining goals other than that of keeping shuttles in the air. They ought only to be launched when putting men into space was absolutely essential to the success of a mission. For other exercises, NASA should revert to the unmanned rockets. Ride's working party suggested that NASA start up an entirely new form of space exploration. She called this project Mission to Planet Earth. NASA could help ecologists by monitoring the earth from a network of new satellites. Undeniably a different and much more down-to-earth perspective than anything NASA's own policymakers were wont to come up with.

This report was given a chilly reception at NASA headquarters. James Fletcher had had something quite different in mind, and his first impulse was to leave the report to gather dust in the archives. But when the American press began to grow restless, he had it made public without committing himself to its recommendations. After that, a disappointed Sally Ride resigned from the astronaut corps.[26]

John Young was second to none when it came to running the gamut of the American space program; as a young man he had flown with Gemini 3 and Gemini 10, later with Apollo 10, and finally, on the Apollo 16 mission in April 1972, he had set foot on the surface of the moon. Thereafter, at the age of fifty, he had taken the space shuttle *Columbia* on its maiden voyage along with Robert Crippen, and later still he had been the commander on STS-9. As we have seen, as chief astronaut, Young did not balk at voicing his criticism of conditions at NASA after the *Challenger* explosion.

Very few people, not even John Young himself, considered it a coin-

cidence when — out of the blue, in April 1987 — he was removed from
the astronaut office and "promoted" away from his former colleagues
and away from active duty.[27]

Ten other astronauts left the space shuttle program voluntarily, although some of them had been planning to retire before *Challenger*.[28]

Not until thirty-two months after the *Challenger* explosion did the American space shuttle *Discovery* take to the skies once more, on September 29, 1988. By this time, alterations had been made to hundreds of
points in the space shuttle system and the development costs had risen
by a further $2.4 billion.[29] Special attention had of course been given to
redesigning the field joints — in line with the proposal made by Hercules
Inc. — in such a way that the joint rotation would no longer occur. In
addition, heating elements had been laid alongside the O-rings to ensure
that they would always retain their elasticity.

In the meantime, quite a number of civilian satellite companies had
resorted to other channels and managed to have their satellites put into
orbit by NASA's foreign competitors. But the Pentagon was standing by,
with plenty of uses for the space shuttle — although they were not prepared to go into detail.

But even all these design alterations could not put an end to NASA's
troubles. In 1989, with new and much more cautious schedules, the
space shuttle should have taken off on nine missions. However, having
learned their lesson from *Challenger*'s fate, the technicians refused to be
pushed, and only five flights were made.[30] A year later, all three space
shuttles were grounded for a while after cracks were observed in the fuel
lines leading to the main engines.[31]

After a fair amount of procrastination, Congress agreed to put up the
money for *Challenger*'s replacement. This new space shuttle, *Endeavor*,
was constructed largely from spare parts taken from the other space shuttles. In this way — running true to form — NASA would be able to make
its grants go further. New spare parts could always be produced later.[32]
Endeavor made its first flight in 1992.

NASA's image was never the same after *Challenger*, and the spaceflight
agency has had a hard time reestablishing itself in the public eye. Even
today the American press will pounce, almost with relish, on every failure
or delay.

46. *The* Challenger *Syndrome*

Challenger exploded because a solid rocket booster had burned through and ignited the Orbiter's fuel supply. From a technical point of view it is as simple as that. The burn-through was caused, as we have seen, by a combination of cold conditions, rubber, and a faulty booster joint design. Still an explanation that makes sense; one which — once the technical causes have been ascertained — could provide a basis for improvement. And in the aftermath, that is just what NASA did.

But this does not explain everything. Why did they fly at all with a joint which they knew to be problematic? Why did the subcontractors not listen to their own engineers? And why, later on, did NASA ignore its subcontractors' warnings? Why did they belittle the risk? And why were the people at the top and the bottom of the various organizations living in such totally different worlds when it came to evaluating the risk involved? It was not just in the realms of space telemetry that NASA suddenly found itself in a "no downlink" situation.

There are no easy answers to these questions. Most of those involved in the decision-making process believed, as we have seen, that they had acted professionally and responsibly, and accomplished the task set for them. They were loyal to their respective bosses, followed established procedures and regulations, and believed that they were working for the best interests of their respective companies. Afterward, NASA's senior management persisted in maintaining that they would never have launched *Challenger* had they known anything about the teleconferences held the night before. Although the House committee which investigated the accident after the Rogers commission definitely did not agree with them on this point. Apropos of the Levels II and I who were not notified, their report states quite clearly:

> The Committee finds no evidence to support a suggestion that the outcome would have been any different had they been told.[1]

This was the same management body that had thundered, again and again, that the space shuttle program would brook no more problems and delays. It was time for everyone to get their asses in gear and "fly out that manifest"!

Personnel at NASA were obviously under pressure during January 1986. Reagan had to give his State of the Union address, which really ought to contain a piece on the attractive high school teacher from Concord, New Hampshire. NASA badly needed this pat on the back from the President; just as it needed the softening touch that McAuliffe would provide amid all the high-tech jargon. And the President would be able to use her as proof of his interest in the field of education.

Then there were the politicians, who — on behalf of the taxpayers — were looking for the cost-effective Space Transportation System which they had been promised. As Senator William Proxmire barked at the "recalled" administrator, James Fletcher, in 1987: in the 1970s the latter had assured Congress on several occasions "that the Shuttle was the greatest thing since sliced bread."[2] Congress was growing tired of NASA's excuses and its endless appeals for increased funding. But the finances could only be improved by stepping up the launch rate — sending the shuttle back up again as soon as possible after landing. None of this was likely to predispose an administration toward exercising caution when it came to making decisions.

Behind the politicians were the American people, who were at long last beginning to take some interest in NASA again — now that an ordinary American was to take part in a mission. And, as mentioned earlier, this interest was a prerequisite for NASA's continued existence, essential if the agency was to have any hope of obtaining its funding.

So, this time around, the media had to be given their pictures on time — after that succession of embarrassing postponements of the previous mission, with a politician on board. Television and reporters were just as vital to the space exploration agency here on earth as oxygen was to the astronauts in space. During the very years when investigative journalism was celebrating some major coups in the United States, NASA seemed remarkably sacrosanct. As one space journalist on *The Washington Post* expressed it later, with some wonder: "It was almost like writing about entertainment . . ."[3]

NASA was also subjected to commercial pressure. If the space shuttle system could not provide a regular flight schedule, then their civilian customers would look elsewhere, and the space shuttle system would gradually be swallowed up by military interests instead — a development that could very easily mean the end of a powerful civilian space administration.

All of these various considerations ran together and merged with the pressure that an enormous spacecraft — all fueled up and ready to go — will always exert on overwrought, overworked decision makers.

But why, one might ask, albeit a little naively, didn't Thiokol stick to its guns? Nothing would have been easier than refusing to sign the final approval certificate, and without that, even NASA could not fly. But was it really that simple? There *were* no launch commit criteria stipulating a benchmark of 53 °F. Besides, Thiokol could already feel its competitors breathing down its neck. Such a situation calls for a fair amount of sangfroid, if one is to openly challenge an omnipotent customer on whom one's living is going to depend in the years to come. Especially when this customer so clearly expresses its almost uncontrollable rage at one's concern and carefully calculated technical reservations.

So Thiokol was in fact scared of NASA, which was scared of Congress, which was constantly having to answer to the American people. And, of course, all of these parties were wary of journalists.

Then, as we have seen, there was the quagmire of jealousy that existed between the various NASA field centers, and that web of tactical bureaucratic rivalries. Over the years, the individual centers had developed into what one observer has described as "quasi-independent baronies."[4] In line with the cutbacks that had been made during the 1980s, NASA had endeavored to carve up new projects in such a way that every center would get a slice of the cake — a policy which, naturally, generated even more confusion and scope for infighting. The physical distances between the different centers only served to reinforce these tendencies.[5]

That innovative management style of which James Webb and others had been justifiably proud — one which made widespread use of professional accountability as a leadership tool — had gradually become permeated, at all levels, by bureaucratic accountability. Whereas, in the past, NASA's managers came from a technical background and had, therefore, great respect for their own technical experts and their judgment, these managers were now being supplanted ever more often by a burgeoning layer of bureaucrats, who had never wielded a monkey wrench and could only manage by means of orders, memos, rules, and regulations. And it is not hard to imagine the kind of commitment and personal judgment that a technician is going to display in such an environment. *If that's the way the boss wants it, then that's what he'll get; it's his responsibility, not mine, thank goodness.*

This new management style can only be described as new because we are talking about NASA. In most other companies, it had been a fact of life for so long that no one ever gave it a thought. At NASA they had taken pride in being different from the run-of-the-mill industrial concerns. But the space administration was now continually being monitored and held politically accountable, to an extent that would have been unthinkable in the 1960s. So, in actual fact, NASA now needed bureaucrats and lobbyists more than it needed people who were used to getting their hands dirty.[6]

And into the midst of this welter of conflicting interests, into this eye at the center of a whirlwind of pressure, early one January morning, they popped Christa McAuliffe. Mercifully, she can have had little idea, as Sonny Carter solicitously settled her in her couch, that she had climbed aboard much more than just a space shuttle.

There is no reason for believing that the people at NASA or at Morton Thiokol were any meaner or more irresponsible than the average employee in any large, modern organization. On the contrary, there is every reason to suppose that, in many instances, they have been far more conscientious, and that their testing and safety procedures have been better than average. Had they been convinced that *Challenger* would come to grief, no one in their right mind would have dreamed of letting it take off. No one at NASA had any desire to find themselves in the glare of the public spotlight — especially not after a thirty-two-month break in their launch schedule. But since they were not sure that anything *would* go wrong, they assumed that everything would be okay. It would appear to be as complicated, and yet as simple, as that.

By all accounts, NASA is in many ways a model organization, when compared with other large, modern companies. But the agency still did not manage to pick up all the warning signals in time; nor could it discern the overall pattern formed by countless tortuous relationships both inside and outside the organization — not until it was too late. Strangely enough, it seems to have been particularly difficult for NASA — with its glorious past and famous can-do attitude — to note the danger signs.

The course of events in the years leading up to the *Challenger* disaster demonstrates, quite conclusively, how difficult it is for unpalatable information to penetrate large organizations. Obviously, their size alone necessitates the sifting and filtering of information traveling upward, to save those on the receiving end, higher up, from drowning in "noise." But at each level this filtering process seems only to facilitate the transmission of information that conforms to or is consonant with the management's goals and ideals. The faintest consonant whisper will always get through, while any conflicting views or dissonant information will have to be shouted from the rooftops, or writ very large, if it is ever to be received. Company directors usually end up hearing nothing but their own echoes, because underlings in an organization will always be very good at finding their bosses' "frequencies."

Even if fortune smiles and such dissonant information does get through, that is not enough. The information also has to have an impact on the recipient—never an easy task for dissonant information. Just try telling a heavy smoker that smoking is bad for his health. We are all good at contradicting, mocking, or suppressing points of view which do not suit us. And the best, most effective acts of suppression are, of course, collective. If, for instance, a whole group of senior executives decide that they cannot afford any more delays, there will be almost no budging them on that point—even when, on closer inspection, they are seen to be acting counter to the organization's interests. Funnily enough, if anything goes wrong they will have great difficulty in remembering that they had ever been warned.

But let us assume that this unpalatable information does get through and, in time, has some impact—and I am quite certain that many people at all levels in NASA were aware that the situation with the SRB joints was serious. The way I see it, if NASA had had a free hand they would not have thought twice about grounding the shuttle and taking all the time they needed to solve the problem. But the point is that NASA did not have a free hand. So, clearly, even when the information does have an impact, external forces can, in fact, be so powerful that they override these warnings, and fingers are crossed and chances taken anyway.[7]

The present Administrator of NASA, Daniel S. Goldin, is evidently well aware of the underlying gravity of such problems, which seem to crop up in all large organizations. In his first speech to his workforce in April 1992, he seemed almost to be pleading with them:

> Let me just digress from the written words here and say, when you have a concern, feel free that NASA is an open system to express those concerns. If you have an idea, before you take that idea forward, why don't you test it, do some peer review. . . . If you make it through peer review, and a consensus builds, take it forward and don't let anybody in the organization stop you, go to your boss, talk about it, see if there's some consensus there, take it as high in the organization as it has to go, and if it has come to my office, I'll stay night and day, I'll stay weekends, but I plan to listen. And I fully expect each of your bosses to encourage you to take it forward and not to stop you. NASA is an open book and I deeply and firmly

believe it from the bottom of my heart and I believe that each of you believes that, so I really, truly want you to participate.[8]

Just to be on the safe side — and as a kind of indirect admission that the above appeal for openness might not be enough — NASA has also introduced an anonymous whistle-blower system.[9]

All the sprawling ramifications of this process only serve to highlight how fluid the boundaries are between a large technological system, which in this case happens to be NASA, and the outside world. There is much more to such a system than crawlers, transfer vans, astronauts, and rockets. National and political ideals also interact with the system and affect its behavior. The archetypal American belief in modern technology as a panacea for all ills and NASA's optimistic faith in its own ability to get the better of every tricky situation gradually merged to form what was, in essence, the crux of the matter. For many years the contemporary belief in limitless and unchecked growth through the bold application of technological resources had been part of the NASA credo. And there were many who felt that it was not only *Challenger* but also this belief that plunged, blazing, into the ocean on January 28, 1986.

There is much to suggest that NASA had evolved into something of a dinosaur. It is almost as though the agency had been programmed to stay right on course for the abyss, as though it were incapable of adapting to a new age with different needs and different values. It seems almost pathetic that NASA's present Administrator, Dan Goldin, should find it necessary — in speech after speech — to refer to a long-dead President: "The real issue is, does America want a space program or not? Do we want to stop what President Kennedy started?"[10]

Those glib references to man's urge to explore, to cross new frontiers; to the salmon that will always swim against the current — not because it is easy, but because it is hard — are now beginning to sound like a tired cliché.

While such statements might have whipped up enthusiasm in the 1960s, nowadays most people find them a bit tiresome and vaguely phony. Even *Aviation Week*, usually so loyal to NASA, has now seen the writing on the wall. In an editorial published on the twenty-fifth anniversary of the first lunar landing, the magazine had this sober comment to make on the Apollo program:

> Now that the Cold War is over, such costly technological muscle flexing will not fly. . . . The lesson NASA should take from this

Apollo silver anniversary is that the United States probably will never again commit the resources to human space flight that it did in the 1960's.[11]

Nevertheless, NASA's planners doggedly persist in proposing spectacularly ambitious projects — each one more inappropriate than the one before. The Mars project apart, the latest has been a costly global warning system to combat heavy meteor showers that could threaten the existence of the human race. NASA would undertake to develop rocket technology capable of intercepting the meteors and "impressing a force" on them, to send them shooting past the earth. Earthly powers do not, however, seem to have any influence over NASA's own course or self-image.

After the accident on Three Mile Island in 1979, the Kemeny Report was published. This report described all of the complex concurrent causes behind this technological systems breakdown. In a foreword to the Danish edition of this report, Professor Ove Nathan wrote:

> In one instant, the accident at Three Mile Island exploded the myth of well-regulated American technology, in competent and responsible hands. Instead, in a flash, the United States and the rest of the world were afforded a glimpse of a universe fraught with vacillation, error, thoughtlessness and bureaucratic infighting.[12]

It was this universe that was exposed once again when *Challenger* exploded and — that same year, in the Soviet Union — when things went badly wrong with the Chernobyl nuclear reactor. Granted, in the latter case, one was dealing both with much less conscientious decision makers and much less advanced Russian technology, but here too social values were inextricably bound up with a technological system. In the Russian case this amounted simply to the notion that such rampant growth in the production of cheap electricity was vital, if the regime was to have any credibility in the eyes of an increasingly dissatisfied populace.[13]

It looks very much as though Charles Perrow was right in saying that, far from being exceptions, such major technological disasters are becoming "normal" wherever systems have grown so vast, complex, interactive, and tightly coupled that even their designers find it hard to figure them out. Perhaps it would be best if, in the future, boards of inquiry did not use the phrase "an unusual combination of mistakes." It may be that the occurrence of such coincidences in large modern organizations is, in fact, inevitable and, hence — sadly — more and more "usual."

In the modern world, we are surrounded by bigger and bigger organizations, whether publicly or privately run, and even relatively harmless slipups within these organizations can trigger off increasingly dramatic breakdowns, purely by virtue of the size of the systems and their complexity. We cannot do without these systems, and we cannot ignore modern technology. The French philosopher Jacques Ellul states that it would be just as absurd for us today to rebel against technology as it would have been for a peasant in the twelfth century to wage war against the trees, the rain, or the wind. Large modern technological systems are a precondition of the twentieth century.[14]

And no matter how much we might like to, we cannot differentiate between good and evil technology, since both good and evil are inextricably bound up with almost all modern technology and with most large systems. So we will just have to live with the intransigent dinosaurs that surround us and try to exert some influence on them, once we have worked out how they "think" and why they react as they do. And we have to know that our own ethical instincts are not simply good enough but all we have to go on.

We should not expect the experts to intervene, nor should we believe that they always know what they are doing. Often they have no idea, having been blinded to the situation in which they are involved. These days, it is not unusual for engineers and scientists working within systems to be so specialized that they have long given up trying to understand the system as a whole, with all its technical, political, financial, and social aspects. In such a situation it is very tempting to regard oneself as just a tiny cog in a greater entity, and to leave it to others to look after it — and if they don't do it, then the system will just have to take care of itself. Often, even the politicians do not understand these technological systems well enough to have any hope of controlling them.[15]

The reality of such systems corresponds all too seldom with our own ideas of order and logic. It is considerably more chaotic, haphazard, and confused than most of us would ever dare to imagine.

A faded commission report from 1973, published after the Skylab 1 accident, contains the following passage:

> The management system developed by NASA for manned space flight places large emphasis on rigor, detail and thoroughness. In

hand with this emphasis comes formalism, extensive documentation and visibility in detail to senior management. While nearly perfect, such a system can submerge the concerned individual and depress the role of the intuitive engineer or analyst. It may not allow full play for the intuitive judgment or past experience of the individual. An emphasis on management systems can, in itself, serve to separate the people engaged in the program from the real world of hardware.[16]

This quote is interesting because it highlights two meanings of responsibility. On the one hand, one has to abide by the rules — "go by the book"; on the other hand, one has to use one's own intuition. Both are vital and neither can be dispensed with, but the commission wonders whether such intensive use of certificates, formalism, and technical documentation — all of which have been introduced to allay fears about the safety aspect — does not submerge personal judgment and thus, strange as it may seem, actually increase the risk.

As a rule, large organizations prefer that their employees not bring their personal ethics and concepts of decency to work. And often — even where they are not involved with technical issues — they will be inspired and impressed by scientific dissection, or reduction, of the world. An employee's vague feeling of unease and intuitive sense for what one can or cannot get away with will rarely be rewarded either by his bosses or by his colleagues. On the contrary. More often than not, keeping such wishy-washy layman's sensitivities at arm's length, and replacing them with a certain operational cynicism, will be viewed as the mark of a professional.

Lewis Mumford once said that Galileo's sin did not consist, first and foremost, in his defiance of the Catholic Church. His real sin lay in the fact that he was the first person to exchange man's holistic approach to life for the narrow, focused segment of it that can make very precise observations of phenomena in a limited period of time and interpret them solely by means of terms such as "mass" and "motion" and "number." Thus, indirectly, he rejected the significance of a wide and detached use of human experience, thereby also diminishing us as people. In his work, Mumford reminds us that, its benefits notwithstanding, this techno-scientific viewpoint, which — with its bias toward accurate factual observation and generalized statements — has been gaining ground in all sectors of modern society, does, in fact, constitute a reduction and deformation of reality.

In *Technics and Civilization*, published in 1934, Mumford writes:

In general, the practice of the physical sciences meant an intensi-fication of the senses: the eyes had never before been so sharp, the ear so keen, the hand so accurate. . . . But with this gain in accuracy went a deformation of experience as a whole. The instruments of science were helpless in the realm of qualities. The qualitative was reduced to the subjective: the subjective was dismissed as unreal, and the unseen and unmeasurable, non-existent. . . . As the outer world of perception grew in importance, the inner world of feeling became more and more impotent.[17]

This voluntary reduction of what Mumford calls human experience as a whole has proved to be astonishingly effective within the world of science, and NASA's incredible successes in space would be unthinkable without it. The problem is, however, that it would be just as correct to say that this also applies to the complicated process which led to the *Challenger* explosion.

Great strength of conviction may be demanded for an individual to speak his or her mind and trust to his own sense of what is right and decent — as Roger Boisjoly and Allan McDonald did on the eve of the *Challenger* launch. You have to set aside all thought for yourself if you are going to take on the organization's logic and its dissection of reality single-handed. The personal cost can be very high. But often the only ones left to stand guard over a system's social ethics will be a few re-sponsible individuals who have suddenly had all they can take. The sys-tems are not going to guard themselves.

So — in the Space Transportation System, as in all other large, complex systems — it takes employees with personal integrity, intuition, and com-mon sense, along with undiminished human experience. Nowadays we can experiment and invent, design and administer on an ever-larger scale — and we do. But once in a while it might be a sobering exercise — no matter what post one occupies in the scheme of things — to give some thought to the events that led to the fate of the space shuttle *Challenger*.

 *Notes*

PREFACE

1. James Gleick, *Genius: The Life and Science of Richard Feynman*, p. 410.
2. Ronald Reagan, *An American Life: The Autobiography*, (New York: Simon & Schuster Trade, 1990) p. 404.
3. See, among others, Russell P. Boisjoly, Ellen Foster Curtis, and Eugene Mellican, "Roger Boisjoly and the Challenger Disaster: The Ethical Dimensions," pp. 217–30.

1. A MAJOR MALFUNCTION

1. This introductory account is based chiefly on material from *Time*, February 10, 1986, supplemented by material from other sources, primarily *The New York Times*, McConnell, and Joels, Kennedy, and Larkin.

2. HEROISM AND NOBLE SACRIFICE

1. MacArthur, ed., *The Penguin Book of Twentieth-Century Speeches*, pp. 448 ff.
2. *Time*, February 10, 1986.
3. *The New York Times*, January 29, 1986.
4. Ibid.
5. Ibid.
6. Ibid.
7. *Time*, February 10, 1986.
8. Ibid.
9. *Aviation Week and Space Technology*, February 3, 1986.

3. SOME USEFUL GROUNDWORK

1. Sir Isaac Newton, *The Mathematical Principles of Natural Philosophy*.
2. Per Holm, *Raketter (Rockets)*, p. 16.
3. *Encyclopaedia Britannica*, "Rocket and Missile Systems."
4. Ronald W. Clark, *Works of Man: A History of Invention and Engineering,*

from the Pyramids to the Space Shuttle, p. 313; Robert M. Powers, *The World's First Spaceship SHUTTLE*, p. 39.

5. *Time*, 60th Anniversary Issue, 1983, p. 91.
6. David Baker, *The History of Manned Space Flight*, p. 9.
7. *Encyclopaedia Britannica*, "Space Exploration."
8. Alfred Fritz, *Med raket ud i verdensrummet (Rocketing into Space)*, p. 61.

4. THE GERMANS

1. Frederick I. Ordway and Mitchell R. Sharpe, *The Rocket Team*, p. 8.
2. Ibid., p. 272.
3. Ibid., p. 274.
4. Ibid., p. 245.
5. Ibid., pp. 283 f.
6. Michael J. Neufeld, *The Rocket and the Reich. Peenemünde and the Coming of the Ballistic Missile Era*, p. 22.
7. Ordway, p. 41, and Fritz, p. 110.
8. Ordway, p. 69.
9. Ibid., p. 302.
10. Ibid., pp. 287 ff.
11. Neufeld, p. 264.
12. Ordway, p. 342.
13. Ibid., p. 346.
14. Ibid., p. 348.
15. Ibid., p. 353.
16. Tom Wolfe, *The Right Stuff*, p. 139.
17. Neufeld, p. 271.
18. Dale Carter, *The Final Frontier: The Rise and Fall of the American Rocket State*, p. 205.

5. THE RUSSIANS ARE COMING!

1. Walter A. McDougall, *The Heavens and the Earth: A Political History of the Space Age*, p. 114.
2. Paul B. Stares, *Space Weapons and U.S. Strategy*, p. 24.
3. McDougall, p. 102.
4. Ibid., pp. 108 ff.
5. Ibid., p. 106.
6. Ordway, p. 380.
7. McDougall, p. 122.
8. Ibid., p. 123.
9. Baker, pp. 26, 51.
10. According to Erling Bjøl, p. 125, it was later revealed that as late as 1961 the Russians had no more than a handful of intercontinental rockets at their disposal and still had no hydrogen bombs for them to carry.

11. McDougall, p. 123 — although Rip Bulkeley has not found anything in his studies to support this scenario.
12. Carter, p. 127.
13. McDougall, p. 128.
14. Baker, p. 28.
15. McDougall, p. 148.
16. Ibid., p. 149.
17. Robert Stone, *The Satellite Sky*, Television documentary, 1990.
18. Wolfe, p. 57.
19. Ibid., pp. 57 f.
20. Rip Bulkeley, *The Sputniks Crisis and Early United Space Policy*, p. 194.
21. Stone's documentary.
22. Bulkeley, p. 5.
23. Stone's documentary.
24. Bulkeley, p. 189.
25. Carter, p. 120.
26. Baker, p. 27. Joseph J. Trento, *Prescription for Disaster: From the Glory of Apollo to the Betrayal of the Shuttle*, p. 8. See also Bjøl in note 10.
27. Stone's documentary.
28. Carter, p. 129.
29. Ibid., p. 157.

6. THE BIRTH OF NASA

1. Carter, p. 123.
2. McDougall, p. 143.
3. Baker, p. 28.
4. Ibid., p. 29.
5. Ibid., p. 30.
6. Trento, p. 13.
7. Ibid., p. 17.
8. McDougall, pp. 173 f.
9. Carter, p. 155.

7. NASA'S PRESIDENT

1. Baker, pp. 30, 50.
2. Trento, p. 15.
3. Wolfe, p. 117.
4. Trento, p. 28.
5. Wolfe, p. 173; Baker, p. 60.
6. Wolfe, p. 174.
7. Stone's documentary.
8. McDougall, p. 210.
9. Carter, p. 138.

10. Stone's documentary.
11. Bjøl, p. 122; Jack Raymond, "Growing Threat of Our Military-Industrial Complex."
12. Raymond, p. 64.
13. McDougall, p. 233.
14. Thomas P. Hughes, *American Genesis: A Century of Invention and Technological Enthusiasm, 1870–1970*, p. 449.

8. "Before This Decade Is Out"

1. Trento, pp. 31 f.
2. Ibid., p. 32.
3. McDougall, p. 390.
4. Wolfe, pp. 185 ff.
5. Baker, p. 86.
6. Ibid.
7. McDougall, p. 148.
8. Baker, p. 88.
9. Carter, p. 139.
10. Ibid., p. 216.
11. Baker, p. 91.
12. Ibid., p. 92.
13. McDougall, p. 390.
14. Baker, p. 93 f.
15. Stone's documentary.

9. The Astronauts

1. Trento, p. 21.
2. Wolfe, pp. 56 f.
3. Ibid., pp. 60 f.
4. Ibid., p. 151.
5. Ibid., pp. 152, 155.
6. Ibid., p. 229.
7. Ibid., p. 160.
8. Ibid., p. 150.
9. Ibid., p. 231.
10. Charles Perrow, *Normal Accidents: Living with High-Risk Technologies*, p. 267.
11. Wolfe, pp. 205 ff.
12. Carter, p. 160.
13. Baker, p. 92.
14. Carter, p. 187.
15. Ibid., pp. 160, 175.
16. The following account appears in Wolfe, pp. 258 ff.

17. Ibid., p. 262.
18. Carter, p. 188.
19. Ibid.; Wolfe, p. 289.

10. NASA's Finest Hour

1. Valentina Tereshkova, June 16, 1963.
2. Stares, p. 258.
3. Wolfe, p. 193.
4. Joseph A. Raelin, *The Clash of Cultures: Managers and Professionals*, p. 60.
5. Ibid., pp. 19 f.
6. Ibid., p. 38.
7. Trento, p. 98.
8. Raelin, p. 64.
9. Ibid., p. 67.
10. Barbara S. Romzek and Melvin J. Dubnick, "Accountability in the Public Sector: Lessons from the Challenger Tragedy," p. 229.
11. Wolfe, p. 194.
12. Trento, p. 54.
13. Carter, p. 202.
14. Trento, pp. 18 f.
15. Carter, p. 196 (John F. Kennedy, October 22, 1963).
16. Trento, p. 61.

11. Jim Webb's NASA

1. James E. Webb, *Space Age Management: The Large-Scale Approach*, p. 169.
2. Ibid., p. 36.
3. Ibid., p. 55.
4. Ibid., p. 59.
5. Ibid., p. 22.
6. Ibid., p. 21.
7. Ibid., p. 25.
8. Ibid., p. 16.
9. McDougall, p. 382.
10. Carter, p. 144.
11. Ibid., pp. 159, 174 f.
12. Ibid., p. 248.
13. Norman Mailer was asked to cover the American lunar program, an assignment which resulted in the book *A Fire on the Moon*, 1970. This quote is from p. 10.
14. This expression comes from nuclear science, where a certain "critical mass" (of, for example, uranium) has to be exceeded to create a chain reaction.
15. Webb, pp. 60–64.
16. Ibid., p. 142.

17. Wolfe, pp. 330 f.
18. Webb, p. 104.
19. Ibid., p. 76.

12. System Error

1. Tim Furniss, *Manned Spaceflight Log*, p. 14; Wolfe, p. 312.
2. Furniss, p. 21.
3. Ibid., pp. 32 f.
4. Ibid., pp. 34 ff.
5. *Voshkod 1* (October 1964).
6. Furniss, pp. 36 ff.
7. The following account is based on Baker, pp. 276 ff.
8. Carter, p. 194.
9. Mike Gray, *Angle of Attack: Harrison Storms and the Race to the Moon*, p. 216.
10. Trento, p. 67.
11. Gray, p. 252.
12. Ibid., p. 241.
13. Trento, p. 68.
14. This account is based on Perrow, pp. 267 ff.; Baker, pp. 98 f.; Wolfe, pp. 235 ff.
15. Theory expanded upon in Perrow, pp. 89 ff.

13. The End of a Decade

1. Trento, p. 74.
2. Webb, p. 23.
3. Trento, p. 73.
4. *Time*, 60th Anniversary Issue, p. 87.
5. *Time*, January 2, 1988, p. 19.
6. Ibid., p. 22.
7. Ibid., p. 14.
8. Carter, p. 165.
9. Trento, p. 83.
10. Mailer, p. 13.
11. Ordway, p. 397.
12. Trento, p. 57.
13. Stone's documentary.
14. Mailer, p. 62.
15. Ibid., p. 60.
16. Ibid., p. 63.
17. Ibid., p. 64.
18. Ibid.

14. THE *EAGLE* HAS LANDED

1. Mailer, p. 153.
2. Carter, p. 176.
3. Mailer, pp. 51 f.
4. Ibid., p. 52.
5. Carter, p. 197.
6. Baker, p. 350.
7. Gray, p. 282.
8. Carter, p. 210.
9. Baker, p. 356.
10. Mailer, p. 89.
11. Baker, p. 361.
12. Ibid.
13. Mailer, p. 104.
14. Trento, p. 84.
15. Ibid., p. 86.
16. Webb, p. 96.

15. SAY AGAIN, PLEASE

1. The following description is based on two sources. Partly on the very detailed account in Baker, pp. 375–88; partly on the more analytical version in Perrow, pp. 271–81.
2. Andrew Chaikin, *A Man on the Moon: The Voyages of the Apollo Astronauts,* p. 294.
3. Perrow, p. 277.
4. The Apollo 13 accident is most fully described in Henry S. F. Cooper, *Thirteen: The Flight That Failed* (New York: DIAL, 1973), and in Jim Lovell and Jeffrey Kluger, *Lost Moon: The Perilous Voyage of Apollo 13* (New York: Houghton Mifflin, 1944).
5. Baker, p. 387.
6. Ibid., p. 377.
7. Ibid., p. 388.

16. AFTER APOLLO

1. McDougall, p. 412.
2. Ibid., p. 138.
3. Ibid., p. 407.
4. Ibid., p. 422.
5. Carter, p. 231.
6. McDougall, p. 420.
7. Richard S. Lewis, *The Voyages of Columbia: The First True Spaceship,* p. 19.

8. Hughes, p. 462.

9. Baker, p. 362.

10. Ibid.

11. Lewis, *The Voyages of Columbia*, p. 20.

12. *Report to the President by the Presidential Commission on the Space Shuttle Challenger Accident*, p. 2.

13. "NASA: Lost in Space," *The World & I*, p. 29.

14. Chaikin, p. 336.

15. See also Hughes, p. 460.

16. McDougall, p. 423.

17. Baker, p. 426.

18. Malcolm McConnell, *Challenger: A Major Malfunction*, p. 9.

17. THE POLITICAL SPACECRAFT

1. Carter, p. 223.

2. Lewis, *The Voyages of Columbia*, p. 18.

3. Ibid., p. 24.

4. John Logsdon, "The Decision to Develop the Space Shuttle," p. 118.

5. Lewis, *The Voyages of Columbia*, p. 23.

6. Nigel Macknight, *Shuttle*, pp. 8 f.

7. Wolfe, p. 215.

8. Stares, p. 129.

9. Lewis, *The Voyages of Columbia*, p. 25.

10. Ibid., p. 27.

11. McConnell, p. 35.

12. Baker, p. 426.

13. Lewis, *The Voyages of Columbia*, p. 44.

14. Logsdon, p. 109.

15. Most of the space shuttle's orbits were made at between 120 and 200 miles above the earth. A few missions did, however, call for orbits to be made over 300 miles out. The details of military missions have not been made public.

16. Logsdon, p. 109.

17. Lewis, *The Voyages of Columbia*, p. 35.

18. McConnell, p. 41.

19. Lewis, *The Voyages of Columbia*, p. 31.

18. AEROSPACE

1. Carter, p. 202.

2. Raymond, p. 56.

3. Ibid., p. 57.

4. Paul A. C. Koistinen, *The Military-Industrial Complex: A Historical Perspective*, p. 16.

5. Eisenhower, in *The Saturday Evening Post*, May 18, 1963; quoted here from Koistinen, p. 13.
6. *The New York Times*, April 23, 1986.
7. Ibid.
8. Koistinen, p. 17.
9. *The New York Times*, April 23, 1986.
10. Carter, p. 218.
11. Quotes from Raymond, p. 54.
12. McDougall, p. 383.
13. Ibid., p. 440.
14. Raymond, p. 63.
15. Carter, p. 221.
16. Raymond, p. 60.
17. *The New York Times*, April 23, 1986.
18. *The New York Times*, May 8, 1986.
19. Trento, p. 132.
20. McConnell, p. 50.
21. Ibid., pp. 46 ff.
22. Ibid., p. 45.
23. Ibid., p. 49.
24. Lewis, *The Voyages of Columbia*, p. 35.
25. McConnell, p. 49.
26. Ibid., pp. 51 ff.
27. Lewis, *The Voyages of Columbia*, p. 35; Macknight, p. 29.
28. Baker, p. 451.
29. Ibid., p. 507.

19. Off the Shelf

1. Society of Automotive Engineers, *The Space Shuttle: Its Current Status and Future Impact*, p. 8.
2. Lewis, *The Voyages of Columbia*, p. 60.
3. Ibid., p. 39.
4. *Current Status*, p. 16.
5. Lewis, *The Voyages of Columbia*, p. 64.
6. Powers, p. 54.
7. *Current Status*, p. 34.
8. Lewis, *The Voyages of Columbia*, p. 31.
9. Ibid., p. 72.
10. Macknight, pp. 16 f.
11. Lewis, *The Voyages of Columbia*, p. 63.

20. Tiles

1. *Current Status*, pp. 12 f.
2. Macknight, p. 19.

3. *Current Status*, p. 14.
4. Lewis, *The Voyages of Columbia*, pp. 88, 90.
5. Ibid., p. 86.
6. Ibid., p. 84.
7. *The New York Times*, April 23, 1986.
8. Lewis, *The Voyages of Columbia*, p. 87.
9. Trento, p. 163.
10. Lewis, *The Voyages of Columbia*, p. 111.
11. Ibid., p. 110.
12. Ibid., p. 113.

21. The Flying Brickyard

1. "Envelope" is actually a slang term for a graph indicating how fast or slow an aircraft is able to fly at different heights.
2. Macknight, pp. 13 ff; Lewis, *The Voyages of Columbia*, pp. 1–14.
3. *Current Status*, p. 10.
4. Macknight, pp. 66 ff.
5. Joseph P. Allen, *Entering Space: An Astronaut's Odyssey*, p. 206.
6. Alfred Lunde, *Romfartsrevolusjonen*, p. 18.
7. Macknight, p. 22.
8. *Current Status*, pp. 11 f.
9. Lewis, *The Voyages of Columbia*, p. 41.
10. Trento, p. 173. Years later, the Russian space shuttle *Buran* flew its first mission — with no astronauts on board — and landed safely.

22. Raw Thrust

1. *Encyclopaedia Britannica*, "Rockets."
2. Richard S. Lewis, *Challenger: The Final Voyage*, pp. 65 ff.
3. Ibid., p. 73.

23. Creative Bookkeeping

1. *Current Status*, p. 70.
2. Trento, p. 111.
3. "NASA'S Challenge: Ending Isolation at the Top," *Fortune*, May 12, 1986.
4. "NASA's Mid-Life Crisis," *The New Republic*, March 24, 1986.
5. Trento, p. 110.
6. Lewis, *The Voyages of Columbia*, pp. 35 f.
7. Trento, p. 118.
8. Howard E. McCurdy, *Inside NASA: High Technology and Organizational Change in the U.S. Space Program*, p. 28.
9. Trento, p. 121.

10. Ibid., p. 102.
11. Ibid., p. 103.
12. Ibid., p. 138.
13. Stares, p. 23.
14. Ibid., p. 65.
15. *Encyclopaedia Britannica*, "Space Exploration."
16. Trento, p. 127.
17. Logsdon, p. 118.
18. Trento, p. 127.

24. FRESH ENERGY

1. Robin Kerrod, *The Illustrated History of NASA*, p. 104.
2. Lewis, *The Voyages of Columbia*, pp. 128 ff.
3. Allen, p. 164.
4. Trento, p. 192.

25. JUST ROUTINE

1. Clark, p. 7.
2. Ibid., pp. 137 f.
3. Lewis, *The Voyages of Columbia*, p. 186.
4. McConnell, p. 28.
5. Ibid., p. 29.
6. Powers, p. 90.
7. H. R. Siepmann and D. J. Shayler, *NASA Space Shuttle: From the Flight-deck*, p. 40.
8. Lunde, p. 21.
9. Siepmann, p. 39.
10. Ibid., p. 27.
11. Trento, p. 201.
12. Allen, p. 113.
13. Furniss, p. 157.
14. Kerrod, p. 160.
15. Trento, p. 250.

26. BACKSTAGE

1. McConnell, p. 29.
2. Perrow, p. 260.
3. The section on the technical problems is based primarily on Furniss, Lewis, McConnell, and Siepmann.
4. Trento, p. 173.
5. McConnell, p. 69.
6. Ibid., p. 74.

7. Trento, p. 25.
8. McConnell, p. 72.
9. Trento, p. 183.
10. Ibid., p. 276.
11. *IEEE Spectrum*, February 1987, p. 45.
12. McConnell, p. 26.
13. Ibid., p. 73.
14. *Aviation Week and Space Technology*, January 13, 1986.
15. Trento, p. 276; Lewis, *Challenger: The Final Voyage*, p. 48.

27. IN THE INTERESTS OF NATIONAL SECURITY

1. McConnell, p. 73.
2. *The New York Times*, February 11, 1986.
3. McConnell, p. 62.
4. Ibid., p. 64.
5. Trento, p. 233.
6. Ibid., p. 169.
7. Ibid., pp. 184 f.
8. William J. Broad, *Star Warriors*, p. 17.
9. Broad, p. 125.
10. Stares, p. 231.
11. Trento, p. 243.
12. Ibid., p. 217.
13. Ibid., pp. 218 f.
14. *The New York Times*, April 23, 1986.

28. TISP

1. *Newsweek*, February 10, 1986, p. 25.
2. Ibid.
3. This was the case, for example, with the astronaut Don Lind in April 1985.
4. McConnell, p. 102.
5. David Shayler, *Shuttle Challenger*, p. 57.
6. *The New York Times*, January 29, 1986.
7. Ibid.
8. Carter, p. 257.
9. Trento, p. 250.
10. Carter, p. 255.
11. Ibid., p. 258.
12. Ibid., p. 259.
13. McConnell, pp. 22 f.

29. One Day to Go

1. Lewis, *Challenger*, p. 123.
2. McConnell, p. 24.
3. Ibid., p. 23.
4. Ibid., p. 25.
5. Lewis, *Challenger*, pp. 52 f.
6. Trento, pp. 170 f., 178 f., 180 f.
7. Lewis, *Challenger*, p. 53.
8. McConnell, pp. 131 ff.
9. Ibid., pp. 137 ff.
10. Lewis, *Challenger*, p. 3.
11. Ibid., p. 9.
12. McConnell, pp. 154 ff.; Lewis, *Challenger*, p. 3.
13. *Time*, February 10, 1986, p. 12.
14. *The New York Times*, February 10, 1986.
15. Ibid., January 28, 1986.

30. Liftoff

1. Lewis, *Challenger*, p. 124.
2. McConnell, pp. 164 f., 217 ff.
3. Ibid., p. 208.
4. Siepmann, p. 13.
5. McConnell, p. 209.
6. *Report*, p. 17.
7. Lewis, *Challenger*, p. 10.
8. *Time*, February 10, 1986, p. 20.
9. Lewis, *Challenger*, p. 10.
10. McConnell, p. 223.
11. Lewis, *Challenger*, p. 11.
12. Ibid.
13. *Time*, February 10, 1986, p. 20.
14. Lewis, *Challenger*, p. 12.
15. Ibid., p. 22; *The New York Times*, January 29, 1986.
16. McConnell, p. 232.
17. *Newsweek*, February 10, 1986, p. 10.
18. Lewis, *Challenger*, pp. 22 f.

31. The Morning After

1. Carter, p. 265.
2. *Newsweek*, February 10, 1986, pp. 19, 21.
3. *Time*, February 10, 1986, p. 16.
4. Ibid., p. 22.
5. *Newsweek*, February 10, 1986, p. 18.

6. Ibid., pp. 16 ff.
7. Ibid., pp. 16, 18.
8. Ibid.
9. Ibid., pp. 17 f.
10. McConnell, p. 9.
11. *Newsweek*, February 10, 1986, p. 18.
12. Ibid., p. 20.

32. COMMISSIONS

1. Lewis, *Challenger*, p. 25.
2. Ibid., p. 27.
3. Ibid.
4. Ibid., p. 30.
5. Ibid., p. 34.
6. Ibid., p. 46.
7. Ibid.
8. *The New York Times*, February 4, 1986.
9. Further information on the members of each commission in *Report*, pp. 202 f.
10. Lewis, *Challenger*, p. 47.

33. RICHARD P. FEYNMAN

1. Gleick, p. 11.
2. *Physics Today*, February 1989, p. 70.
3. Ibid., p. 23.
4. "Mr. Feynman Goes to Washington," in Richard P. Feynman, *What Do You Care What Other People Think?*, p. 88.
5. Ibid.
6. Ibid., p. 89.
7. Ibid., p. 90.
8. Ibid., p. 91.

34. DISCORD

1. "Mr. Feynman," p. 95.
2. Lewis, *Challenger*, p. 47.
3. Ibid., p. 106.
4. Ibid., p. 102.
5. Ibid., p. 105.
6. "Mr. Feynman," pp. 101 ff.
7. *Report*, p. 97.

35. REMEDIAL ACTION—NONE REQUIRED

1. *Report*, p. 206.
2. *IEEE Spectrum*, February 1987, pp. 39 f.

3. *Report*, p. 120.
4. Ibid., pp. 121 f.
5. Ibid., p. 122.
6. Ibid., p. 123.
7. Ibid.; *IEEE Spectrum*, p. 40.
8. *Report*, p. 236.
9. Ibid., p. 238.
10. Ibid., p. 125; *IEEE Spectrum*, p. 42.
11. *Report*, p. 241.
12. Ibid., p. 124.
13. William H. Starbuck and Frances J. Milliken, "Challenger: Fine-Tuning the Odds Until Something Breaks," p. 325; Henry S. F. Cooper, "Letter from the Space Center," p. 86.
14. Starbuck, p. 325.
15. *Report*, p. 128.
16. Ibid., pp. 133 f; *IEEE Spectrum*, p. 43.
17. *Report*, p. 134.
18. Ibid., p. 136.
19. Ibid.
20. Ibid., p. 137.
21. *IEEE Spectrum*, p. 43.
22. *Report*, pp. 137 f.
23. "Letter from the Space Center," pp. 89 f.
24. McConnell, p. 177.
25. *Report*, p. 249.
26. McConnell, pp. 179 f.
27. *Report*, p. 139.
28. *IEEE Spectrum*, p. 45.
29. *Report*, pp. 141, 255.
30. McConnell, p. 180.
31. *Report*, pp. 141, 252.
32. Ibid., pp. 141 f.
33. Ibid., p. 143.
34. Lewis, *Challenger*, p. 188.

36. FEYNMAN'S EXPERIMENT

1. "Mr. Feynman," pp. 98 f. Later it was revealed that Kutyna had learned this from a source among the NASA astronauts, whom he did not want to jeopardize.
2. *Report*, p. 137.
3. "Mr. Feynman," pp. 107 f.
4. Lewis, *Challenger*, p. 99.
5. Ibid., pp. 99 f.

37. TELECONFERENCES

1. *Report*, pp. 82 f.
2. McConnell, p. 20.
3. *Report*, p. 85.
4. McConnell, p. 19.
5. *Report*, p. 85.
6. McConnell, pp. 105 f.
7. *Report*, p. 86.
8. Ibid., p. 87.
9. Ibid., p. 88.
10. McConnell, p. 108.
11. Ibid., p. 109.
12. Ibid., p. 115.
13. *Report*, p. 89.
14. Ibid.
15. Ibid., p. 90.
16. Ibid., p. 96.
17. Maureen Hogan Casamayou, *Bureaucracy in Crisis: Three Mile Island, the Shuttle Challenger, and Risk Assessment*, p. 53.
18. Russell P. Boisjoly, p. 222.
19. Ibid., pp. 92 f.
20. *Report*, p. 94.
21. Lewis, *Challenger*, p. 118.
22. McConnell, p. 187.
23. *Report*, pp. 145 f.
24. *Report*, p. 95.
25. Ibid., p. 93.
26. Ibid., p. 95.
27. Lewis, *Challenger*, p. 120.
28. *Report*, p. 97.
29. Lewis, *Challenger*, p. 116.
30. Roger M. Boisjoly, "Ethical Decisions: Morton Thiokol and the Shuttle Disaster." Speech given at MIT, January 7, 1987. Quoted from Russell P. Boisjoly, p. 223.
31. Lewis, *Challenger*, p. 114.
32. *Report*, p. 98.
33. Ibid., p. 100.
34. Ibid., pp. 100 f.
35. Ibid., p. 101.
36. Lewis, *Challenger*, p. 121.
37. Ibid., p. 126.

38. ROCKWELL'S WARNING

1. *Report*, pp. 114 f.
2. McConnell, pp. 210 f.
3. *Report*, p. 115.
4. Ibid.
5. *Report*, p. 117.
6. Lewis, *Challenger*, p. 134.
7. Ibid.
8. Ibid., p. 132.

39. THE REACTION

1. Trento, pp. 4 f.
2. *Time*, March 17, 1986.
3. *The New Republic*, March 24, 1986, p. 12.
4. *Time*, March 3, 1986.
5. *Aviation Week and Space Technology*, February 17, 1986.
6. Lewis, *Challenger*, p. 180.
7. Ibid.
8. Ibid., p. 184.
9. Ibid., p. 191.
10. "Mr. Feynman," p. 141.
11. Lewis, *Challenger*, p. 188.
12. Ibid.

40. SLOW MOTION

1. *Report*, pp. 19 ff., 37 ff.
2. Ibid., p. 70.

41. EVALUATING THE RISK

1. "Mr. Feynman," p. 134.
2. Ibid., pp. 134 f.
3. Ibid., p. 135.
4. Lewis, *Challenger*, p. 24.
5. Appendix F, "Mr. Feynman," p. 165.
6. Ibid., p. 167.
7. Ibid., p. 166.
8. Ibid., p. 168.
9. Ibid., p. 171.
10. Ibid., p. 172.
11. Ibid., pp. 176 f.
12. Ibid., p. 177.

13. Ibid., p. 178.
14. Ibid., pp. 178 f.

42. CRISIS IN CONSCIENCE

1. Webb, p. 157.
2. See, for example, Michael Brody, "NASA's Challenge: Ending Isolation at the Top."
3. Fendrock, p. 115.
4. Ibid., p. 116.
5. Ibid., p. 119.
6. Ibid.
7. Ibid.
8. Ibid., p. 115.
9. Ibid.
10. *Report*, p. 104.

43. RECOMMENDATIONS

1. *Report*, p. 208.
2. *The New York Times*, April 23, 1986.
3. *Aviation Week and Space Technology*, May 19, 1986.
4. *The New York Times*, May 31, 1988.
5. "Mr. Feynman," p. 148.
6. Ibid., p. 151.
7. Lewis, *Challenger*, pp. 218 f.
8. Ibid., p. 219.
9. *Aviation Week and Space Technology*, June 16, 1986.
10. Ibid.

44. TWO MINUTES AND FORTY-FIVE SECONDS

1. *Time*, May 12, 1986.
2. Lewis, *Challenger*, p. 178.

45. AFTERMATH

1. *The New York Times*, May 8, 1986, and April 24, 1986.
2. Lewis, *Challenger*, p. 220.
3. Ibid., p. 221.
4. Ibid., p. 231.
5. Ibid., p. 222.
6. Ibid.
7. *Aviation Week and Space Technology*, February 9, 1987.

8. Diane Vaughan, "Autonomy, Interdependence, and Social Control: NASA and the Space Shuttle Challenger," p. 249.

Later, the sum the company would receive for the projected SRBs was, however, increased. In 1987 a number of congressmen criticized NASA for signing contracts amounting to $1.8 billion with Morton Thiokol (*The New York Times*, April 20, 1987).

9. "Morton Thiokol: Reflections of the Shuttle disaster," *Business Week*, March, 14, 1988, p. 38.

10. *The New York Times*, June 17, 1986.

11. *Aviation Week and Space Technology*, July 21, 1986.

12. *The New York Times*, January 21, 1987.

13. *Aviation Week and Space Technology*, January 5, 1987.

14. *The New York Times*, February 18, 1988.

15. *Aviation Week and Space Technology*, February 2, 1987.

16. Ibid., April 27, 1987.

17. *The New York Times*, September 3, 1988.

18. Ibid., January 28, 1987, and May 31, 1988.

19. Cooper, "Letter from the Space Center," p. 84.

20. Trento, p. 1.

21. "Letter from the Space Center," p. 92.

22. Ibid., p. 93.

23. Ibid., p. 95.

24. *Aviation Week and Space Technology*, January 26, 1987.

25. Ibid., February 2, 1987.

26. Extracts from Ride's analysis appear in *Historic Documents of 1987*, pp. 649–69.

27. Chaikin, *A Man on the Moon*, pp. 572 ff.

28. "Letter from the Space Center," p. 84.

29. *Britannica Book of the Year 1989*, pp. 304 f.

30. *Britannica Book of the Year 1990*, p. 320.

31. *Politiken* (Danish daily newspaper), July 24, 1990.

32. *Aviation Week and Space Technology*, August 10, 1987.

46. The *Challenger* Syndrome

1. Quoted from Casamayou, p. 32.

2. Ibid., p. 68.

3. Ibid., p. 65.

4. Michael T. Charles, "The Last Flight of Space Shuttle Challenger," p. 160.

5. As a symbolic touch, a system has subsequently been introduced whereby the identification badges worn by all staff do not identify their affiliation within the organization; they simply indicate that they are NASA personnel. Casamayou, p. 89.

6. Concerning accountability models, see Romzek and Dubnick.

7. See Maureen Hogan Casamayou's brilliant analysis of this situation in *Bureaucracy in Crisis.*
8. Daniel S. Goldin, "Address to Employees," April 1, 1992.
9. Casamayou, p. 92.
10. "NASA: Lost in Space," p. 25.
11. *Aviation Week and Space Technology,* July 18, 1994, p. 90.
12. Ib Lindberg, *Sådan gik det til (That's How It Happened).* The Kemeny Report on the Inquiry into Three Mile Island.
13. Hughes, p. 470.
14. Ibid., pp. 451 f.
15. Ibid., p. 452.
16. Quoted from Russell P. Boisjoly, p. 224.
17. Lewis Mumford, *Technics and Civilization,* p. 49.

 *Bibliography*

BOOKS

Aaen, Harold. *Rummets Udforskning (Space Exploration)*. Frederikshavn: Dafolo, 1991.

Allen, Joseph P. *Entering Space: An Astronaut's Odyssey*. London: Stewart, Tabori & Chang, 1984.

Baker, David. *The History of Manned Space Flight*. New York: Crown, 1981.

———. *The Rocket. The History and Development of Rocket and Missile Technology*. London: New Cavendish Books, 1978.

Bilstein, Roger E. *Stages to Saturn: A Technological Survey of the Apollo/Saturn Launch Vehicles*. Washington: Scientific and Technical Information Branch, National Aeronautics and Space Administration, 1980.

Bjøl, Erling. *Politikens USA Historie*. Copenhagen: Politikens Forlag, 1988.

Britannica Book of the Year. Chicago: Encyclopaedia Britannica Educational Corp., 1987–90.

Broad, William J. *Star Warriors*. New York: Simon & Schuster, 1985.

Bulkeley, Rip. *The Sputniks Crisis and Early United States Space Policy: A Critique of the Historiography of Space*. Bloomington: Indiana University Press, 1991.

Carter, Dale. *The Final Frontier: The Rise and Fall of the American Rocket State*. New York: Routledge, Chapman & Hall, 1988.

Casamayou, Maureen Hogan. *Bureaucracy in Crisis: Three Mile Island, the Shuttle Challenger, and Risk Assessment*. Boulder, Colo.: Westview Press, 1993.

Chaikin, Andrew. *A Man on the Moon: The Voyages of the Apollo Astronauts*. New York: Viking Penguin, 1995.

Clark, Ronald W. *Works of Man: A History of Invention and Engineering, from the Pyramids to the Space Shuttle*. New York: Viking, 1985.

Encyclopaedia Britannica. Chicago: Encyclopaedia Britannica Educational Corp., 1980.

Encyclopedia of Physical Science and Technology. San Diego: Academic Press, 1987.

Feynman, Richard P. *What Do You Care What Other People Think?* New York: Bantam, 1988.

Fritz, Alfred. *Med raket ud i verdensrummet (Rocketing into Space)*. Copenhagen: Jul. Gjellerups Forlag, 1958.

Furniss, Tim. *Manned Spaceflight Log*. London: Jane's Publishing Inc., 1986.

Gleick, James. *Genius: The Life and Science of Richard Feynman*. New York: Pantheon, 1992.

Gray, Mike. *Angle of Attack: Harrison Storms and the Race to the Moon*. New York: Norton, 1992.

Historic Documents of 1987. Washington: Congressional Quarterly, 1988.

Holm, Per. *Raketter (Rockets)*. Copenhagen: Gyldendal, 1974.

Hughes, Thomas P. *American Genesis: A Century of Invention and Technological Enthusiasm, 1870–1970*. New York: Viking Penguin, 1989.

Joels, Kerry M., Gregory P. Kennedy, and David Larkin. *The Space Shuttle Operator's Manual*. New York: Ballantine, 1982.

Johannesen, Erik. *Midt i en Jettid (Living in the Jet Age)*. Viby J.: Centrum, 1987.

Kerrod, Robin. *The Illustrated History of NASA*. London: Prion, 1986.

Koistinen, Paul A. C. *The Military-Industrial Complex: A Historical Perspective*. New York: Praeger, 1980.

Lewis, Richard S. *The Voyages of Columbia: The First True Spaceship*. New York: Columbia University Press, 1984.

———. *Challenger: The Final Voyage*. New York: Columbia University Press, 1988.

Lindberg, Ib, Jorgen Flindt Pedersen, and Arne Herlov Petersen. *Sadan gik det til (That's How It Happened)*. The Kemeny Report on the Inquiry into Three Mile Island. Copenhagen: Albatros, 1979.

Lunde, Alfred. *Romfartsrevolusjonen (The Space Travel Revolution)*. Oslo: Gyldendal, 1986.

MacArthur, Brian, ed. *The Penguin Book of Twentieth-Century Speeches*. New York: Viking Penguin, 1993.

Macknight, Nigel. *Shuttle*. Nottingham: Motorbooks International, 1985.

Mailer, Norman. *A Fire on the Moon*. Boston: Little, Brown, 1970.

McConnell, Malcolm. *Challenger: A Major Malfunction*. New York: Doubleday, 1987.

McCurdy, Howard E. *Inside NASA: High Technology and Organizational Change in the U.S. Space Program*. Baltimore: Johns Hopkins, 1993.

McDougall, Walter A. *The Heavens and the Earth: A Political History of the Space Age*. New York: Basic Books, 1985.

Mumford, Lewis. *Technics and Civilization*. New York: Harcourt Brace Jovanovich Inc., 1934.

Neufeld, Michael J. *The Rocket and the Reich: Peenemünde and the Coming of the Ballistic Missile Era*. New York: Free Press, 1995.

Ordway, Frederick I., and Mitchell R. Sharpe. *The Rocket Team.* New York: Cromwell, 1979.

Perrow, Charles. *Normal Accidents. Living with High-Risk Technologies.* New York: Basic Books, 1984.

Powers, Robert M. *The World's First Spaceship SHUTTLE.* Harrisburg: Stackpole, 1979.

Raelin, Joseph A. *The Clash of Cultures: Managers and Professionals.* Boston: Harvard Business School Press, 1986.

Report to the President by the Presidential Commission on the Space Shuttle Challenger Accident. Salem: Ayer Co., 1986.

Sapolsky, Harvey M. *The Polaris System Development: Bureaucratic and Programmatic Success in Government.* Cambridge: Harvard University Press, 1972.

Shayler, David. *Shuttle Challenger: Aviation Fact File.* Englewood Cliffs, N.J.: Prentice Hall, 1987.

Siepmann, H. R., and D. J. Shayler. *NASA Space Shuttle: From the Flightdeck.* London: Motorbooks International, 1987.

Society of Automotive Engineers. *The Space Shuttle: Its Current Status and Future Impact.* San Diego: Univelt, 1978.

Stares, Paul B. *Space Weapons and U.S. Strategy.* London: Croom Helm, 1985.

Stone, Robert. *The Satellite Sky.* Television documentary, 1990.

Trento, Joseph J. *Prescription for Disaster: From the Glory of Apollo to the Betrayal of the Shuttle.* New York: Crown, 1987.

Webb, James E. *Space Age Management: The Large-Scale Approach.* New York: McGraw-Hill, 1969.

Wolfe, Tom. *The Right Stuff.* New York: Farrar, Straus & Giroux, 1981.

Yeager, Jeana, and Dick Rutan. *Voyager.* New York: Knopf, 1987.

ARTICLES

Baker, David. "Science Crashed with Challenger." *New Scientist,* no. 1545 (1987).

Bell, Trudy E., and Karl Esch. "The Fatal Flaw in Flight 51-L." *IEEE Spectrum,* February 1987.

Boisjoly, Russell P., Ellen Foster Curtis, and Eugene Mellican. "Roger Boisjoly and the Challenger Disaster: The Ethical Dimensions." *Journal of Business Ethics,* Vol. 8 (1989).

Brody, Michael. "NASA's Challenge: Ending Isolation at the Top." *Fortune,* May 12, 1986.

Charles, Michael T. "The Last Flight of Space Shuttle Challenger," in Uriel Rosenthal, ed., *Coping with Crises.* Springfield, Ill.: Charles C Thomas, 1989.

Cooper, Henry S. F. "Letter from the Space Center." *The New Yorker*, November 10, 1986.

Fendrock, John J. "Crisis in Conscience at Quasar." *Harvard Business Review*, March–April 1968.

Goldin, Daniel S. (April 1, 1992). "Address to Employees Transcript," [Online]. Available: spacelink.msfc.nasa.gov.

———. (March 14, 1991). "Keynote Address Transcript, Goddard Memorial Symposium," [Online]. Available: spacelink.msfc.nasa.gov.

Lambright, W. Henry. "The Augustine Report, NASA, and the Leadership Problem." *Public Administration Review*, Vol. 52 (March–April 1992).

Logsdon, John. "The Decision to Develop the Space Shuttle." *Space Policy*, May 1986.

McCurdy, Howard E. "NASA's Organizational Culture." *Public Administration Review*, Vol. 52 (March–April 1992).

"NASA: Lost in Space." *The World & I* (July 1994).

"NASA's Mid-Life Crisis." *The New Republic*, March 24, 1986.

Raymond, Jack. "Growing Threat of Our Military-Industrial Complex." *Harvard Business Review*, May–June 1968.

Romzek, Barbara S., and Melvin J. Dubnick. "Accountability in the Public Sector: Lessons from the Challenger Tragedy." *Public Administration Review*, Vol. 47 (May–June 1987).

Shafritz, Jay M. "An Indictment of NASA's Merit System." *Public Administration Review*, Vol. 52 (March–April 1992).

Starbuck, William H., and Frances J. Milliken. "Challenger: Fine-Tuning the Odds Until Something Breaks." *Journal of Management Studies*, Vol. 25, no. 4 (July 1988).

Turner, Barry A. "The Organizational and Interorganizational Development of Disaster." *Administrative Science Quarterly*, Vol. 21 (September 1976).

Vaughan, Diane. "Autonomy, Interdependence, and Social Control: NASA and the Space Shuttle Challenger." *Administrative Science Quarterly*, Vol. 35 (June 1990).

Werhane, Patricia H. "Engineers and Management: The Challenge of the Challenger Incident." *Journal of Business Ethics*, Vol. 10 (1991).

 *Permissions*

We are grateful for permission to reprint the following:

From "McAuliffe's Goal: Humanize Technology of the Space Age," Associated Press, *The New York Times*, January 29, 1986, copyright © 1986 by Associated Press, reprinted by permission of The Associated Press.

From *Aviation Week & Space Technology*, reprinted by permission of *Aviation Week & Space Technology*.

From "The Fatal Flaw in Flight 51-L" by Trudy E. Bell and Karl Esch, *IEEE Spectrum*, February 1987, copyright © 1987 by IEEE, reprinted by permission of IEEE.

From "Roger Boisjoly and the Challenger Disaster: The Ethical Dimensions" by Russell P. Boisjoly, Ellen Foster Curtis, and Eugene Mellican, *The Journal of Business Ethics*, Vol. 8, 1989, copyright © by Kluwer Academic Publishers, reprinted by permission of Kluwer Academic Publishers.

From *Star Warriors* by William J. Broad, copyright © 1985 by William J. Broad, reprinted by permission of Simon & Schuster.

From *The Sputnik Crisis and Early United States Space Policy: A Critique of the Historiography of Space* by Rip Bulkeley, copyright © 1991 by Rip Bulkeley, reprinted by permission of Indiana University Press.

From *The Final Frontier* by Dale Carter, copyright © 1988 by Dale Carter, reprinted by permission of Routledge, Chapman and Hall.

From *Man on the Moon* by Andrew Chaikin, copyright © 1994 by Andrew Chaikin, reprinted by permission of Viking Penguin, a division of Penguin Books USA Inc.

From "Letter from the Space Center" by Henry S.F. Cooper, *The New Yorker*, November 10, 1986, copyright © 1986 by Henry S.F. Cooper, reprinted by permission of *The New Yorker*.